The ASCENT of MAMMALS

Also by John Reilly

The Ascent of Birds: How Modern Science is Revealing their Story

Greetings from Spitsbergen: Tourists at the Eternal Ice 1827–1914

Spitsbergen's Early Postcards: an Annotated Catalogue

The ASCENT of MAMMALS

How DNA Discoveries are Rewriting Our Story

JOHN REILLY

First published in 2026 by
Pelagic Publishing
20–22 Wenlock Road
London N1 7GU

www.pelagicpublishing.com

The Ascent of Mammals: How DNA Discoveries are Rewriting Our Story

Copyright © text, figures and images 2026 John Reilly

The moral rights of the author have been asserted by him in accordance with the Copyright, Designs and Patents Act 1988.

All rights reserved. Apart from short excerpts for use in research or for reviews, no part of this document may be printed or reproduced, stored in a retrieval system, or transmitted in any form or by any means, electronic, mechanical, photocopying, recording or otherwise, now known or hereafter invented, without prior permission from the publisher.

https://doi.org/10.53061/OLRH7565

Without limiting the exclusive rights of any author, contributor or the publisher, any unauthorised use of the contents of this publication to train generative artificial intelligence (AI) is expressly prohibited. Pelagic Publishing also exercises its rights under Article 4(3) of the Digital Single Market Directive 2019/790 and reserves the entirety of this publication from the text and data mining exception.

A CIP record for this book is available from the British Library

ISBN 978-1-78427-632-4 Hardback
ISBN 978-1-78427-633-1 ePub
ISBN 978-1-78427-634-8 PDF

EU Authorised Representative: Easy Access System Europe – Mustamäe tee 50, 10621 Tallinn, Estonia, gpsr.requests@easproject.com

Cover image: *Out on a Limb* © 2025 Kevin Hayler. All rights reserved, DACS

Typeset in Adobe Caslon Pro by S4Carlisle Publishing Services, Chennai, India

10 9 8 7 6 5 4 3 2 1

Printed in the Czech Republic by Finidr

'The one most adaptable to change is the one that survives.'
Charles Darwin (1809–1882)

'Nothing in biology makes sense except in the light of evolution.'
Theodosius Dobzhansky (1900–1979)

'A curious aspect of the Theory of Evolution is that everybody thinks he understands it.'
Jacques Monod (1910–1976)

To a special mammalian family:
Janette, Philip, Mark, Rachel, Jess, Oliver, Harriet and Sienna

Contents

Acknowledgements	ix
List of Illustrations	xi
Timeline	xiii
Geological Ages	xv
Prologue: *Linnaeus's legacy*	1

PART ONE: MONOTREMES — 13

1. The Platypus's Story: *Monotremes oviparous, ovum meroblastic* — 15

PART TWO: MARSUPIALS — 33

2. The Monito del Monte's Story: *The marsupial diaspora* — 35
3. The Marsupial Mole's Story: *Austral doppelgängers* — 54
4. The Tasmanian Tiger's Story: *De-extinction* — 75

PART THREE: EUTHERIANS — 83

5. The Aardvark's Story: *Evolutionary distinctiveness* — 85
6. The Hyrax's Story: *Aquatic origins* — 98
7. The Elephant's Story: *Admixtures, ratchets and retrogenes* — 105
8. The Sloth's Story: *Regressive evolution and pseudogenes* — 117
9. The Solenodon's Story: *EDGE scores, venom and the K–Pg event* — 128
10. The Camel's Story: *High latitudes and domestication* — 142
11. The Whale's Story: *Loss of gene function* — 150
12. The Buffalo's Story: *Adaptability and domestication* — 160
13. The Giraffe's Story: *Comparative genomics* — 172
14. The Horse's Story: *A bushy phylogeny* — 181
15. The Bear's Story: *Inter-species gene flow* — 191

16	The Cat's Story: *Dispersals and bottlenecks*	202
17	The Bat's Story: *Powered flight and echolocation*	211
18	The Rat's Story: *Extreme evolution*	223
19	The Lemur's Story: *Sweepstake dispersal*	239
20	The Tarsier's Story: *Speciation genes*	251
21	The Howler Monkey's Story: *Trade-offs, reinforcement and duplications*	258
22	The Gibbon's Story: *Jumping genes*	269
23	The Gorilla's Story: *Ghost admixtures*	276
24	The Bonobo's Story: *Vicariance, neoteny and genetic fossils*	284
25	The Human Story: *Palaeogenomics and adaptive introgressions*	296
	Epilogue: *The descent of mammals*	310
	Glossary	313
	Dramatis Personae	321
	Notes	326
	Bibliography	360
	Index	363

Acknowledgements

There are few things I enjoy more than exploring new and exciting scientific discoveries and sharing their significance with others. As with my earlier book, *The Ascent of Birds*, the current project has not only been challenging and ambitious, but also immensely rewarding. Over the past three years, I have had the opportunity to immerse myself in the latest research that is reshaping our understanding of mammalian evolution. I hope that my efforts to distil these advances for a general readership will help foster a broader appreciation of how modern science is rewriting our 200-million-year story, and how DNA holds clues to some of evolution's most enduring puzzles.

I could not have written this book without the generous support of many individuals. First and foremost, I am immensely grateful to the scientists who generously shared their time and expertise. Their willingness to respond to my questions, share relevant publications and offer feedback on draft sections of the manuscript has been invaluable. I am also thankful to the professional wildlife guides who enabled me to observe many of the book's dramatis personae in their natural environments – experiences that have deepened my understanding of the extraordinary diversity and adaptability of mammals. My sincere thanks also go to the photographers who kindly and generously provided the striking images that have enriched the text. Those who have helped include (in alphabetical order): Alex Bäcker, Ben Bluhm, Nicolás Chimento, Nathaniel Dominy, Christopher Emerling, Tim Flannery, Rikki Gumbs, David Hoddinott, Jason Horn, Amiyaal Ilany, Shoji Kawamura, Christopher Kirk, Martin Kuhlwilm, Asato Kuroiwa, Tomas Marques-Bonet, Amanda Melin, Pete Morris, Joe Nunez-Mino, Linda Ongaro, Svante Pääbo, Andrew Pask, Eduardo Patrial, Harvinder Pawar, Tom Peat, Tom Rich, Kenneth D. Rose, Roberto Nespolo Rossi, Beth Shapiro, David Thybert, Samuel Turvey, Carrie Veilleux, Mike Watson, Michael Westbury, Janine Williams. Acknowledgement of those who have offered support does not imply their endorsement of the views expressed in this book; I alone am answerable for them. Similarly, any errors of fact or omissions are entirely my responsibility.

I am also grateful to Nigel Massen and his team at Pelagic Publishing for their steadfast support. Nigel believed in this project from the beginning and offered much-needed encouragement during moments of uncertainty. My sincere thanks also go to Hugh Brazier, whose sharp editorial eye and meticulous attention to detail have once again spared me many a blush. I am also appreciative of his tactful hints that some of my explanations for complex and challenging concepts may not have been as cogent as I had assumed. Special acknowledgement also goes to David Hawkins for skilfully overseeing the production process from manuscript to finished book.

I would also like to thank my son Philip for his skill, patience and thoughtful input in producing the figures that accompany the text. Finally, and most importantly, I am indebted to my wife and travelling companion, Janette, for her constant support and good-humoured tolerance of the many hours spent researching and writing.

Illustrations

FIGURES

1.1	Evolution of the mammalian placenta and middle ear	18
1.2	Evolution of the mammalian jaw joint and middle ear	24
2.1	Gondwanan origin of stem therians	38
2.2	Phylogeny and biogeography of marsupials	48
3.1	Phylogeny of Australian marsupials	55
5.1	Phylogeny of Mammalia	86
5.2	Phylogeny of Afrotheria	88
9.1	Phylogeny of Laurasiatheria	129
9.2	Models for the diversification of placental mammals	140
10.1	Phylogeny of Cetartiodactyla	144
10.2	Historical dispersal of the camelid family	145
12.1	Phylogeny of Bovidae	161
15.1	Hybrid origin of the Asiatic black bear	196
17.1	Comparison of a human upper limb with a bat wing	212
18.1	Phylogeny of Euarchontoglires	226
19.1	Phylogeny of primates	241
21.1	Sweepstake dispersal of New World monkeys	260
24.1	Map showing distribution of chimpanzees and bonobo	285
25.1	Origin and dispersal of *Homo sapiens*	304

PLATES

1. Morganucodon
2. Platypus
3. Short-beaked echidna
4. Tim Flannery
5. Monito de monte
6. Koala
7. Thylacine and dingo

8	Fat-tailed dunnart
9	Aardvark
10	Lowland streaked tenrec
11	Rock hyrax
12	African savanna elephant
13	Brown-throated sloth
14	Hispaniolan solenodon
15	Arabian camel
16	Common hippopotamus
17	Grey whale
18	Muskoxen
19	African buffalo
20	Northern (Rothschild's) giraffe
21	Plains zebra
22	Polar bear and brown bear
23	Jaguar
24	Bat's wing
25	Brown rat
26	Spiny mouse
27	Naked mole-rat
28	Amami spiny rat
29	Ring-tailed lemur
30	Western tarsier
31	Mantled howler monkey, Nancy Ma's night monkey and golden-headed lion tamarin
32	Lar gibbon
33	Mountain gorilla
34	Bonobo
35	Svante Pääbo

Timeline

The *Ascent of Mammals* is the story of our 200-million-year history, from the Early Jurassic to the Holocene. But the beginnings can seem unimaginably distant to us. To better appreciate the time scale, let us imagine the Earth's 4.5-billion-year history condensed into a single 24-hour day. In this scenario, life first appeared as simple cells (prokaryotes) at five in the morning,, with photosynthesis evolving 2–3 hours later. Complex cells (eukaryotes) formed at 11 a.m., while multicellular life forms only appeared at the 18th hour. Continuing this imagined storyline, mammals have graced the planet for less than an hour, with modern humans spanning the final few seconds – a blink of an eye in the history of life. This book explores the rise of mammals, from the earliest egg-laying monotremes to the primates, including *Homo sapiens*, events that occurred during the last 60 minutes of the Earth's day-long history.

Years before present	Major event
4.54 billion	Formation of Earth
3.8 billion	Emergence of life on Earth
600 million	Last common ancestor (LCA) of all vertebrates
400 million	Tetrapods first venture onto land
233 million	Endothermy
231–66 million	Age of dinosaurs
225 million	*Brasilodon*
205 million	*Morganucodon*
187 million	Monotremes diverge from stem therians
140 million	Pangaea splits into two supercontinents
125 million	*Sinodelphys szalayi*
125 million	Metatherians diverge from eutherians
85 million	Laurasiatheria and Euarchontoglires
76 million	Solenodon
66 million	K–Pg mass extinction
64 million	Hyracoidae
63 million	Bats
60 million	Earliest known elephant, *Eritherium*

59.7 million	Monito del monte
58 million	Divergence of sirenians from proboscideans
56 million	Palaeocene–Eocene Thermal Maximum (PETM)
56 million	*Teilhardina* – earliest known crown primate genus
55 million	Divergence of platypus and echidna
48 million	*Indohyus*
40 million	Stem New World monkeys (platyrrhines)
40 million	Caviomorphs arrive in South America
39 million	Crown cetaceans
38 million	Honey possum
33 million	Crown New World monkeys (platyrrhines)
30–25 million	Apes diverge from Old World monkeys
26 million	Vampire bats
25 million	Titis, sakis and uakaris
18 million	Earliest known bovid, *Eotragus*
16.9 million	Middle Miocene Climatic Optimum (MMCO)
15 million	Quolls
12 million	Howler, spider and woolly monkeys
10 million	Gorilla
8 million	Divergence of African and Asian buffalo
8 million	Macropodidae descend from trees
6 million	LCA of chimpanzees, bonobos and *Homo sapiens*
6–5 million	Tree-kangaroos diverge from rock-kangaroos
5–3 million	Formation of the Panamanian isthmus
4.58 million	LCA of genus *Panthera*
3.65 million	Jaguar
3.47 million	Snow leopard and tiger
2.57 million	Leopard and lion
2 million	LCA of chimpanzees and bonobos
1.7 million	LCA of extant uakaris
1.6 million	Emergence of human speech
400,000	Polar bear
400,000–40,000	Neanderthals and Denisovans
300,000	*Homo sapiens*
160,000–60,000	Dispersals of *Homo sapiens* out of Africa
45,000	*Homo sapiens* reach Australia
16,000	*Homo sapiens* arrive in North America
12,000	*Homo sapiens* colonise South America
12,000–10,000	Agricultural Revolution
10,500	Domestication of cattle
7,000	Domestication of camels

Geological Ages

Era	Period	Epoch	Mya
Cenozoic	Quaternary	Holocene	0.01
		Pleistocene	2.5
	Neogene	Pliocene	5.3
		Miocene	23
	Palaeogene	Oligocene	33.9
		Eocene	56
		Palaeocene	66
Mesozoic	Cretaceous	Late	
		Early	
			145
	Jurassic	Late	
		Middle	
		Early	201

K–Pg Boundary (at 66 Mya)

Mya, million years ago

PROLOGUE

Linnaeus's legacy

God created; Linnaeus organised.
Carl Linnaeus (1707–1778)

Under the scorching midday sun, with our shadows all but gone, we drifted out on the shallow waters of the Laguna San Ignacio. A white, over-exposed desert harshness defined our horizon, while towards the southwest, the Pacific Ocean lay reassuringly azure under a cloudless sky. Once our panga's outboard fell silent, my fellow travellers settled into an easy reverie, lulled by the rhythmic lap of water and the occasional call of a distant seabird.

Laguna San Ignacio is one of several secluded inlets on the western shore of the Baja California peninsula, a place that remains a timeless haven of peace and tranquillity. Our group had gathered here for one reason: to experience a close encounter with one of nature's true leviathans, the grey whale. These mammals use the secluded waterway in the winter months to mate, give birth and care for their newborn calves before migrating to remote feeding grounds off Alaska. A few individuals even migrate as far as Sakhalin Island in Russia, the longest-known migration of any mammal.[1] Laguna San Ignacio and several other Baja coastal inlets are now recognised as critically important for the species' survival.[2] Thankfully, with the support of the Mexican government, the local population has ensured that these waterways have become federally protected marine sanctuaries, a development crucial for the future of the grey whale. Already, these measures have yielded results, and help to explain the lagoon's popularity with today's naturalists and eco-tourists.

As we floated beneath the cloudless sky, it was not long before we noticed a female whale nudging her inquisitive calf towards our panga. With mounting excitement, we leaned over the gunwale and watched as the youngster surfaced and gently brushed against the side of the boat. My scientific training counted for little, and any resolution to avoid anthropomorphising vanished instantly. Indeed, I quickly became convinced that the calf was directly communing with me alone, gazing with her rheumy eyes, questioning who I was and what I was doing in her remote marine hideaway. Like those around me, I couldn't resist the temptation to stroke the creature and poke its skin. The texture was unexpected, like a smooth, well-insulated wetsuit – soft but spongy – with little evidence of the barnacle 'hitchhikers' that adorned her mother. To my amazement, the calf pushed its head further out of the water, seemingly for greater interaction, and appeared to relish the moment's intimacy. Without warning, the young whale vanished beneath our boat, only to resurface on the opposite side. Then, with its double blowhole open, it exhaled with a mighty whoosh, drenching our excited group with a fine mist of seawater and nasal secretions. It was a magical experience, a humbling, one-of-a-kind moment, a wildlife encounter that resulted entirely from innate calf curiosity rather than any human encouragement or harassment. To this day, my impressions are just how human the infant was – curious, playful, and confiding. And I do not doubt we were sprayed on purpose and that the calf may even have smiled afterwards.

The whale encounter in Laguna San Ignacio remains one of my most memorable wildlife experiences. The calf's human-like behaviour intrigued me and made me ponder our evolutionary relationships. Although we are both mammals, how have we evolved and adapted to such different environments? Furthermore, what exactly are mammals, and when did they first appear? What are the evolutionary stories of all the other 6,400-plus species of living mammal – a class that includes such diverse forms as the platypus, kangaroos, sloths, armadillos, anteaters, elephants, tenrecs, bats, cats, rats, lemurs, apes and humans? Their unparalleled diversity is reflected in the range of sizes, from the 2-gram bumblebee bat to the blue whale, some 80 million times larger. Finally, how do all the myriad species relate to each other, and to us, and how did we all evolve from a common ancestor? These are a few of the many questions I shall attempt to answer in this book, an account that explores aspects of our shared heritage that are both profoundly important and little known until recently.

Mammalian taxonomy

The Greek philosopher Aristotle (384–322 BCE) undertook the very first taxonomic foray, attempting to classify creatures according to their behaviour and physiological similarities and differences. Such musings presaged the concept of *Mammalia*, since Aristotle recognised a distinct group of air-breathing, viviparous 'quadrupeds' with characteristic dentition and digital form. Furthermore, he stressed that 'the viviparous are such as man, and the horse, and those animals which have hair; and of the aquatic animals, the whale kind as the dolphin and cartilaginous fishes.' He also commented that all such creatures possessed teeth, and their varying arrangements could aid identification. Concerning digital structure, he observed that 'some are many-cloven, as the hands and feet of man', while 'some are many-toed, as the lion, the dog, the panther'. Others are 'bifid or have hoofs instead of nails, as the sheep, the goat, the elephant, the hippopotamus', while yet more have 'undivided feet, as the solid-hoofed animals, the horse and ass'.[3] Aristotle then used these groupings to show a ladder of the simplest organisms moving upwards to the highest, his *scala naturae*, or 'Great Chain of Being'.

Aristotle's nascent ideas relating to the classification of life lay undeveloped for centuries, only to be replaced by hearsay and mythology until well after the Middle Ages. Later attempts often appeared arbitrary or even whimsical, and it was not until 1735 that taxonomy was formalised by the ambitious Swedish botanist Carl von Linné (1707–1778), better known as Carolus Linnaeus. The young Swede was obsessed with collating the natural world and devised a binomial system in which every organism is assigned a unique two-part scientific name, the genus and the species, that remains in use today.[4] He also arranged all living creatures in a hierarchy of family relationships, from kingdom and class to order and genus and ultimately to the species itself. Such ideas, mostly completed while Linnaeus was still a student, were published in a short work entitled *Systema Naturae* in 1735. However, it was not until the larger tenth edition, published in 1758, that he introduced and defined Mammalia, a class that includes humans and other milk-producing animals. Not surprisingly, Linnaeus agonised over the inclusion of humans, as the belief that we are just another animal was highly controversial in the eighteenth century.

Nevertheless, Linnaeus became convinced that human beings, which he named *Homo sapiens* ('man of wisdom'), belonged with the apes. Previously, we had always been at the centre of the natural world, and it's hard for us to imagine how revolutionary Linnaeus's ambivalence was. It was only to appease the church that he placed us in splendid isolation as the sole member of our genus, albeit within the Primate order. Despite the predictable ecclesiastical outcry, Linnaeus's approach paved the way for other naturalists and scientists to investigate the similarities further, laying the foundations for our understanding of human evolution (Chapter 25).

Mammalia means 'of the breast', and although mammals have many other features, it was suckling that Linnaeus highlighted to define our group. In doing so, he made the female mammary gland the symbol of our biological class. Strangely, of his six major divisions – Mammalia (mammals), Aves (birds), Amphibia (reptiles and amphibians), Pisces (fish), Insecta (insects and arachnids) and Vermes (worms) – it was only Mammalia that he classified using a sexual characteristic, and even then, one restricted to the female sex.[5]

So why did Linnaeus choose *Mammalia* over alternative and more inclusive terms such as *Pilosa* (the hairy ones) or *Triossauria* (those with three ear bones) that apply to both sexes? One suggestion proposed by Londa Schiebinger, Professor of History of Science at Stanford University, is that external pressures, especially social ones, influenced Linnaeus greatly.[6] She argues that the Swede, a practising physician and father of seven children, 'venerated the maternal breast' in an era when the benefits of breastfeeding were starting to be appreciated. Linnaeus also championed the struggle against wet nursing and stressed the benefits of women breastfeeding their own babies. Such views coincided with political realignments underlining women's public power and attaching a new value to women's domestic roles. His efforts were not in vain, as breastfeeding in the late eighteenth and early nineteenth centuries became much more acceptable across Europe as attitudes to the welfare and health of mothers and children changed.

It was never Linnaeus's intention to imply any evolutionary relationship between his different groupings. His classification was merely an attempt to understand the natural world to reflect God's logic. Indeed, he regarded himself as the second Adam and is quoted to have said, 'God created; Linnaeus organised.' It is insightful that his *opus magnum*'s frontispiece depicts the author in the Garden of Eden, applying names to an array of freshly created creatures. Later, speculation about species' malleability arose as European explorers and collectors returned with a wealth of animals that revealed nature's bewildering diversity. It was a debate that would culminate a century later in the controversial ideas of Charles Darwin and Alfred Russel Wallace, in which species are not fixed or static but evolve over geological time.

Mammalian diversity

All mammals share specific traits besides their milk-producing mammary glands – a large forebrain, warm-bloodedness and a high metabolic rate, a single lower jawbone or dentary, and three middle-ear bones (auditory ossicles) for conducting sound. Other features, while occurring in most species, are not universally present, such as viviparity (monotremes lay eggs), molar teeth designed for chewing food (absent in baleen whales and anteaters) and insulating hair or fur (absent in pangolins). The evolutionary success of these innovations is reflected in the myriad of mammalian lifestyles: jumpers, runners, burrowers, swimmers, tree dwellers, gliders and flyers. Every imaginable kind of food sustains one species or another: meat, fish, krill, vegetation, fruit, nuts, ants, termites, and even nectar and pollen. They have also successfully colonised all seven continents, including the harshest environments, whether oceans, deserts, tropical forests or mountains.

But how many species of mammal are alive today? This question might seem straightforward, but it is not. A recent publication lists 6,399 species but concludes it is likely to be a significant underestimate. Given the current rate of increase, approximately 25 species per year for the last 15 years, it is predicted that there will be 7,342 mammal species recognised by 2050 and 8,590 by 2100. Indeed, with the increasing use of more sophisticated DNA analysis techniques, these projections may be on the low side.[7] Nevertheless, the message is clear: Linnaeus's Mammalia contains a considerably greater species diversity than is widely recognised.

But how are scientists identifying so many new species? A few discoveries are evident, even in the field, as they differ morphologically from any known taxa. Examples include a blossom bat (*Syconycteris* sp.) that feeds on rainforest nectar, discovered in the Foja mountains of West Papua in 2010, and the Vangunu giant rat found in the Solomon Islands in 2017. However, molecular analysis enables phenotypically cryptic but genetically divergent evolutionary lineages to be recognised. Since 2004, over half of all new species have resulted from taxonomic splits based partly or solely on genetic data. For example, Bryan Carstens from Ohio State University, working with his then graduate student Ariadna Morales, revealed in 2018 that the little brown bat, widespread across North America, is, in fact, five separate species.[8] Further work by the same group suggests that many cryptic species await discovery and are likely to be hidden in predictable places, especially if small-bodied with large, climatically variable ranges, such as bats, moles, rats and shrews.[9]

Given the wealth of available phylogenetic data, it has become almost impossible to fully visualise the complexities of the mammalian tree of life, let alone one for every known living organism, as it would require a vast sheet of paper or multiple computer screens. However, the development of an interactive website called OneZoom has made the task easy. The program allows the interrogation of a wealth of data by laying it out in ever smaller bubbles using a fractal structure and a zooming interface so that the computer never runs out of space. Each leaf on the tree represents a different species, and the branches show its relatedness to other species and how it has evolved over millions of years. Each leaf is also colour-coded, from green to red, to convey the species' vulnerability to extinction, although many leaves remain grey due to our lack of knowledge. Type in your species of interest, from platypus to human, and within seconds of hitting the search button you will have travelled along the tree's complex network of branches to your target animal. I encourage readers to explore this valuable resource (OneZoom.org), as it is free and fun and makes the mammalian tree of life understandable in a way that only years of study could have done in the past.

A molecular approach

The present work is a collection of 25 self-contained chapters or stories covering the evolution, dispersal and speciation of all the world's 6,400-plus known species that constitute Linnaeus's Mammalia – from the egg-laying monotremes and the pouch-bearing marsupials of the austral continents to the ubiquitous placentals, including the large-brained primates. Human evolution is covered in the final story, with the same level of coverage as other key mammals, in line with Linnaeus's realisation that we are but one of many wondrous species. However, given Mammalia's wealth of diversity, I have had to be ruthless in my selection of species discussed. Therefore, the responsibility for any omission or commission lies with me alone. Many of the issues discussed are still subjects of intense and sometimes passionate debate, with experts holding many disparate and conflicting views. For the sake of brevity and readability, I have resorted to consensus science and restricted the discussion of alternative hypotheses.

I have not dwelt on our earliest ancestors, the so-called pre-mammals – pelycosaurs, therapsids and cynodonts – as their accounts are covered comprehensively in several recent publications.[10] While fossils provide unique clues about the existence and morphology of the wealth of past species, I have only dealt with a small selection, limiting the discussion to those that help us understand the evolution, distribution and adaptations of the species alive today. To further explore the impressive variety of these ancient bones and the stories they tell, I refer readers to popular accounts by the American palaeontologists Donald Prothero and Steve Brusatte.[11]

In contrast, genetics or the genetic code will be this book's leitmotiv, a Rosetta Stone that has enabled the deciphering of many facets of mammalian evolution. I will focus on the recent cascade of novel and often unexpected insights from such studies, mind-blowing advances that show no signs of slowing. Indeed, since the success of the Human Genome Project in 2003, the cost and speed of sequencing DNA have improved so much that the Earth BioGenome Project has already embarked on the next 'moon-shot for biology', the genomic sequencing of 1.5 million eukaryotic, or multicellular organisms.[12] By 2020, the Zoonomia Consortium had already published the genomic sequences for 240 mammalian species, including representatives from most major families.[13]

But to appreciate the code's evolutionary secrets, we need to understand the structure of DNA and how the complex 'molecule of life' replicates – features first revealed in a famously brief paper by Watson and Crick in 1953, prosaically entitled 'A structure for deoxyribose nucleic acid'.[14]

The two scientists deduced that each DNA molecule possesses two strands that intertwine to form a double helix. Each strand has a backbone of alternating sugar and phosphate groups, and one of four possible bases, or nucleotides – A (adenine), T (thymine), C (cytosine) and G (guanine) – is attached to each sugar and faces inwards. Using cardboard cut-out models, they realised that an A on one strand could only bond or pair to a T on the other. In contrast, a C could only bond to a G. Crucially, the formation of these so-called base pairs (A-T and C-G) means that the order on one DNA strand must precisely specify the order on the other strand. Watson and Crick's momentous discovery immediately suggested how DNA must replicate. When a new DNA molecule is required, the two strands of the helix split in two, like the zip on a jacket, and each half acts as a template, facilitating the production of a mirror image or complementary strand of itself.

The enzyme DNA polymerase is essential for the molecule's replication, and it does so, one base at a time, using the single strands as templates. While highly accurate, this process occasionally suffers from errors. Such mistakes or mutations can arise from external insults, such as radiation or chemicals, or from an intrinsic error in the genetic replication process. The effects may vary from single base substitutions to multiple base changes, including random breaks, duplications and deletions. Mutations can be silent and harmless or catastrophic and fatal. However, a few, especially if they involve genes encoding critical proteins, may provide a survival advantage by altering the animal's morphology, physiology or behaviour. These changes are then subjected to natural selection and, if advantageous, are more likely to be passed on to the next generation and subjected to further selection pressures. Most modifications occur during the normal reshuffling processes of reproduction and account for the genetic uniqueness of every individual. Indeed, without such random genetic mutations, there would be no germline heterogeneity and no opportunity for adaptation and eventual speciation. In other words, mutations are the raw materials of evolution.[15]

While Watson and Crick's paper was a Nobel Prize-winning breakthrough, they had yet to determine how DNA's genetic information leads to protein synthesis. It turns out that a species' DNA acts as an instruction manual or genetic blueprint for its unique development. The order of a gene's four bases – A, C, G and T – is not random but occurs in a precise sequence of triplets, or codons, each of which codes for one of the 20 amino acid units that make up proteins. These linear codes are transcribed into RNA (ribonucleic acid) and then translated within cellular factories, or ribosomes, to produce specific proteins. A gene's base order dictates the amino acid sequence of the corresponding protein, and any base change or mutation will alter the complementary RNA sequence and produce a protein with a different amino acid composition. Such changes may lead to proteins with new functions, potentially with dramatic consequences for evolution.

A detailed knowledge of the genetic code of mammals (or any other group of organisms) has enabled evolutionary biologists to probe aspects of their distant past. As we will discuss, comparing the base sequences of different species allows the construction of their family tree or phylogeny. Previously, less robust phylogenies were constructed using the anatomical similarities and differences of living creatures (comparative morphology), fossil records and embryological clues. DNA changes can also help determine the timing of a divergence or speciation event and provide clues for a species' ancestral dispersal route when coupled with palaeontological and geological evidence. Lastly, sequence data from expressed genes, those stretches of DNA that code for proteins, and their genetic switches, can reveal mechanisms underpinning a taxa's phenotypic evolution and adaptation.

So let us look at each of these areas in turn.

Phylogenetic trees can be constructed by lining up and comparing the order of DNA's four bases from the same genomic region from different species. Since an animal's whole genome must be copied before it is inherited, and as DNA replication is imperfect, changes will inevitably occur. Should these changes, or mutations, survive, they will accumulate over time, and the gene sequence will slowly drift away from its original configuration. Indeed, as highlighted above, if genes are changeless, taxa cannot evolve. This observation implies that different species must be closely related if only a few differences exist in their base sequences. In contrast, the presence of many changes implies a more distant relationship.

By analysing DNA from a cohort of species and documenting their differences, it is possible to trace their line of descent back to a distant shared ancestor. Although this approach cannot provide definitive answers, it does reveal the most likely hypothesis based on statistical probability. The first DNA-based genealogies became available in the late 1990s and early 2000s, showing that many long-accepted kinships were fictitious, mere illusions of anatomical convergence. For example, the near-identical anatomical features of many 'ant-eating' mammals are no longer thought to imply common ancestry but instead reflect independently derived adaptations for survival in similar ecological niches, a condition known as convergent evolution. Unlike palaeontology and comparative anatomy, DNA analysis can often provide sufficient clues to untangle convergence from common ancestry. As a result, many branches of the mammalian tree of life have had to be pruned and regrafted, including those for bats, pangolins, tenrecs, and many hoofed and insectivorous species.

The analysis of DNA sequences can also provide clues to the timing of evolutionary events.[16] As specific genetic changes occur at a steady rate per generation, scientists use them to estimate the time that species diverged from a common ancestor. Such genetic changes, or mutations, accrue like the ticks on a stopwatch and provide palaeontologists with a 'molecular clock'. These evolutionary timepieces have become increasingly sophisticated, thanks to improved DNA sequencing, better computational algorithms, and a greater knowledge of the factors contributing to genetic change. The potential for error can be further reduced by calibrating molecular-based phylogenies using accurately dated fossil records.[17] By applying such techniques, geneticists have constructed increasingly robust timelines using the growing database of mammalian DNA, both extant and ancient. Unexpectedly, they reveal that species can evolve long before the dates of their first fossils, suggesting an unrecognised antiquity for many lineages.

Integrating data from genetic phylogenies and molecular clocks with palaeogeography provides significant insights into the continental origins of various mammalian groups and their probable dispersal routes. For instance, it is now established that the earliest mammals, known as monotremes, evolved in Gondwana and dispersed to Australia before the fragmentation of this southern supercontinent. Subsequent declines in sea levels facilitated their migration northwards to New Guinea. In contrast, marsupials originated in the northern hemisphere and undertook an extensive dispersal journey through South America and Antarctica to reach Australasia. Correlating molecular dates with geological events has also underscored the importance of the tectonic separations of Africa, Central and Southern America, Europe and Asia in the early evolution of placental mammals. Moreover, molecular clocks corroborate that transoceanic dispersals were crucial in the early history of New World primates and Madagascar's endemic mammals.

By comparing the sequence of genes from different species, it is possible to determine what distinguishes one from another at the molecular level. This approach, termed comparative genomics, has proven invaluable for studying evolutionary change. It highlights genes conserved or common among species and those that differ and contribute to an organism's uniqueness. Today, the availability of near-complete and error-free long-reading sequences enables precise comparisons that were not possible a decade ago, facilitating the assembly of 92–96 per cent of genomes into chromosomes.[18]

We will also investigate how specific genetic processes have played a role in evolution, including gene duplication (lactation), genomic imprinting (live birth), retroviral genes (placenta), retrogenes (elephants), pseudogenes and regressive evolution (sloths), gene loss (cetaceans), and inter-species gene transfer (*Homo sapiens*). Additionally, we will explore how mutations that influence gene function – by turning them on or off or modifying their expression levels – can significantly impact embryonic development and evolution. While most human genes are similar to those of other primates, differences in gene expression are responsible for many of our key phenotypic traits, such as larger brains, bipedalism, and changes in facial and limb structure.

The last common ancestor (LCA)

Our story of the *Ascent of Mammals* begins at an unknown location on Pangaea, a supercontinent that contained nearly all the landmasses we recognise today. Approximately 200 million years ago, in the Late Triassic/Early Jurassic, the last common ancestor (LCA) of all living mammals scuttled amid the continent's horsetails, ferns, and fallen cycad leaves. However, the species is a theoretical one. It doesn't have a scientific name, and we can only guess at its morphology and behaviour, for no fossils are known. Indeed, the LCA of mammals may remain hidden forever.

Twenty-five million years earlier, however, a population of 'near' mammals, the nocturnal *Brasilodon*, emerged from burrows to forage while their cold-blooded predators awaited the sun's return. Like us, the shrew-sized *Brasilodon* grew two sets of enamelled teeth, a deciduous or milk set that developed during early life and a second and permanent set in adulthood.[19] Reptiles, in contrast, generate teeth throughout life to replace damaged ones. The replacement of deciduous by adult teeth, or diphyodonty, is a complex and unique mammalian trait. Milk teeth are poorly formed, irregularly spaced, and fail to occlude properly, features that enable breastfeeding and preserve space for the adult teeth as the jaw grows. However, their presence implies the coexistence of other mammalian traits: a hard palate (you can't suck and breathe simultaneously without one), placentation, warm-bloodedness, fur, lactation, suckling, and parental care. It is impressive what palaeontologists can infer from just three fossilised jaw bones!

However, the complete set of mammalian characteristics didn't evolve all at once but accrued piecemeal over tens of millions of years of trial and error. Hence, identifying the precise point at which 'near' mammals became 'true' mammals is an unrealistic endeavour likely to remain out of palaeontological reach.

While no actual fossil evidence for the LCA exists, its chromosomal composition has been deduced. In 2022, a global team of scientists computationally reconstructed its genomic organisation and showed that it possessed

40 chromosomes (38 autosomal and two sex chromosomes). Furthermore, nine whole chromosomes or chromosome fragments had the same order of genes as found in modern birds.[20] Intriguingly, over 1,000 blocks of the LCA's genes were located on the same chromosome and in the same order as found in all mammals today, from monotremes to humans – areas that contain genes critical for normal embryonic development. This remarkable finding reveals that many mammalian genes have remained stable for over 300 million years of evolution. In contrast, the intervening regions with more repetitive sequences were prone to breakages, rearrangements and duplications that have driven mammalian development.

To differentiate 'near' mammals from modern mammals, scientists use wonderfully arcane nomenclature – 'mammaliamorphs', 'basal mammaliaforms', or even 'non-mammalian mammaliaforms' – descriptions only understood by the cognoscenti. However, to emphasise the need for such terminology and the difficulty of identifying the first 'true' mammal, let me introduce the genus *Morganucodon* ('Glamorgan tooth', named after the region of Wales where remains were first discovered) (Plate 1).

Like *Brasilodon*, these diminutive creatures were nocturnal and insectivorous, but they appeared slightly later, around 205 million years ago.[21] They laid small, leathery eggs and had venomously spurred feet like the modern platypus. Their genus was clearly widespread throughout Pangaea, as fossils are known from China, Europe and North America. In 2011, a team from the University of Austin in Texas, led by Timothy Rowe, deduced its lifestyle using high-resolution X-ray computed tomography (CT) of its 2.5-centimetre-long skull. The resultant three-dimensional images allowed the construction of a virtual 'endocast' of its underlying brain. The scans revealed enlarged areas for smell (prominent frontal olfactory lobes), touch (from body hair) and motor coordination.[22] Such observations led Rowe and his colleagues to conclude that the morganucodons had whiskers, lived underground in burrows, and foraged amongst the leaf litter at night, looking for insects and grubs.[23]

Despite having a mammalian skeleton and teeth and possessing heightened olfactory and tactile senses, the morganucodons still retained a degree of reptilian physiology and anatomy. They were not as warm-blooded as modern mammals and had a longer lifespan, like reptiles.[24] Furthermore, their lower jawbones still retained features found in their non-mammalian ancestors. Intriguingly, the three ear ossicles of the middle ear, a *sine qua non* of modern mammals, evolved later and on at least two separate occasions – once in the early monotremes and once in the common ancestor of the marsupials and placentals.[25] Therefore, a common ancestor of all extant mammals with three middle-ear bones cannot have existed. Nevertheless, although the LCA of every mammal alive today may never be found, the morganucodons and *Brasilodon* were undoubtedly very similar.

Let us return to endothermy so as to underscore the stepwise evolution of mammalian characteristics. In recent times, scientists have pinpointed the moment when our ancestors transitioned to warm-bloodedness within the pre-mammals (mammaliamorphs) that inhabited the Late Triassic 233 million years ago. The discovery was made by studying the intricate canals within the inner ear, responsible for transmitting balance-related information to the brain. In cold-blooded animals, the endolymph fluid within the tubes behaves like syrup, necessitating wider tubes to function. Conversely, the endolymph is more fluid in endothermic mammals and accommodates narrower tubes. Researchers from Lisbon University used the diameter of these tiny tubes as a proxy for body temperature, inferring that pre-mammals exhibited a sharp increase in temperature of 5–9 degrees Celsius 233 million years ago.[26]

From our vantage point, these pocket-sized, nocturnal insect-eaters may seem puny and inconsequential in a world dominated by the mighty dinosaurs. However, appearances can be deceptive. Their novel combination of traits would be their golden ticket, enabling their descendants to survive the end-Cretaceous apocalypse that would befall the Earth some 150 million years in the future.

From the Jurassic world of Pangaea, our hypothetical LCA of all mammals evolved along a wondrous path of contingency and inevitability, convergence and divergence, and trial and error. As we will see, it was a journey shaped by continental drift, multiple climatic upheavals, changing ocean currents and extinction events, one that has led a population of small furry creatures to become the intelligent, self-aware primates reading this page. I hope the present story of our 200-million-year ascent and how science has revealed it will open your eyes to the wonders of mammalian evolution and its resultant diversity.

PART ONE

Monotremes

CHAPTER 1

The Platypus's Story

MONOTREMES OVIPAROUS, OVUM MEROBLASTIC

The gene's description of ancestral worlds is overlain by modifications and detailed refinements scripted since the animal was born.
Richard Dawkins

Captain John Hunter, the second governor of New South Wales, should never have been appointed to his post. He spent too much time observing and collecting flora and fauna and not enough attending to his administrative duties. As a self-taught naturalist, Hunter devoted many hours to his obsession, exploring the hinterland around Sydney. Indeed, he is best remembered for one such foray in 1797, in the company of an indigenous hunter, to a muddy lagoon near Hawkesbury, north of the settlement. From the bank, he watched as the local man patiently tried to spear an odd-looking creature that repeatedly surfaced to breathe. The Englishman's curiosity was piqued – the animal was unlike anything previously seen by a European (Plate 2).

Hunter was a member of several learned English societies and was eager to hear their opinion. And so, once in possession of a decent specimen, he preserved it in a barrel of alcohol and shipped it to the Literary and Philosophical Society in Newcastle-upon-Tyne in 1798. At first, its members thought it was a joke and that the animal was a hoax, given that the specimen possessed fur, lacked conspicuous mammary glands, sported a bill, and had webbed feet and no teeth. Furthermore, Hunter informed his English colleagues that the animal foraged in water and, according to native lore, laid eggs and nested like a bird. One can imagine the society members guffawing in amusement, smug in the knowledge that none of God's creations could exhibit such an enigmatic mélange of traits – mammalian, reptilian, avian and piscine. Hunter knew better but still thought its origins resulted from 'a promiscuous intercourse between the different sexes of all these different animals.'

A second specimen reached the English doctor-turned-zoologist George Shaw at the British Museum. Despite writing 'it naturally excites the idea of some deceptive preparation by artificial means', Shaw accurately described the animal in 1799, giving it the name 'duck-billed platypus'.[1] It was a remarkable paper, as its conclusions relied on a single dried skin with a desiccated and hardened bill, unlike the pliable soft bill of the living animal.

As the colonisation of eastern Australia increased and settlers moved further afield, platypus sightings became more common, quelling any lingering doubt about their authenticity. The creature inhabited creeks and lakes in various habitats, from Queensland's tropical rainforests to Tasmania's glacial tarns. It swam effortlessly through the water, propelled by its webbed front feet with its hind feet and tail acting as brake and rudder. When feeding, it closed its eyes, nose and ears and probed with its bill for insect larvae and other small invertebrates amongst the water's muddy sediments, where it could feed for two to three minutes at a time. The colonists also noted that the females occasionally ventured onto dry land to dig burrows. These could be of two types: a resting place for both sexes and a more extended, complex structure where females raised their young. Although awkward on land, the platypus is adept at excavating tunnels using its strong forepaws and powerful claws. The female then lined her nesting chamber with wet leaves, twigs and vegetation transported between its hind feet and tail. The males, unlike females, possessed spurs on their ankles, conduits for potent venom used to see off rivals during the breeding season.

A full anatomical description had to await the arrival of a whole animal rather than simply a skin. In 1802, the distinguished English surgeon Sir Everard Home received two such specimens – a male and a female – and immediately set about their dissection. With great skill and patience, he discovered that the female reproductive organs are of a 'very extraordinary nature'. There were 'no nipples, and no regularly formed uterus', findings more in keeping with lizards than mammals. Furthermore, he discovered that the reproductive organs and ureters opened into the rectum in both sexes. Such anatomical features implied that the platypus possessed a cloaca (derived from the Latin *cloaca* for public sewer or drain), a multipurpose orifice used for defecating, urinating and reproduction, 'as in the bird'. He also reported an unusual interclavicular bone near the shoulder, now termed a furcula – the 'wishbone' of domestic fowl. Finally, after noting that the trigeminal nerves supplying the face were 'uncommonly large', Home made the astute deduction that the bill must be a major sensory organ.[2]

A few years before the arrival of the first platypus skin in England, George Shaw had received another puzzling specimen to study, also from Australia. The animal in question was first documented on 9 February 1792 by George

Tobin, one of Captain Bligh's officers on HMS *Providence*. He wrote that, during a banquet, they ate a most unusual creature:

> *A kind of sloth about the size of a roasting pig with a proboscis two or three inches in length ... on the back were short quills like those of a porcupine ... The animal was roasted and found of a delicate flavour.*[3]

Unbeknown to the naval officers, they had dined on an echidna, a species new to science (Plate 3). Later that year, George Shaw formally described the animal in his *Naturalist's Miscellany*, a multi-volume production that illustrated all newly discovered fauna worldwide.[4] He recorded that it was 'amongst the most curious and interesting quadrupeds yet discovered'. It possessed a long tubular snout, with small nostrils at the tip, lacked teeth, and had a cylindric, long extensile tongue. The whole upper body and tail were 'thickly coated with strong, sharp spines, of considerable length', while the head, legs and underparts were 'thickly coated with strong, close-set, bristly hair'.[4] Shaw gave it the common name 'porcupine ant-eater', believing it was a cross between the common porcupine and the giant anteater. While its taxonomy was confusing, he eventually lumped it with the anteaters. However, unlike Home, Shaw was not an anatomist and failed to notice that the echidna possessed a single rear orifice like the platypus.

It was only later that the close anatomical relationship between the platypus and the echidna was recognised, resulting in the French zoologist Etienne Geoffroy Saint-Hilaire formally placing them together as the Monotremata (meaning 'one-holed animals', referring to the fact that they both have a cloaca). However, Hilaire could not assign the animals to any zoological series, and it would take many years before their place on the 'tree of life' was determined (Figure 1.1).

Oviparous or viviparous?

Early anatomical studies failed to answer one of the nineteenth century's most hotly debated zoological questions. Did monotremes lay eggs, as the indigenous peoples attested, or did they give birth to live young? The puzzle needed solving, as the answer had fundamental implications for mammalian evolution.

Many contemporary scientists couldn't countenance the idea of an egg-laying or oviparous mammal, as this would give credence to evolutionary theory and the transmutation of species. As a result, the French anti-evolutionist Henri de Blainville predicted viviparity, or live birth, and insisted

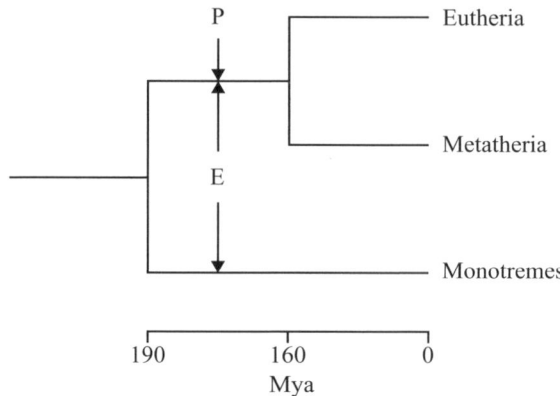

Figure 1.1 Evolution of the mammalian placenta (P) and the three middle-ear ossicles (E). Note the independent origin of the mammalian middle ear in the therian and monotreme lineages.

that no one would ever find eggs. In contrast, early evolutionists, including Jean-Baptiste Lamarck and Etienne Geoffroy Saint-Hilaire, were convinced that the platypus's anatomy implied oviparity and that it must lay eggs. To add to the often-vitriolic debate, the petulant English anatomist Richard Owen suggested ovoviviparity, meaning that the absence of eggs was due to their dissolution within the female's reproductive tract.

The argument was finally settled in 1884, some 86 years after the first platypus skin had arrived in England. William Hay Caldwell, an ambitious young Scottish zoologist, had sailed for Australia the year before to solve the problem in grand style. With the aid of more than a hundred locals, he killed and dissected a vast number of monotremes. Despite labouring for almost a year, he finally recorded, 'I shot a [platypus] whose first egg had been laid; her second egg was in a partially dilated os uteri [the mouth of the uterus]'. With the mystery solved, Caldwell hurriedly announced his discovery with the most famous telegram in zoological history, wiring 'Monotremes oviparous, ovum meroblastic' to the annual meeting of the British Association for the Advancement of Science in Montreal. After the chairman had conveyed the message, the delegates stood and cheered. Caldwell's minimalist, four-word cable, unintelligible to the non-specialist, carried a wealth of meaning for the cognoscenti: the platypus lays eggs (monotremes oviparous), and the early stages of its embryo are yolky and resemble those of birds and reptiles, not mammals (ovum meroblastic).[5] Caldwell's discovery was timely, for it coincided with the 25th anniversary of the publication of Darwin's *Origin of Species*. Science had advanced, and the heretical ideas of evolution and intermediacy had been accepted by many.

Although recognised as true mammals since the 1830s, the evolutionary origins of monotremes remained unclear. Indeed, it was not until the 1970s and 80s that palaeontologists discovered fossils that showed the platypus and echidnas to be rare survivors of an ancient lineage that diverged from all other mammals around 187 million years ago. In other words, monotremes lie phylogenetically between the egg-laying reptiles and the therians (marsupials and placentals), which helps explain many of their unusual features – oviparity, egg-teeth, cloaca, interclavicle bone, reptile-like gait and ankle spurs. As a result, monotreme genomes have given scientists a unique window into the evolutionary innovations underpinning the reptile–mammal transition, including clues to the lifestyle of the last common ancestor of mammals.[6]

Let us start by exploring the evolution of two features that characterise all living mammals: the mammary gland and the three-boned middle ear.

'A scarcely nutritious fluid'

Monotremes are unique since, although they lay eggs, they possess mammary glands to nourish their offspring – an anatomical feature first recognised by Richard Owen many years before the confirmation of their oviparity. However, lactation is not by way of nipples but by the secretion of milk through numerous pores on the female's lower chest and abdomen. Monotreme hatchlings emerge from their eggs with the help of a transient egg-tooth that shares many similarities with reptilian teeth – a reflection of their shared evolutionary heritage.[7] Baby platypuses, or puggles, suck milk from their mother's fur, whereas young echidnas feed from two milk-producing areas within a temporary pouch on the female's abdomen. All monotremes are born immature and helpless – or altricial, in scientific jargon – and remain dependent on their mother's milk for up to 3–6 months.

The evolution of complex traits, including the mammary gland, is often raised as an argument against Darwinism. How can such an innovative organ arise from a small and seemingly insignificant beginning? Darwin's contemporary, St George Jackson Mivart, forcefully promulgated such concerns in his book *On the Genesis of Species* (1871):

> *Is it conceivable that the young of any animal was ever saved from destruction by accidentally sucking a drop of scarcely nutritious fluid from an accidentally hypertrophied cutaneous gland of its mother?*

In other words, Mivart could not accept that the many intermediate stages required between species lacking mammary glands and those possessing a complete milk-dispensing ductal system could be viable. To Mivart, intermediate stages didn't appear plausible, and he offered one example after another in support of what he viewed as Darwin's fatal flaw. Mivart's insightful criticism, one of the few concrete objections to the theory of natural selection, forced a response from Darwin in the form of an additional chapter in the sixth edition of his *Origin of Species*. Using evidence from the animal kingdom, Darwin wrote that intermediate stages of organ development, such as sensitivity to light in the case of the eye, had a proven utility. He emphasised that if you lack an eye, you are blind, but the presence of even a few rudimentary photoreceptors might provide a survival advantage. Indeed, no matter how slight the benefit, the genes for primitive structures are more likely to be inherited by future generations, enabling further changes and improvements.

Although Darwin was wrong about the origin of the mammary gland (he thought they evolved from glands found in the brood pouches of some fishes), he correctly argued that earlier forms of complex organs may not have had the same function as they do today, and that functional shifts are 'an extremely important means of transition'. Darwin supported his defensive argument with several convincing examples. Flotation bladders of fish gave rise to the lungs of terrestrial animals. Skull sutures in human neonates that enable birth through the pelvic girdle arose in pre-mammals as a simple by-product of cranial development, and feathers, which evolved for insulation or sexual display, were repurposed for flight.

The charismatic and erudite biologist Stephen Jay Gould and his colleague Elisabeth Vrba, from Yale University, termed this phenomenon 'exaptation' (derived from the Latin *ex*, meaning by reason of previous form, and *aptus*, fit) and emphasised its evolutionary importance.[8] Indeed, the more scientists looked, the more widespread and pivotal exaptation appeared. Readers, therefore, should not be surprised to learn that this process underpins our story of the evolution of the mammary gland and lactation.

As hinted, the mammary gland did not arise *de novo* but evolved from a pre-existing structure with a different function. According to Olav Oftedal, a nutritional ecologist at the Smithsonian Institution in Washington DC, this was the humble sweat gland, or, more precisely, the apocrine gland.[9] This structure forms part of a cutaneous complex in mammals, a widespread triad containing a single hair that the apocrine gland sweats along and a sebaceous gland that secretes sebum to help waterproof the skin. Oftedal's hypothesis also helps explain the association of hairs and mammary ducts in platypuses and echidnas, with the hairs forming tracks along which milk escapes. Furthermore, marsupial development has a transient association between mammary ducts and hair, a feature lost in placental mammals due to active repression mechanisms.[10] Interestingly, the secretory cells in both apocrine and mammary glands are surrounded by a single layer of smooth muscle cells. As its function changed, it is not hard to imagine the muscle layer in the apocrine gland undergoing modification to facilitate milk ejection. Darwin was right when he argued that evolutionary innovations have antecedents that extend deep in time. In other words, nothing ever begins when you think it does.

The evolution of the mammary gland in therian mammals, including humans, involved further exaptation, although not from pre-formed structures but from the hijacking and recycling of genes already operating before the emergence of mammals. These 'architect' or *Hox* genes control body plans during embryonic development, including organ and limb formation. In turn, complex networks regulate and control the *Hox* genes. After the divergence of the monotremes, a component of this ancient regulatory network was repurposed and used for a different function – to switch on the development of the mammalian bud that elongates and sprouts into the underlying tissue to form the complex ductal system of the mammary gland and its associated nipple.[11]

The mammary gland's function is to produce milk – but where did this nutritious substance come from? Oftedal believes that the forerunner was the increased abdominal sweat production required to prevent leathery eggs from drying out – a process that evolved well before the existence of true mammals. Indeed, fossil evidence indicates that more than 200 million years ago, some Triassic pre-mammals produced a nutrient-rich milk-like secretion.[12] The stimulus for its production was the leaky eggshells of the increasingly warm-blooded animals. For unknown reasons, the mammal lineage never developed hard calcified eggshells like birds, and their eggs were always at risk of desiccation. Pre-mammals overcame the problem by releasing extra secretions during incubation, and it's not hard to imagine such fluids becoming more nutritious and evolving as a food source.

Unlike mammals, bird and reptile embryos obtain all their nourishment from egg yolk, whose ingredients derive from a single set of proteins, the vitellogenins. These large molecules are released from the mother's liver and transport the essential components – amino acids, lipids, phosphorous and calcium – to the developing egg. Therian embryos, however, depend on the placenta, a complex organ which allows nutrients to cross from mother to fetus, and then, once born, milk from the mammary gland.

To understand how this transition evolved, David Brawand and colleagues, working in the Centre for Integrative Genomics at the University of Lausanne, compared the vitellogenin (*VTG*) genes from monotremes, marsupials and placentals with those of the chicken (an egg-laying, milkless control).[13] They confirmed that chickens have three functional *VTG* genes but noted that monotremes possess only one, while therian mammals have none. Although all three *VTG* genes are still detectable in mammals, they have become pseudogenes, useless DNA sequences that no longer produce protein. In effect, DNA mutations have sequentially turned off mammalian *VTG* genes during evolution, with the first occurring around 170 million years ago, the second 70–90 million years ago, and the last 30 million years ago. Given that the amount of protein produced is proportional to the number of functioning genes, these findings help explain why monotreme eggs contain some yolk but not as much as the eggs of reptiles and birds.

But nutrients lost from eggs must be found elsewhere. David Brawand's team found that the platypus's genome contains three genes for casein, the primary milk protein. Caseins bind calcium and functions similarly to vitellogenin by providing essential nutrients for growing young. Further studies indicated that casein genes arose in the common ancestor of all mammals between 310 and 200 million years ago, well before the evolution of the placenta. But where did casein genes come from? The answer is from tooth-enamel genes that duplicated, with the extra copies evolving later to produce caseins. Indeed, the tooth-enamel genes lie adjacent to the casein gene cluster in all mammals, including the monotremes.[14] Therefore, mammals were producing milk before they stopped laying eggs. It seems that lactation reduced the nutritional needs of the egg and contributed to its abandonment in marsupial and placental mammals in favour of the placenta. The result was that the genes associated with egg production mutated, becoming inactive pseudogenes without affecting the fitness of the mammalian lineages.

The evolution of life-supporting milk involved a dramatic increase in the secretion of calcium salts and protein. However, these modifications posed a problem: calcium phosphate is highly insoluble and, left unchecked, would precipitate out and block the mammary glands. Furthermore, two casein proteins found in milk tend to aggregate and produce toxic fibrils, or amyloid protein, compounding the problem. According to Carl Holt and John Carter from the University of Glasgow, evolution overcame these obstacles by co-secreting two ancestral caseins that can form tiny hollow microspheres or micelles. These soluble structures then sequestrate the troublesome caseins and salts, a fact that enabled the evolution of a more nutritious fluid without endangering the mother's fertility.[15]

Monotreme lactation is inherently non-sterile, as the lack of nipples means milk is expressed directly onto the animal's furry abdomen while it is underground in a germ-laden burrow. Since their young are born immature, blind, hairless and immunologically naïve, they are especially vulnerable to infection. However, selection pressure over millions of years has solved the problem. In 2014, scientists from Deakin University in Australia discovered that monotreme milk contains a unique protein with potent antimicrobial activity against a wide range of bacteria.[16] In collaboration with scientists from CSIRO, Australia's national science agency, protein crystals were successfully grown in the laboratory and subjected to X-ray studies to determine their structure.[17] What the team discovered surprised and excited everyone. The protein, called monotreme lactation protein (MLP), has a strange shape, probably related to its function, that has never previously been seen in over 100,000 known proteins. Its numerous folds form a series of tight ringlets, which led to the molecule being dubbed the 'Shirley Temple' protein after the young actress's hairstyle. According to Janet Newman, the lead scientist, the protein is so unusual that 'it's like being in a florist and seeing a completely new flower.' Another monotreme milk protein, echidna antimicrobial protein (EchAMP), has an action complementary to MLP and protects against spore-forming microbes in the soil.[18] Natural selection, it seems, has modified the monotreme's genome to protect their offspring until weaning with a potent combination of antimicrobial proteins. Interestingly, because these milk constituents are especially effective against bacteria that cause human skin and urinary tract infections, they may offer a new approach to the global threat of antibiotic resistance.

It is worth pondering further on the issues of calcium and protein precipitation in milk and the infection risk of suckling hatchlings underground. If pre-mammals had not solved these two developmental constraints, it is doubtful that our reproductive strategy of placentation and lactation could have evolved.

Ears and jaws

Birds and reptiles have only one middle-ear bone, the columella or stapes, that links the eardrum to the inner ear or cochlea. Although the stapes became increasingly thread-like and more sound-sensitive over time, there was a limit. To enhance performance further, mammals evolved a middle ear with three small articulating bones or ossicles. Vibrations in the tympanic membrane (eardrum) are picked up by the malleus and transferred to the stapes via the incus before being conducted to the cochlea (Figure 1.2). The role of the ossicles is to provide a lever effect that amplifies the sound and improves the impedance matching between the air waves and those of the fluid-filled cochlea. The fluid vibrations then trigger minute hairs, whose movements are converted to electrical impulses, which are transmitted to the brain by the auditory nerve for further processing. The result is the perception of sound. But where did the mammals' extra middle-ear bones come from?

Fossil data, comparative anatomy and developmental biology have provided the answer. The lower jaw of reptiles and pre-mammals consists

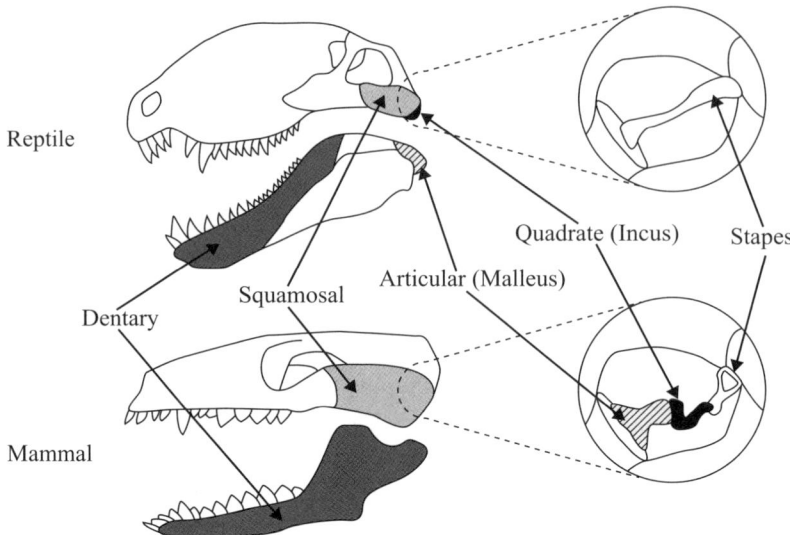

Figure 1.2 Evolution of the mammalian jaw joint and middle ear. Two reptilian skull bones, the quadrate and articular, shrank and moved posteriorly to become the incus and malleus of modern mammals.

of several bones: the large dentary in the front that carries the teeth, and several smaller bones that form the primary jaw joint between the mandible and skull. During the reptile–mammal transition, the dentary enlarged and moved back to create a separate hinge or secondary jaw joint. The result was that the smaller bones became redundant. Freed from their jaw duties, they migrated to the middle ear to become the malleus and incus, assisting the stapes in transmitting sounds more efficiently (Figure 1.2). However, the malleus remained connected to the jaw by several thin ligaments. This functional transformation provides one of the most iconic examples of exaptation in vertebrate evolution – the repurposing of jaw and skull bones as auditory bones. Remarkably, this spectacular evolutionary event occurred independently in the monotreme and therian lineages, suggesting a strong selection pressure for sensitive hearing.[19]

Recent evidence indicates that different pressures drove these changes. The first was the advantage of chewing and grinding more efficiently, whether the motion was side-to-side, up-and-down, or back-and-forth. Developing a single, solid, muscular jaw joint provided better-coordinated mastication and more efficient access to the array of novel foods that emerged during the Jurassic and Cretaceous. They included leaves, fruits, seeds and roots of flowering plants (angiosperms) and their associated wealth of pollinating insects. The second selection pressure was the extended range of hearing for higher-pitched sounds that aided the detection of insects, especially at night.

The enhanced hearing also encouraged communication and, with it, increased social interactions. Until the evolution of the mammalian ossicles, ear flaps or pinnae were likely absent, as they are mainly helpful in detecting and determining the direction of high-frequency sounds.

In placental animals, the skeleton of the secondary joint is fully formed before birth. However, monotremes and marsupials are born at a much earlier embryonic stage and before the adult joint exists. Until recently, it has been unclear how the neonates of these species move their jaws to feed on their mother's milk. Dr Neal Anthwal, a research associate at King's College, London, has spent his career probing the mammalian ear's evolution and development (evo-devo). By studying newborn monotremes, he discovered that they use their middle-ear bones to support and articulate the lower jaw with the skull until the formation of the adult joint is complete. It's only much later in life that the connection between the middle ear and the mandible is lost. Interestingly, the transitory coexistence of a primary and secondary joint in monotremes mirrors the adult anatomy of the pre-mammals, including the morganucodons.[20]

Anthwal believes that placental mammals may also use their ear bones to support the jaw during early embryonic development. This observation reinforces the evolutionary association of the jaw and ear. It is not surprising, therefore, that trauma to our jaw joint can cause dislocation of the ear ossicles (remember that the malleus remains attached to the mandible via ligaments). Similarly, rare developmental abnormalities of the jaw can affect the ear, such as in Treacher Collins syndrome, where the jaw fails to form and the ear bones attempt to take their place.

The development of the auditory ossicles had far-reaching effects, as they increased the mammalian ears' capacity for further modification, or 'evolvability'. According to an Austrian research team from Vienna, headed by Philipp Mitteroecker, the evolution of the three ear ossicles increased the genetic, regulatory and developmental complexity of the mammalian ear, which, in turn, markedly increased the number of 'dials' and 'knobs' for natural selection to tweak.[21] As we will discuss, this contributed to the evolutionary success and adaptive diversification of mammals that has led to the vast diversity of ecological and behavioural niches observed today – from the high-frequency echolocation of flying bats to the ultra-low frequencies used by marine mammals.

Teeth, venom and sex

In 2020, an international team of researchers led by biologists from the University of Copenhagen mapped the complete platypus and most of the echidna genomes for the first time.[22] Unlike past studies, they also allocated

96 per cent of the DNA sequences to individual chromosomes. The result is that we now have a much better understanding of how some of the bizarre features of monotremes evolved.

First, let's consider their teeth, or rather their absence, since adult monotremes, unlike most mammals, are edentulous. As puggles grow and lose their vestigial teeth, the resorbed tissues are replaced by two horny plates that mash their insect and crustacean prey. Adult echidnas also lack teeth but pulverise their insect diet between the bottom of their mouth and tongue, using spines on the tongue to improve food processing. How did the edentulous state in monotremes evolve, given that their ancestors possessed a full set of teeth?

The answer lies hidden in their DNA. Four of the eight genes involved in the development of tooth enamel are missing from platypus and echidna genomes. Another two genes were later lost from the echidna lineage after it diverged from the platypus. It appears that the four genes were knocked out in a distant monotreme ancestor approximately 120 million years ago when a dietary change would have provided strong selection pressure for their loss. The remaining genes, coding mainly for dentine, a tissue that lies beneath the enamel, were likely retained for egg and vestigial tooth development.

Monotremes also have fewer bitter taste receptors than most mammals, owing to a reduced need to detect harmful plant toxins.[23] A similar reduction is observed in pangolins, which suggests convergent evolution due to both lineages having similar insectivorous diets. Another bizarre consequence of changing diet is that the monotreme's oesophagus is connected directly to the intestine. They have no stomach, nor any of its powerful acids and protein-digesting enzymes. As a result, monotremes have lost many genes implicated in the function and action of gastric juice, an event that occurred, as with tooth-enamel genes, during the time of their common ancestors.

Venom was a widespread defensive mechanism during the 'dark age' of mammalian history when our ancestors were small, and dinosaurs ruled the Earth.[24] Indeed, it has been suggested that poisonous spurs may be a defining trait of the first mammals before being lost in the extant species. Today, venom is only present in the platypus unless one includes the toxic saliva of the solenodons (Chapter 9) and the slightly venomous bites of some shrews. Even in the platypus, however, its function has changed. During the breeding season, the male platypus releases venom from a hollow spur on the hind legs connected to a pair of crural glands – or modified sweat glands – on the abdomen. At other times, the glands atrophy, as the toxin's primary use is for fending off rival males rather than for defensive purposes. Even so, the poison can kill a dog, and humans accidentally injected suffer from immediate pain and swelling resistant to morphine-like drugs. Echidnas also have spurs and glands, but they secrete scent, not venom, and use it for chemical communication during the breeding season.[25] The echidna's change in lifestyle, from aquatic to terrestrial, and the development of a spiny coat, made the synthesis of metabolically demanding venom redundant, resulting in an altered function for its crural glands.

Analysis of the platypus's genome has revealed the origins of this most unusual of mammalian features. In conjunction with a team of international colleagues, Camilla Whittington from the University of Sydney identified three genes that code for the significant components of monotreme venom.[26] All three genes owe their origins to the duplication of genes that coded originally for a set of antimicrobial proteins called beta-defensins. Present in all vertebrates, defensins protect cell surfaces from microbial colonisation and may even have a role in protecting the platypus's immature offspring from infection. Intriguingly, snakes have also duplicated their defensin genes to produce toxins, suggesting that there may only be a limited number of protein structures that can act as templates for venom.[27]

Gene duplication is a major force in evolution, one worth considering in further detail, as without it, the ability of any species to adapt to a changing environment would be severely limited. Duplications usually occur during the production of sex cells, or gametes, when chromosomes fail to divide equally. Although they're mistakes, if they persist, they can be fortuitous. According to the Japanese biologist Susumu Ohno, the resultant duplication creates redundancy, and redundancy provides scope for innovation.[28] Since one gene copy retains its original function, the second copy can be tinkered with and altered to produce a new function. Nature rarely starts from scratch, since the chance of producing a novel protein with the required function *de novo* is vanishingly small.

Let me expand on this crucial point. Since proteins comprise a linking chain of amino acids, and as any one of 20 different amino acids can occupy each position, the number of possible structures for a relatively short protein of only 100 amino acids in length is enormous. Indeed, it's hyper-astronomical (10^{130}), or greater than the total number of hydrogen atoms in the universe. Generating even a shorter sequence from scratch, like the 42-amino-acid-long DLP-2 (defensin-like peptide-2) in platypus venom, would take the species longer than the solar system has existed, even if it tried a new sequence every second. As Andreas Wagner, Professor of Evolutionary Biology at the University of Zurich, puts it, 'There are a lot of different ways to arrange molecules. And not nearly enough time.'[29] Instead, the ancestral mammals duplicated their defensin genes and modified the extra ones until a selective advantage evolved. The chance of success is not as improbable as one might imagine, since many different mutations can produce a protein with a similar three-dimensional structure and almost identical function. Surprisingly, whether the protein is to be a regulator, an enzyme or a transporter like haemoglobin, the number of solutions is too large to count. In other words, gene duplications provide vital shortcuts that life uses to solve adaptive challenges.

While gene duplication plays a significant role in the story of mammalian evolution and the evolution of other life forms, it should be noted that duplications occur as random events devoid of any inherent foresight or purpose. Rather than being driven by a predetermined end goal, evolutionary change emerges because chance mutations generate novel phenotypic traits, which are subject to refinement through natural selection. However, this raises a fundamental question that remains unresolved. Is gene duplication actively tolerated because of the advantages that innovation confers, or is it merely another instance of cellular error that manages to evade the quality control mechanisms responsible for maintaining the integrity of DNA replication?

Another monotreme oddity solved by genome analysis is how their sex is determined. All other mammals, including humans, have two sex chromosomes that determine sex, the X and Y chromosome system in which XY is male and XX is female. It turns out that male monotremes have five pairs of XY chromosomes (X1Y1, X2Y2, etc.), which pair with each other head-to-tail and form a chain during meiosis and sperm production. Interestingly, their X chromosomes show no homology with the human X chromosome, having more in common with the avian Z chromosome (birds have a ZZ/ZW system). Furthermore, it has been deduced that all 10 sex chromosomes formed a ring structure in the earliest monotremes, which fragmented over millions of years to give rise to the multiple small X and Y chromosomes.[30] Why this happened and the reason for so many sex chromosomes remain a mystery. Nevertheless, the importance of the finding is that the monotreme sex chromosomes seem more akin to those of birds than humans, and challenge the idea that their sex chromosomes evolved separately.

Electroreception

Early observers were perplexed by the ability of the platypus to find food in deep stygian waters without apparent use of any of the known senses. Their nostrils form a tight water seal, while their eyes and ears lie in a muscular groove that pinches shut when diving. Could they have a sixth sense? Sir Everard Home's first description of the animal in 1802 contained a prescient statement, ignored for nearly two centuries. He noted that the platypus's trigeminal nerve, which innervates the face, is 'uncommonly large' and postulated that 'the sensibility of different parts of the bill is very great.'

At the end of the twentieth century, the distinguished Australian neurobiologist Jack Pettigrew revealed that the bill contains an array of mechano- and electro-sensory receptors, with an orderly spatial mapping in the brain. As in other species, a region called the somatosensory cortex deals with sensory input. Different areas of the cortex receive sensory information from various body parts, and the relative size of these cortical areas reflects the species' major tactile organ. The bill dominates this area in the platypus, whereas in humans it's the hand and lips.[31]

Pettigrew realised that it was the combination of the bill's many electrical and mechanical receptors that enabled the animal to detect its underwater prey. When hunting, platypuses move their heads from side to side, just as we would use a metal detector, to track down the telltale signs of their prey's muscle contractions and the resultant water disturbance. Signals from both types of receptors reach the brain, where the animal forms a detailed three-dimensional 'fix' of its target.[32] Furthermore, as the electrical waves arrive first, the time lag between the two signals enables the platypus to calculate the distance to its goal, just as we judge the distance of a storm by the time gap between thunder and lightning. In contrast, echidnas have few receptors, as combined electrical and mechanical detection is of limited value in terrestrial environments.

Electroreception is an ancient sense that evolved in the common ancestor of all vertebrates approximately 600 million years ago. Later, most groups lost the ability to detect electricity, including frogs, reptiles, mammals and most fish. Given its effectiveness in detecting underwater prey, why such a loss should have happened remains to be determined. Even stranger is that a few species independently re-evolved electroreception, including electric fish, the Guiana dolphin, and the platypus and echidnas. To understand why this might have happened in the monotremes, we need to consider their origins.

Enter the echidnas

Professor Tim Flannery (Plate 4) is one of Australia's best-known scientists. Dubbed the 'Indiana Jones of mammalogy', his insatiable thirst for knowledge and widespread expertise has resulted in many careers: adventurer, anthropologist, author, environmentalist, explorer, Harvard University professor, museum curator, palaeontologist, zoologist. In his early years, he undertook many arduous expeditions to the world's most inhospitable and remote places, including New Guinea and the Solomon Islands, where he discovered dozens of new species – the most impressive being the black-and-white tree-kangaroo, one of the last 'unknown' large mammals.[33]

As one might expect, two years of COVID lockdown didn't interfere with Flannery's scientific drive and ambition. With the unexpected liberation from routine academia, he had time to address several unsolved problems of monotreme evolution. Where did they originate from, why are they restricted to Australia and New Guinea, and where are the fossil echidnas? The last question was the most intriguing, as the oldest fossils are less than 2 million years old, despite echidnas having been around for over 50 million years.

In collaboration with his colleague Professor Kristofer Helgen, Flannery started by examining all the available fossils and related evidence on monotremes to get a clearer picture of their history.[34] The result was the reclassification of the fossilised jawbone of the smallest and oldest known monotreme, *Teinolophos trusleri*, a small shrew-like animal that lived in southeast Australia approximately 130 million years ago. Like extant monotremes, it possessed electroreceptors in its snout, shown by the marked groove in its dentary that housed the trigeminal nerve and related tissue.

While *T. trusleri* emerged from its burrow to forage, its territory in modern southeast Australia formed part of Gondwana and lay within the Earth's southern polar circle. The terrain was covered by dense polar forests, with periods of snow and low temperatures, especially during the 3–6 months of the austral winter. Such conditions would have favoured the emergence of mammals with electro-sensitive snouts to aid them in finding prey in the leaf litter and soil of Gondwana's dark and dank forests. Indeed, this ancestral trait is present in all the monotremes alive today. Flannery argues that this scenario explains why the early monotremes didn't disperse, even though they evolved when the southern continents were all joined together. The animals were adapted to high-latitude environments and only spread to other habitats after developing strategies to survive warmer temperatures.

Gondwana hosted many monotreme species, ranging from southern Patagonia to Australia, but by 100 million years ago, only one had evolved an aquatic lifestyle – the ancestral platypus.[35] Then, 66 million years ago, the Earth suffered a dramatic mass extinction. At the end of the Cretaceous, an asteroid strike off the coast of Mexico, coupled with massive volcanic activity in India, resulted in the extinction of around 75 per cent of all species, including dinosaurs and most monotremes. Against the odds, the terrestrial, semi-aquatic platypus survived. Its austral location and the more robust photosynthetic food chains in river systems than those on the land and in the oceans would have helped it weather the global apocalypse.[36]

Once the Earth's ecosystems had recovered and the dinosaurs were long gone, the echidna lineage diverged from the platypus-like monotremes. Indeed, embryonic echidnas possess facial features that reveal their snout evolved from a 'platypus-like' bill. However, Flannery and Helgen were puzzled by the large gap in the echidna fossil record, which appears restricted to rocks less than 2 million years old. Where did the early echidnas hide before diversifying to give rise to the four extant species?

After several brainstorming sessions, Helgen suggested New Guinea as being the most likely answer. Millions of years ago, proto-New Guinea remained connected to Australia by a land bridge, allowing floral and faunal interchange.[37] Eventually, the restless jostling of the Earth's crust wrenched New Guinea from Australia, and its wildlife, including its platypus-like monotremes, became isolated.

Flannery and Helgen's novel hypothesis envisages that the evolution and diversification of echidnas occurred amongst the mountains of western New Guinea in the Vogelkop or Bird's Head Peninsula area. This highland region started to form about 12 million years ago due to the tectonic collision of the north-moving Australian plate with the Pacific plate. The resultant Arfak mountain range in the west had few rivers but extensive humid rainforests. Without predators or competition, these early monotremes would have strayed from the rivers for food and eventually evolved a terrestrial lifestyle. They became insectivorous, with less need for electroreception, and developed a protective spiny coat. Echidnas then spread to give rise to New Guinea's four extant species – the western long-beaked echidna of the Vogelkop and Foja mountains, the eastern long-beaked echidna of the central highlands, the critically endangered Sir David's long-beaked echidna of the Cyclops Mountains (named after Sir David Attenborough), and the short-beaked echidna of southern and eastern New Guinea.[38]

At the beginning of the last ice age, about 2.6 million years ago, when sea levels fell, the short-beaked echidna walked across to Australia to become the continent's most widespread native mammal. If Flannery and Helgen are correct, the gap in the echidna fossil record reflects a 35-million-year sojourn amid the rainforests of New Guinea.

Ancestral but not primitive

The five extant monotremes – a platypus and four echidnas – are the last survivors of an austral tribe that extends back to the Middle Mesozoic, or the 'Age of Reptiles'. That they exist at all is both extraordinary and fortunate. They have provided us with a unique insight into the reptile–mammal transition and have helped us build a picture of the mammals' last common ancestor: a small, furry, insectivorous, egg-laying animal that leaked milk from its abdomen. Furthermore, their genomes have enhanced our understanding of how the mammary gland and lactation evolved – features that Linnaeus chose to define our class.

Finally, I should dispel the mantra that monotremes are primitive because of their reptile-like features. While some traits are undoubtedly bizarre, they merely reflect lineage antiquity and are not inefficient or undeveloped.[39] Early lineage branching would have given the platypus and the echidnas more time and scope to hone their unique ancestral adaptations. As a result, they are no less advanced than any other mammalian group. Any promulgation to the contrary stems from the specious belief that *Homo sapiens* occupies the uppermost twig of the 'tree of life'. We do not. All species reside on twigs of equal importance. Evolution is non-directional, with no plan or regard for the future, with the only goal to exist and reproduce. Indeed, if monotreme primitivity were true, they would not have survived the arrival of another class of mammals, the marsupials.

PART TWO

Marsupials

CHAPTER 2

The Monito del Monte's Story

THE MARSUPIAL DIASPORA

Everything is the way it is because it got that way.
Sir D'Arcy Thompson (1860–1948)

We were in the rainforests of Waigeo, an island off the west coast of New Guinea, to see one of the most sought-after birds in the world, the riotously feathered Wilson's bird-of-paradise. Crouched behind a strategically placed blind, we anxiously waited for a male to show himself. After what seemed an eternity, a thrush-sized gem, splashed with a palette of colour – iridescent green, cobalt blue, chartreuse yellow, jet black and deep crimson – dropped to the ground and nervously tended his terrestrial display court. Amidst the dank foliage and crepuscular gloom, he entertained us with a short but energetic display of acrobatics and fluorescence. However, whether the female was as enamoured as we were remained unclear. Once his morning's performance was over and he had disappeared deep into the forest, we retraced our steps along the trail's sinuous windings, musing about how fortunate and privileged we had been.

Suddenly, our guide stopped and pointed to a small creature clinging to a branch high above our heads. This cute, furry animal with large saucer-shaped eyes, a cross between a cat, a monkey and a fluffy toy, was a Waigeo cuscus, a marsupial found nowhere else in the world. It was quietly resting amongst the canopy, as cuscuses do, after foraging throughout the night for leaves, nectar and fruit. Still euphoric from our avian encounter, I confess that the cuscus received scant attention. We quickly moved on to our next target as more avian delights awaited us. To this day, I regret my somewhat cursory and perfunctory encounter. Only later, while researching this book, did I learn about the cuscus's remarkable 170-million-year evolutionary trek – a worldwide diaspora that took its distant ancestors from Australia across five continents via interlinking terrestrial bridges before returning to Australia and Indonesia, where they remain today.

The story of the cuscus and its many marsupial cousins deserves telling, as their past journeys and diversifications rival any from the mammalian world. However, to explore their epic dispersal, we need to know where they first evolved, and to do that, we need to know about teeth.

Gondwanan origins

As explained in the previous chapter, the evolution of a dentary-only mandible in early mammals significantly improved their ability to grind and chew. However, achieving this efficiency necessitated an innovative tooth design that ensured precise alignment between the upper and lower teeth. The result was the functionally complex 'tribosphenic' molar, an innovation so advantageous that modifications of its basic design characterise most therians alive today. The technical term tribosphenic (from the Greek words *tribo* for grinding and pounding and *sphen* for shearing and cutting) was coined in 1936 by the American palaeontologist George Gaylord Simpson, one of the founders of the 'modern synthesis' of evolution.[1] Compared to the morganucodons, whose teeth consisted of a simple chain of three or four peaks, tribosphenic molars possess peaks and valleys of various shapes and sizes that slot together while chewing, like a pestle and mortar. While the initial selective pressure was the need to cut insect exoskeletons more efficiently, tribosphenic teeth made more versatile feeding strategies possible and underpinned early mammalian diversification. Indeed, without the evolution of these microscopic dental modifications, there would be no insectivorous shrew opossums, leaf-chewing giraffes, meat-slicing lions, fish-eating seals, or omnivorous primates like us.

Seeing as teeth and their often-associated fragmented jaws are the commonest fossil finds of early mammals (skeletal bones were usually too fragile to survive), the tiny tribosphenic molar looms large in the field of therian scholarship. As a result, scientific papers are peppered with an arcane array of terms, including protocones, paracones, preparacrista, paraparacrista, trigonids and talonids. Thankfully, our narrative does not demand familiarity with such morphological minutiae, only that the myriad of dental grooves, peaks and ridges helps palaeontologists to identify fossils as therian and to elucidate the sequence and timing of evolutionary changes that occurred in mammalian molars.

For most of the twentieth century, early therian finds were restricted to Laurasia, and palaeontologists widely assumed that stem therians evolved in the northern hemisphere. Then, in 1997, after 23 years of fruitless searching, a husband-and-wife team of palaeontologists, Thomas Rich and Patricia Vickers-Rich, found a fragile, 16-millimetre-long jaw of *Ausktribosphenos nyktos* from the Early Cretaceous rocks of southeast Australia.[2] The species, so named because it was 'an Australian Cretaceous tribosphenic mammal that lived by night', was the oldest therian fossil from the southern hemisphere, dating from 115 million years ago. This small insectivorous creature, estimated 8.5 centimetres long, lived amongst Australia's polar dinosaurs in the region's cool temperate climate. In the palaeontological world, size doesn't matter, and the evolutionary implication of these comically meagre teeth, the size of rice grains, was profound. Was it possible that the ancestors of all marsupials and placentals evolved in the southern rather than the northern hemisphere?

The remote possibility that science might have gotten mammalian evolution upside down, literally and metaphorically, encouraged fossil hunters to scour the southern continents for more specimens. Within two years, a team headed by the American palaeontologist John Flynn discovered a jaw fragment with three tribosphenic molars from Madagascar. This specimen, from approximately 170 million years ago, was older than any equivalent fossil from the northern hemisphere.[3] Later, more therian jaws and teeth were discovered in Jurassic deposits from Gondwana, in modern-day Patagonia and India.

Eventually, this paltry but very precious collection of teeth and jaws, amassed over 20 years from across Gondwana, and spanning 70 million years of evolution, revealed their collective secret. In 2022, Tim Flannery (whom we met in the last chapter) and his colleagues re-examined all the available specimens using the latest methodology.[4] They concluded that the earliest fossils, retrieved from sediments in Patagonia, Madagascar and India, pre-dated those from the northern hemisphere by up to 50 million years. Furthermore, the youngest fossils, from 110–126-million-year-old Cretaceous deposits in Australia, have features in common with those from the northern hemisphere. In other words, stem therians must have evolved in Gondwana, not Laurasia.

This novel interpretation, presented in dry scientific detail, appeared in the Australian journal *Alcheringa*, with the team's epiphanic leaps tempered by empirical caution. Even so, with such a high-profile and provocative conclusion, one so at odds with conventional wisdom, one might have expected an avalanche of critical correspondence. But as Flannery told me, 'While it has been widely read, nobody has published a refutation or confirmation – more of a wall of silence.' He added, 'It may take some time to find full acceptance among northern hemisphere researchers.' Critics might argue that the finding of similar tribosphenic teeth from both hemispheres results from convergent evolution and that further fossil finds are required for clarification. However, the erosion of northern-centric certainty is not new in evolutionary biology. In 1991, a young Melbourne graduate, Les

Christidis, working with fellow Australian Richard Schodde, revealed that all the world's songbirds, including our familiar thrushes, sparrows and warblers, evolved from Australasia and not Eurasia.[5]

During the Mid-Cretaceous, the nocturnal, furry ancestors of all living therians dispersed northwards to reach Asia from Australia, using volcanic island arcs, or rifted continental crust, as 'stepping-stones' to cross the Tethys Ocean (Figure 2.1).[6] It was a journey that took millions of years. Generation upon generation shared their lives with thunderous titanosaurs, ostrich-sized dromaeosaurs and groups of feathered dinosaurs, some sporting primitive wings. They skirted lakes, surrounded by horsetails and early flowering plants, that brimmed with turtles, fish and croaking frogs. Moths would have hovered above in the fading light to feed on fern spores and conifer pollen while taking the occasional sip of dew. Dragonflies with two-foot wingspans likely danced across their burrows, while stinging hornets sought their unlucky victims. At the same time, pterosaurs, masters of the sky, soared above, searching for birds and small dinosaurs amid the vast tracts of woodlands of towering cypresses, cycads and ginkgos.[7]

While reptiles patiently awaited the morning's sun, our ancestors emerged from their underground burrows to search for mates, defend territories, and feed on the wealth of insects that evolved during the Cretaceous. Many moths, spiders, wasps, flies and beetles appeared alongside the emerging angiosperms and functioned as their pollinators in return for nectar and fruit. Selective

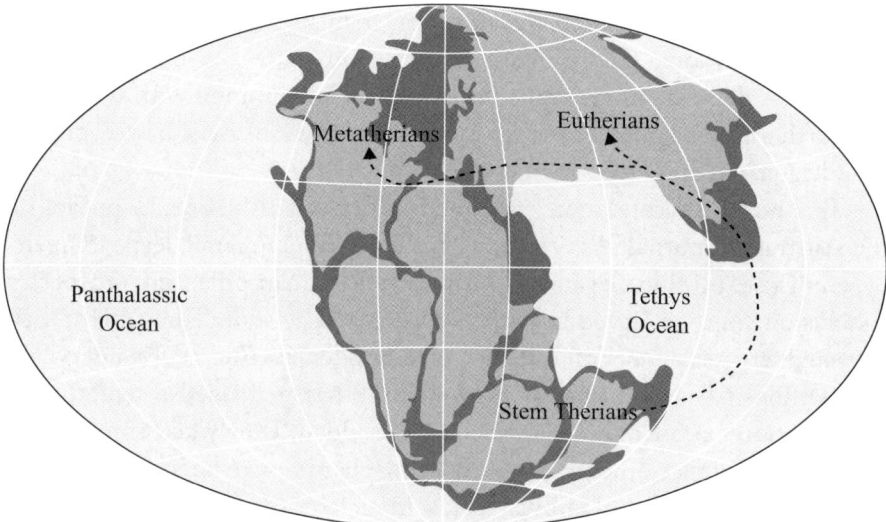

Figure 2.1 Gondwanan origin of stem therians. The early therians evolved in Gondwana and crossed via the Tethys Ocean to Laurasia prior to the metatherian–eutherian split approximately 125 million years ago.

pressure forced the early mammals to exploit this novel resource as dinosaurs had already monopolised the more bountiful herbivore and carnivore niches. But insects are difficult to digest. Their exoskeletons are composed of a rigid carbohydrate polymer called chitin, and it was only because of five functional genes inherited from their non-mammalian ancestors that the early therians could survive on an insectivorous diet. These genes code for proteins called chitinases, which specifically break down dietary chitin into its component sugars and allow access to the insects' soft interiors.[8]

Our modern ant and termite specialists, including aardvarks, certain armadillos, and insectivorous tarsiers, all retain the five chitinase genes. In contrast, the genes of modern carnivores and herbivores contain either knock-out mutations or are deleted. Even animals that would never ingest chitin, like lions, polar bears, and horses, still retain evidence of chromosomal 'molecular fossils' – pseudogenes that betray their insectivorous ancestry. However, insects are small, require effort to catch and are not very nutritious, so many are needed to fulfil a mammal's energy requirements. As a result, our early ancestors remained small and nocturnal, biding their time until the dinosaurs unexpectedly vacated the more nutritious niches.

Early mammals were well adapted to a nocturnal lifestyle, with large eyes and sensitive retinas. However, unlike their reptilian ancestors, they didn't require colour vision and relied mainly on scent to find their way around. Over time, the lack of selective pressure for visual acuity led to the loss of two of their four types of colour-sensitive retinal pigments, or opsins.[9] The resultant state, dichromacy, persists in all living mammals except for primates, including *Homo sapiens* (a topic we'll return to later). Indeed, your cat, dog or horse cannot perceive red and green objects too well because of these genetic changes that occurred 150 million years ago.

During the Cretaceous, therians evolved in other ways that still affect our anatomy. The spur-like ankle structure that delivered venom in all the pre-mammals was lost, a feature retained only by the platypus.[10] An alteration in gait soon followed. Early mammals had a sprawling limb posture like reptiles and monotremes, resulting in decreased ventilation when walking and running. Because of the lateral bending of their spines, reptilian lungs are alternatively compressed during locomotion, shunting stale air from one lung to the other rather than making room for fresh air.[11] Therians solved this pulmonary limitation, known as Carrier's constraint, after a one-off evolutionary event that changed their leg orientation to produce an upright or parasagittal stance.[12] As a result, mammals could run and breathe simultaneously and gain the advantage of superior locomotor stamina.[13]

The ear also changed in several crucial ways. Like the monotreme lineage, therians independently evolved three ear ossicles that boosted their hearing.

Furthermore, the inner ear, which converts sound waves into electrical impulses, evolved to increase the resolution of sound frequencies. Its bony canal, or cochlea, changed from a simple curved structure through intermediate stages to a tightly coiled, snail-like spiral.[14] These anatomical modifications allowed better innervation and blood supply as the organ occupied a smaller space within the skull. A suite of genes, co-opted during the Early Jurassic, led to these bony and neural adaptations – which persist in mammals today.[15]

Viviparity

However, the most profound evolutionary change that occurred is also the least understood – the development of live birth or viviparity. Before the divergence of marsupials and placentals, the early therians had replaced oviparity with viviparity. But why the switch, when reptiles and pre-mammals did very well laying eggs, and live birth isn't inherently superior? One answer relates to the downsizing of mammals that occurred during the 'Age of Dinosaurs.' While their small size was beneficial for evading predators, the resultant miniaturisation of eggs meant shorter incubation times, less nourishment for the embryo, and increased pressure for parental provisioning of 'fetus-like' hatchlings. Conversely, if the young develop within the mother, they would be more mature and less vulnerable once born. Overall, the small size of early mammals likely provided the critical selective pressure for their dramatic change in reproductive strategy. Furthermore, their warm-bloodedness provided a more stable and predictable environment for fetal development than those available to egg-layers, and allowed mammals access to a greater variety of habitats.[16]

Fossil evidence fails to reveal how this change occurred. However, one can imagine a likely scenario: over time, mammalian eggs would have been retained for longer and longer periods until, eventually, they became implanted in the mother's uterus, obviating any need for calcified shells. Instead of egg yolk, mammals evolved the placenta. This organ allows nourishment and oxygen to cross from the mother while ferrying waste products from fetal to maternal blood. Since the placenta is composed of soft tissues, palaeontology cannot help us understand how the organ evolved. However, as marsupials and eutherians are both placentals (confusingly, marsupials do have placentas, albeit small, transient ones), the parsimony principle, which states that the simplest solution is the most likely, implies that the placenta must have emerged in a common ancestor but only after the monotremes had diverged.

The phenomenon of genomic imprinting also supports a single origin for mammalian placentation. According to standard Mendelian genetics, organisms inherit two copies of an autosomal gene, one from their mother and one from their father. However, scientists have discovered that for specific genes involved in placental reproduction, only one is active or 'switched on'. Which copy is expressed depends on the parent of origin – some are only active if originating from the father, while others are only functional if from the mother. An affected gene is 'imprinted' during the formation of germ cells (sperm or egg) by the process of methylation. The latter is a chemical reaction that fixes small molecules, called methyl groups, to specific bases in the relevant gene. The result is altered gene expression, with the imprinted copy becoming inactive. In other words, if the allele inherited from the father is imprinted, it is silenced, and only the maternal gene is expressed. In contrast, only the paternal gene is active if the maternal allele is imprinted.

But what does this have to do with placental evolution? It seems that the placental expression of only a maternal or paternal gene coding for a small range of proteins is essential for a successful pregnancy. Indeed, abnormalities of imprinted genes are associated with rare human diseases characterised by abnormal fetal growth. While poorly understood, imprinted genes may help balance the genetic priorities of mother and father (parental conflict hypothesis) or improve fetal development and maternal provisioning and care (co-adaptive hypothesis).[17] Whatever the explanation, the important point is that genomic imprinting only occurs in therians (eutherians and, to a lesser extent, marsupials) and not in the oviparous monotremes. Since comparative genomics has revealed that two of the known imprinted regions are conserved in marsupials and eutherians, placentation likely had a single origin, probably in an ancestor 125 million years ago.[18] In other words, viviparity has only evolved once in mammals despite having evolved many times in vertebrates (lizards and snakes). However, the original therian placenta has been modified many times by selection pressure to become the most anatomically varied organ in modern mammals.

Recent studies of lizard species that are both oviparous and viviparous, such as the yellow-bellied three-toed skink, suggest that the transition to live birth may not have been as genetically complex as one might imagine.[19] Many of the same genes are involved in the production of both shelled eggs and viviparous embryos, with the main difference being the order and magnitude of their activity. Thus, the transition between oviparity and viviparity was likely a matter of changes in gene expression rather than the evolution of a new raft of structural genes. In other words, mammalian reproduction is another classic example where changes in gene regulation can result in major morphological and physiological adaptations over a relatively short period of evolutionary time.

Until recently, most experts regarded marsupial reproduction as an evolutionary stepping-stone between the egg-laying monotremes and the placental mammals. Indeed, the long-standing scientific bias is reflected in their clade names – Prototheria (first beast), an obsolete term for monotremes, Metatheria (middle beast) and Eutheria (true beast). However, a team led by Professor Anjali Goswami from London's Natural History Museum has challenged this view.[20] By comparing the CT scans of skulls from placental mammals and marsupials in various stages of maturation, the researchers concluded that the developmental strategy of placental mammals, rather than marsupials, is closer to their common ancestor. In other words, marsupials have evolved far more than placental mammals following their divergence. This finding has implications for the evolution of mammalian reproduction. According to Goswami, the developmental strategy of marsupials doesn't represent a 'primitive' or intermediate evolutionary state but is a derived state honed by the sieve of natural selection. It suggests that early therian mammals likely had a precocial reproductive strategy, similar to modern eutherians, with a long-lived placenta.

If Goswami is correct, then the extreme prematurity of marsupial neonates is a derived reproductive strategy rather than a transitional phase between monotremes and eutherians. Although this finding may seem counterintuitive, it does support a recent study of the extinct multituberculates, a lineage of therians that diverged after the monotremes and before the marsupials. By examining the fine structure of their fossilised leg bones, researchers now believe that multituberculates had a prolonged gestation and gave birth to mature offspring that were weaned early, like placentals, but not marsupials.[21] Not for the first time, therian evolution has been turned upside down – marsupials evolved from eutherians and not the other way round. Let's pause for a moment and allow this revelation to sink in, because it's a dramatic scientific *volte-face*. The eutherian lifestyle came first and was later modified by metatherians. As a result, the etymology of the word Metatheria, or 'middle beast', has become an inaccurate descriptor for the marsupial clade.

How the early therian placenta evolved is unclear, although retroviruses may have played a role. On rare occasions, organisms can inadvertently acquire DNA, and hence genes, that did not initially belong to them. This situation can arise following a retroviral infection when the invading viral RNA is reverse-transcribed into double-stranded DNA and inserted into the host's genome. Should the retrovirus-derived gene be incorporated into the host's germ-cell DNA (eggs or sperm or the cells that produce them) and tolerated, it will be inherited and passed on from generation to generation. Most assimilated genes will be silenced, usually by methylation, and will have no downstream effects. However, on rare occasions, a long time after their integration and many mutations later, inserted genes can be activated and produce novel proteins that benefit the host. Natural selection then ensures that the gene's product remains part of the animal's protein repertoire or proteome.

One such gene, *PEG10*, was acquired by early therians and recommissioned for placental growth. Its novel protein has several functions, including controlling placental cell proliferation and developing feto-maternal blood vessels. Indeed, genetically modified mice, in which the *PEG10* gene is rendered inactive, fail to develop normal placentas and have stillborn babies. Therefore, the domestication of *PEG10* and the development of the placenta were critical events in the evolution of therian viviparity.[22] However, *PEG10* is only one of many retroviral genes that eutherians co-opted to produce the panoply of placental anatomies seen today.

Marsupial divergence

Around 140 million years ago, Pangaea began to split into two separate supercontinents, Gondwana in the south and Laurasia in the north. Those therians left behind on Gondwana became extinct, and their genomes made no further contribution to mammalian evolution. In contrast, those that reached Laurasia spread rapidly throughout the continent, reaching Europe and North America, with many becoming arboreal. Minor alterations to their bones, including spine, legs, feet and claws, resulted in greater agility and allowed them to grasp and walk along branches like some of today's treeshrews, opossums and dormice. The selective pressure to venture into these new ecological niches was protection from predators and a supply of insects that remained out of reach of their terrestrial cousins.

In northeast Laurasia (modern-day China), therians would have encountered a temperate climate, alternating between semi-arid and mesic, with heavy seasonal rains. Luckily for our story, a few animals became buried by ash and sludge from nearby volcanic eruptions. Together with an array of dinosaurs, birds and plants, their deaths resulted in what palaeontologists term the Jehol Biota formation, one of the best-preserved continental ecosystems in the Earth's history. Found in Liaoning Province, these exquisite fossils, dating from a 10-million-year period (130–120 million years ago), have revealed how early birds and flowering plants diversified and feathers evolved. Crucially, the Jehol Biota has also provided unprecedented insights into early therian evolution, competition and success. Among its remarkable discoveries is a crow-sized *Microraptor*, a species of feathered dinosaur, feasting on its last meal of a hapless therian.[23]

At some point, the viviparous tribosphenic therians diverged into two clades: the metatherians and eutherians. (Remember, metatherians include all modern-day marsupials and their extinct relatives, while eutherians encompass all placental mammals and their nearest fossilised kin.)

But when did the metatherian–eutherian dichotomy occur? Today, the two clades are easy to distinguish. As we have already discussed, marsupials give birth to highly immature young because of their short-lived placentas and, as a result, their maturation typically requires completion within a brood pouch, a flap of skin usually located on the abdomen. Marsupial male and female reproductive tracts also differ from those of eutherians. Females have two uteri and two vaginas, and before birth, they develop a birth canal between them, the so-called median vagina. Males have a split or double penis in front of the scrotum, with the two ends corresponding to the females' two vaginas. Marsupials also have a different brain anatomy, lacking a corpus callosum, the white matter tract connecting the right and left cerebral hemispheres. Instead, they possess a simple network of nerve fibres. Indeed, according to Rodrigo Suárez and his team from the Queensland Brain Institute, the metatherian cerebral connections or 'connectome' originated at least 80 million years before the emergence of the eutherian corpus callosum.[24]

The problem facing palaeontologists, however, is that these soft-tissue clues are not preserved. At the same time, the discriminatory features that do fossilise, including various aspects of dental and skeletal anatomy, may be difficult or impossible to discern. If that wasn't enough, early therian fossils often appear chimeral, exhibiting characteristics of both clades. In other words, the first metatherians and eutherians can look remarkably alike, and their fossils are challenging, even for experts, to tell apart.

Take the case of *Sinodelphys szalayi*, a mammal which lived 125 million years ago. In 2003, the Chinese American palaeontologist Zhe-Xi Luo and his team reported this small opossum-like tree-dweller as the earliest metatherian.[25] Its nearly complete skeleton was exquisitely preserved in shale and contained impressions of fur and some carbonised soft tissues. However, in 2018, another Liaoning fossil from northeast China, described by Shundong Bi and colleagues, resulted in *Sinodelphys* being reinterpreted as a primitive or basal eutherian.[26] Furthermore, *Sinodelphys* cannot have been an 'oddball' or a one-off, as the arboreal *Eomaia scansoria* ('dawn mother that climbs'), a purported eutherian from the same location, also exhibited metatherian features.[27]

To date, the best contender for the earliest eutherian is a mammal named *Juramaia sinensis* ('Jurassic mother from China'), also from the Liaoning Province. Despite consisting of an incomplete skeleton, the spread-eagled, roadkill-like fossil was found in 160-million-year-old rocks and appears to push back eutherian evolution by 35 million years – a time corroborated by molecular-clock studies.[28] However, doubts have been raised regarding the fossil's stratigraphic provenance and its age. Therefore, 125 million years ago currently reflects a conservative, fossil-informed minimum age of the divergence between metatherians and eutherians, a date I have adopted throughout this book.

The fact that palaeontologists struggle to pinpoint the precise time and place of one of the most important divergences in mammalian history is not surprising. There would have been subtle gradations of change, evolutionary experimentation, and extensive shuffling of genes before establishing identifiable lineages. Today, the oldest metatherian fossils are not from China but from 110-million-year-old rocks from Montana and Wyoming in western North America. Most of these basal metatherian lineages, and there were many, became extinct, some alongside the dinosaurs 66 million years ago.[29]

As discussed, metatherians evolved increasingly short-lived placentas and immature or altricial offspring after diverging from their therian ancestors. Today, for example, the kangaroo's newborn is the size of a jellybean after a gestation of only a month and requires nursing for up to six months. Survival is only possible because the embryo's forelimbs, shoulders, hands and stomachs evolved to grow faster than the rest of its body so that, once born, it can crawl through its mother's fur and reach the milk-producing teats. Metatherian milk also changed dramatically in composition to compensate for the loss of placental support. Recently, Michael Guernsey and colleagues from Stanford University have shown that metatherian lactation co-opted some of the genes used by the therian placenta to achieve this. In other words, the eutherian placenta and the metatherian mammary gland have converged on a similar set of genes to nourish their respective offspring.[30]

The defining feature of marsupials is the skin pouch enclosing the teat area – hence their name (from the Latin *marsupium*, pouch). However, this structure is not present in all South American marsupials, being absent or poorly developed in the small and medium-sized species. This observation suggests that the pouch was not present in the earliest metatherians and that the reproductive cycle of their ancestors must have contained a nesting phase. Furthermore, phylogenetic analysis indicates that, while extant South American marsupials have a variable number of teats, ranging from 4 to 19, the ancestral or plesiomorphic state was only four.

Why natural selection favoured the metatherian reproductive strategy remains unclear. Short-lived placentas and their associated altricial births may be best suited for periods of environmental instability. The long gestation of eutherians means that during periods of limited resources, both mother and offspring may succumb. In contrast, the marsupial strategy is lower risk as mothers can abandon their babies at an earlier stage to increase their chance of survival and have the opportunity to try again later. This may explain why only marsupial mammals survived Australia's harsh conditions.

South American marsupials

By the end of the Cretaceous, one clade, the Marsupialiformes, had dispersed across North America, reaching the northern slopes of Alaska and into Europe and Asia.[31] Indeed, frequent exchanges between the continents may have occurred. They became more diverse and outnumbered their eutherian contemporaries while increasing in size, up to 5.2 kilograms, and dietary diversity. Some remained insectivores, while others became omnivores, frugivores, carnivores or scavengers, consuming a range of soft to hard food sources.

A second clade, one that would produce the crown marsupials, radiated southwards and arrived in South America from North America via the Greater Antilles or Aves Ridge. This volcanic arc straddled the eastern Caribbean Sea in the Late Cretaceous, sometime after 75 million years ago.[32] The land bridge, formed by fluctuations in tectonic activity and sea-level changes, was temporary and is now wholly submerged save for the Venezuelan islands of Aves and La Banquilla. It remains to be seen whether a single migratory event provided the genetic source for the whole radiation of South American metatherians, including all the living species. Some palaeontologists argue that several lineages took part, possibly in several dispersal episodes. What is clear, however, is that the metatherians remaining in Laurasia faced devastating consequences from the end-Cretaceous meteor impact – a catastrophic event that played a significant role in their eventual extinction.[33]

Once in South America, the small opossum-like newcomers entered a world of diverse and untapped ecological habitats – tropical and subtropical forests, mangroves, swamp forests and montane rainforests to the north, with temperate and mixed forests to the south. The Andes had yet to form, and there was no blocking of the equatorial easterlies. Within a few million years, the earliest marsupials had evolved into myriad forms and sizes, facilitated by the high diversity of available niches.

One of the first species to evolve, the rat-like *Pucadelphys*, deserves mention. In 2011, a remarkable trove of 35 individuals was found in Bolivia, consisting of many intact skulls and skeletons.[34] All the animals appeared to have died simultaneously, possibly during a flash flood or other natural catastrophe, indicating that they belonged to a single group. Some skeletons were fully grown females, smaller and with less robust anatomy. Others were fully-grown males, with larger heads and bodies and pronounced canine teeth. A few were juveniles. Since all the fossils were closely packed together, occurring in an area of less than a square metre, *Pucadelphys* must have lived communally. Furthermore, the strong sexual dimorphism implies a mating system characterised by male–male competition and polygyny. It's also likely that *Pucadelphys* huddled together for protection against predators or competitors and that communal living provided a selective advantage with faster reproduction rates.

In contrast, most of today's marsupials are primarily solitary creatures, with the main exception being the kangaroo lineage. In other words, sometime during their evolution, the majority of marsupials abandoned gregarious behaviour, though gaps in the fossil record prevent us from knowing the time and causes of the switch. However, the Bolivian fossils reveal that social interactions occurred early in the Palaeocene and that solitary behaviour was not a primitive or plesiomorphic character for all metatherians.

Many South American species occupied ecological niches that eutherians filled in the northern hemisphere. They included the Sparassodonta (from the Greek *sparassein*, to tear, and *odontos*, tooth), an extinct order of carnivores that ranged from civet-sized to leopard-sized animals. These metatherians showed many similarities with placental carnivores, having sharp molars and lengthened canines, with the teeth of the *Thylacosmilus* being longer than those of *Smilodon*, the 'sabre-toothed tiger'. However, *Thylacosmilus* and its placental equivalent were not closely related, and their phenotypic resemblance resulted from convergent evolution. The 60 species of Sparassodonta were some of the continent's oldest metatherians and were the apex predators of the ancient South American ecosystems from Colombia to Patagonia. With their specialised meat-eating molars, Sparassodonta remained dominant until various environmental and climatic upheavals caused their demise.[35]

The predatory Sparassodonta were not the only examples of convergent evolution within the early metatherians. Consider the marsupial *Borhyaena*, a slouching 90-kilogram ground-dwelling predator that roamed South America in the Early to Mid-Miocene and looked and acted like an African hyena. Another species, *Cladosictis*, a small, sleek metatherian that hunted eggs and small animals in low undergrowth, resembled the placental otters. Then there was *Necrolestes*, the 'thief of the dead', a subterranean mole-like creature that fed on invertebrates. Last but not least, *Microtragulus*, a small bipedal metatherian with ever-growing cheek teeth, hopped about and occupied ecological niches like the extant jerboas and kangaroo rats. Indeed, it's argued that the absence of modern rodents filling such habitats in South America is due to *Microtragulus* and its relatives occupying them until recently. Since the *Microtragulus* lineage extended into the Pliocene, scientists think it might be possible to extract ancient proteins, like collagen, from their fossils to explore their relationship to living marsupials in greater detail. While the convergent

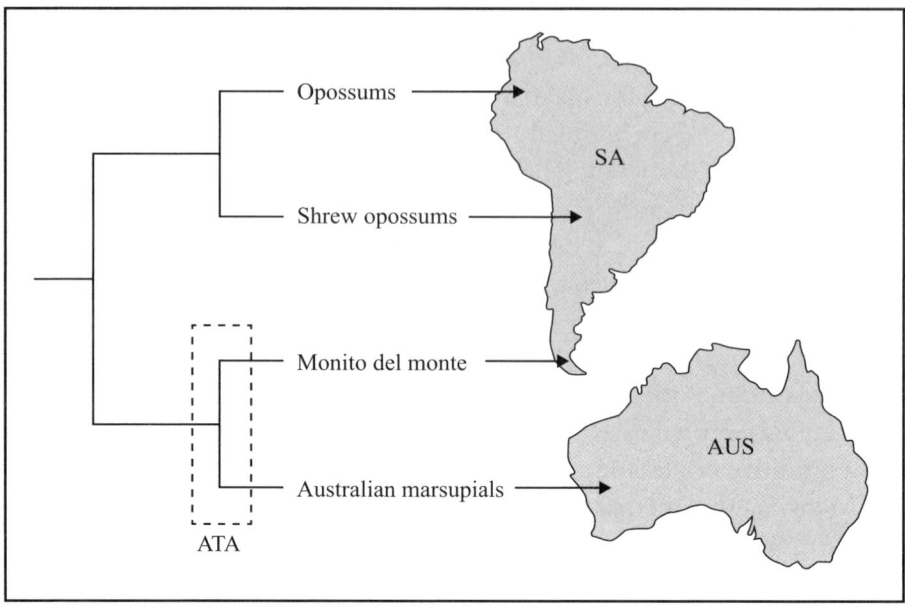

Figure 2.2 Phylogeny and biogeography of marsupials. Whole-genome analysis has revealed that the monito del monte is the sister species to all the Australian marsupials. ATA, Antarctica; AUS, Australia; SA, South America. Modified from Feng et al. (2022).[36]

evolution of these extinct South American metatherians was striking, it was merely a warm-up act, a trial run for the dramatic evolutionary events that would later unfurl in Australia (Chapter 3).

The impressive diversity of South American metatherians waxed and waned due to climatic variations, especially mean annual temperatures and rainfall. However, sometime after the massive K–Pg extinction event, the last common ancestor of all extant marsupials (Marsupialia) lived alongside the many now-extinct taxa. This species, which will forever remain unknown, gave rise to seven orders, of which the first three populated South America and Antarctica, namely the South American opossums (Didelphimorphia), the shrew opossums (Paucituberculata), and the monito del monte, the sole representative of the Microbiotheria (Figure 2.2). These three South American orders comprise over 130 species, accounting for 10 per cent of the continent's terrestrial mammals. New taxa continue to be described at an impressive rate, with 24 named in the last 25 years, and many more await discovery. Most live in tropical and subtropical forests and are small to medium-sized species primarily active at night.[37]

The South American opossums are the most diverse clade, consisting of nearly 100 species. Their most recent common ancestor lived during the Oligocene, sometime between 32 and 24 million years ago. It was small, less than 200 grams, lacked a pouch, sported a prehensile tail, and inhabited rainforests. Furthermore, it was an arboreal or scansorial species, as were most of its descendants by the end of the Oligocene. Later, some taxa, including the ancestors of the short-tailed opossums and the brown four-eyed opossum, abandoned life in trees and became terrestrial by the Early or Middle Miocene. Unlike most other South American marsupials, which have a uniform, pale-brown coat, the short-tailed opossum clade contains several brightly coloured species with reddish furs or marked dorsal stripes, such as the Touan short-tailed opossum and the northern three-striped opossum. Patterned coats may reduce the risk of predation within forests, and seem to have evolved after such species changed to a predominantly diurnal lifestyle.[38]

Subsequent diversifications occurred mainly in the vast area of Amazonia, where the Didelphimorphia remained arboreal or scansorial.[39] However, several lineages ventured out into the dry forests, shrublands and open grassy habitats that had developed by the end of the Miocene. They included the ancestors of the Chacoan pygmy opossum, a recently described species from Argentina that is now a favourite household pet because of its small size, clean nature and omnivorous diet.

Several Mexican endemics, including the gray mouse opossum, diverged 20 million years ago from the clade of South American mouse opossums and crossed the seaway separating North from South America to become dry-forest specialists. They evolved an unusual mating behaviour, hanging upside down from their tails, with the male holding tightly onto the female's neck with his jaws.

After the Middle Miocene Climatic Optimum, 16.9 million years ago (discussed at the end of this chapter), temperatures fell, global sea levels dropped, and the Andean mountains rose to form a scoliotic spine down the continent's western rim, creating many novel habitats. Among the early vertebrates seeking new opportunities were birds, reptiles, fish and mammals, including the omnivorous white-eared opossum. Because of selection pressures and genetic isolation at altitude, the pioneering population of opossums evolved denser fur, with longer outer or guard hairs and a larger body to combat the alpine climate. They became a new species, the Andean white-eared opossum, that today inhabits the Andes from Venezuela to Bolivia.[40]

A unique feature of South American marsupials, especially the opossums, was a mass extinction event during the Miocene, around 11 million years ago, due to predation and possible geographical stresses. While ancient opossums would have lived in equilibrium with their endemic predators, such as boas, 'terror birds' (phorusrhacids) and sparassodonts, they would have been unprepared for the arrival of placental carnivores from North

America. During this period, procyonid carnivorans, the ancestors of coatis and racoons, arrived by overwater dispersals and significantly culled the marsupial species, especially the terrestrial taxa.[41]

During the Pliocene, about 3 million years ago, the North and South American landmasses were reunited after the volcanic isthmus of Panama rose from the sea floor to form a land bridge. Millions of years of separation ended, allowing terrestrial and freshwater fauna to migrate in both directions – the so-called Great American Biotic Interchange. While most species moved southwards, two marsupial species crossed to North America, the Virginia opossum and the southern opossum. The Virginia opossum arrived late, around 800,000 years ago, but rapidly expanded its range northward through Central America to North America. Faced with increasingly temperate environments, the tropical species adapted by evolving an increased body size, shorter extremities and less coat pigmentation the further north they colonised. These phenotypic changes occurred in less than 15,000 years and reflected the high selection pressure faced by a mammal with a relatively slow metabolism and low body temperature.[42]

The order Paucituberculata, or shrew opossums, was the second order of marsupials to evolve, and today it consists mainly of predators that feed on insects, earthworms and small vertebrates (Figure 2.2). Their collective name is not particularly appropriate, for they lack the appearance of shrews, do not have a shrew-like lifestyle, and are not even technically opossums. The seven extant species are the only survivors from a once-abundant Oligocene dynasty.[43] Today, they are restricted to inaccessible cold and humid environments dotted along the Andes from west Venezuela to southern Chile. Shrew opossums' lower incisors have a single cusp, or protruding bump, on the crown, unlike most mammalian teeth. This diagnostic dental trait gives the order its name, Paucituberculata (Latin for 'few bumps').

The third and last of the South American marsupial orders, the Microbiotheria, consisting of the monito del monte, warrants a detailed discussion. But before doing so, we must take a trip to the frozen continent of Antarctica.

Trans-Antarctic dispersal

A long-standing puzzle in the story of mammals is how the South American marsupials reached Australia. We can discount flight and ocean crossings, as no marsupial evolved the necessary physical adaptations to achieve such journeys. As with monotremes, the existence of Gondwana, the ancient supercontinent that connected South America, Antarctica and Australia, is key. In other words, the early marsupials went by foot across Antarctica to reach Australia before the southern supercontinent's break-up.

Seymour Island, near the tip of Graham Land on the Antarctic Peninsula, provided the first tangible evidence of a Gondwanan dispersal. Despite its remote location, this ice-free island has been a gold mine for fossil hunters. Indeed, it has come to be regarded as the Rosetta Stone of Antarctic palaeontology due to the unparalleled window its fossils have provided into the region's biogeographical history.

Despite nearly a century of collecting in the area, it was not until 1984 that the first marsupial fossil was recovered from a species that belonged to an extinct family.[44] Fifteen years later, more informative fossils were unearthed from the same site, the La Meseta Formation of sedimentary rocks. Dating from the middle Eocene, around 40–45 million years ago, this collection of small 'opossum-like' species represents a number of insectivorous and frugivorous taxa.[45] At the time, Antarctica was ice-free, although with a cooling climate. As a result, the flora changed, with conifers giving way to forests dominated by southern beech (*Nothofagus*), increasing the available arboreal niches and facilitating marsupial radiation.[46]

Once Gondwana started to fragment, marsupials had already reached Australia, most likely during the Early to Middle Palaeocene (Figure 2.2). At the same time, monotremes entered South America, although they didn't survive, in contrast to those that reached Australia. However, by the Eocene, cooler conditions, freezing winters and mountain glaciation across Antarctica posed increasing difficulties for dispersal, although coastal and lowland routes remained available. The formation of the Tasmanian Passage approximately 50 million years ago would have curtailed dispersal, and once open waters formed along the whole of southern Australia, no further migration was possible.[47] Those animals left behind were ultimately doomed, as once Antarctica broke free and commenced its drift southwards, the climate deteriorated, resulting in the inhospitable, ice-covered continent of today. Frozen to death, the continent's mammals disappeared under a tomb of ice, leaving only their fossils for discovery.

But how many species crossed to Australia, and what were they like? The answer to these questions has recently been found in the most unlikely of places – hidden within the billions of nucleotide bases of the diminutive monito del monte's genome.

A relict species

The enigmatic monito del monte, Spanish for 'little mountain monkey', named after its simian-like prehensile tail, is a small marsupial native to the temperate rainforests of southern Chile and Argentina (Plate 5). The species inhabits mainly old-growth *Nothofagus* and bamboo forests, often at higher altitudes than most other small mammals. They are well adapted for climbing trees, using their tails, large paws and opposable thumbs. When food is scarce, and temperatures fall in winter, they hibernate and survive on fat stores deposited in their tails. They sleep in nests made from water-resistant leaves and mosses, built in the shelter of overhanging rocks, fallen tree trunks and tree roots.

However, the monito del monte (genus *Dromiciops*) was a puzzle for biogeographers, as morphological, genetic and fossil evidence suggested it was more closely related to Australian than South American marsupials. For example, the ankle bone articulation, as well as brain, sperm and chromosome structures, suggested an Australian origin. The only logical explanation was that the monito del monte had evolved in eastern Gondwana (present-day Australia) and had then undertaken a return journey to South America before Gondwana's break-up.[48] It would take a novel genetic approach to resolve this conundrum.

In 2010, a team from Münster University, headed by Maria Nilsson, reported a study of marsupial evolution using retroposons, or 'jumping genes'.[49] These widespread features are repetitive DNA sequences inserted randomly into the genome after being copied, or 'reverse transcribed', from RNA molecules, typically from viruses. Any new insertion is unique to that individual and will be inherited unchanged from the time of insertion. In effect, retroposons are 'molecular fossils' that scientists can track over evolutionary timescales to determine species' relationships.

Using this technique, the German scientists found a clear genetic separation between the South American and Australasian marsupials. Contrary to expectation, the monito del monte emerged as the sister taxon of all living Australasian marsupials and not, as previously thought, deeply embedded within them (Figure 2.2). Furthermore, retroposon analysis revealed that only one species made it across Antarctica to Australia and that no reverse dispersals occurred. Despite the anatomy and the fossil record, the monito del monte's gene structure provided the answer: it evolved and remained in South American Gondwana.[50]

Recently, confirmation came after a Chinese-led team created a draft genome of the monito del monte. Using supercomputers and complex algorithms to compare several marsupial genomes, they deduced that South America's relic marsupial diverged 59.7 million years ago and before the founder species reached Australia.[51]

The monito de monte is an extant member of the Microbiotheria, an order that arose during the Palaeogene in Gondwana. Fossils from at least 14 species have been identified from Bolivia to Antarctica, dating from a time when climatic conditions allowed for a rapid diversification. However, all but one went extinct in the Middle Miocene, leaving *Dromiciops* alone to inhabit the temperate forests of southern South America. The realisation that this small marsupial is a relict species, the sole living representative of a group long thought to be extinct, was a scientific breakthrough. Importantly, the species provides an unparalleled 'window to the past', given its 60-million-year lineage, minimal morphological innovation, and requirement for the same ecological niche as its Gondwanan ancestors.

But the monito del monte may not be alone after all. Recently, a closely related species was found to have evolved through geographical isolation, or vicariance. During the Miocene, the Earth's gradual cooling suffered a brief reversal 16.9 million years ago, during the Middle Miocene Climatic Optimum or MMCO. This short period of global warming was due to volcanic carbon dioxide release from across large swathes of western America. The eruptions were of such magnitude that ash and fumes reached the stratosphere and covered the landscape to a depth of 100 metres for hundreds of kilometres. Global temperatures increased by 4 degrees Celsius, melting polar ice caps, raising sea levels and causing massive flooding of southern South America. Chile became split into two islands, each containing an isolated population of ancestral monito del monte, complete with their Gondwanan-type habitats of *Nothofagus*. When water levels fell, the isolated populations had become two different species – the newly identified Pancho's monito del monte in the north and monito del monte in the south.[52] Both species are 'key species' as they are important seed dispersers and vital for their habitat's survival. Given their ecological role and the scientific insight they offer, the two vulnerable species warrant urgent protection, especially considering the current rate of their habitat's destruction and fragmentation.[53]

As highlighted, sophisticated genetic studies indicate that all Australian marsupials are monophyletic, or more closely related to each other than to any other marsupial. This finding implies that they all evolved from a single founder species. Interestingly, 'one-off' colonisations of isolated landmasses leading to significant adaptive radiations are not unique in our story of mammals. As we will see, transoceanic sweepstake dispersals of only a few individuals gave rise to the New World primates, the South American caviomorph rodents (guinea pigs, capybaras, chinchillas and agoutis) and the tenrecs and lemurs of Madagascar. Nevertheless, it is remarkable that only one species founded Australia's entire empire of marsupials, given that South America, Antarctica and Australia were all connected via Gondwana for millions of years.

It is now time for us to travel to Australia, to see what happened when the single monito del monte-like species arrived on that continent.

CHAPTER 3

The Marsupial Mole's Story

AUSTRAL DOPPELGÄNGERS

Wherever there is a design that is highly successful in a broad range of similar environments, it is apt to emerge again and again, independently.
Daniel Dennett (1942–2024)

The monito del monte-like founder population reached Australia approximately 60 million years ago, before the continent broke free from Antarctica and drifted on its incremental journey northwards. However, the new arrivals did not have the continent's rainforests to themselves, for their ancient cousins, the early therians (soon to face extinction), and the egg-laying monotremes were already in residence. Once isolated, the vast continent became a gigantic natural experiment that culminated in an array of unique and extraordinary marsupial taxa. Indeed, within the first 3 million years, the ancestral species had diversified to produce the four orders we recognise today: marsupial moles (Notoryctemorphia); bandicoots and bilbies (Peramelemorphia); Tasmanian tiger or thylacine, numbat, dunnarts, Tasmanian devil and quolls (Dasyuromorphia); and the koala, wombats, gliders, honey possums, ringtail possums, kangaroos, wallabies and potoroos (Diprotodontia).

Approximately 334 taxa evolved from the single founder species in Australia and parts of Southeast Asia, leading to the development of some of the best-known marsupials on Earth. However, the rapid pace of their diversification has made it challenging for phylogeneticists to determine the exact sequence of evolutionary events.[1] Consequently, their phylogeny is better represented as a network rather than a traditional branching tree. Rapid diversification often generates ambiguous and inconclusive genetic signals, influenced by evolutionary processes like hybridisation and incomplete lineage sorting (explained below). The most robust phylogeny, published by a team from the University of Sydney led by David Duchêne, was derived using thousands of DNA sequences from representatives of most marsupial

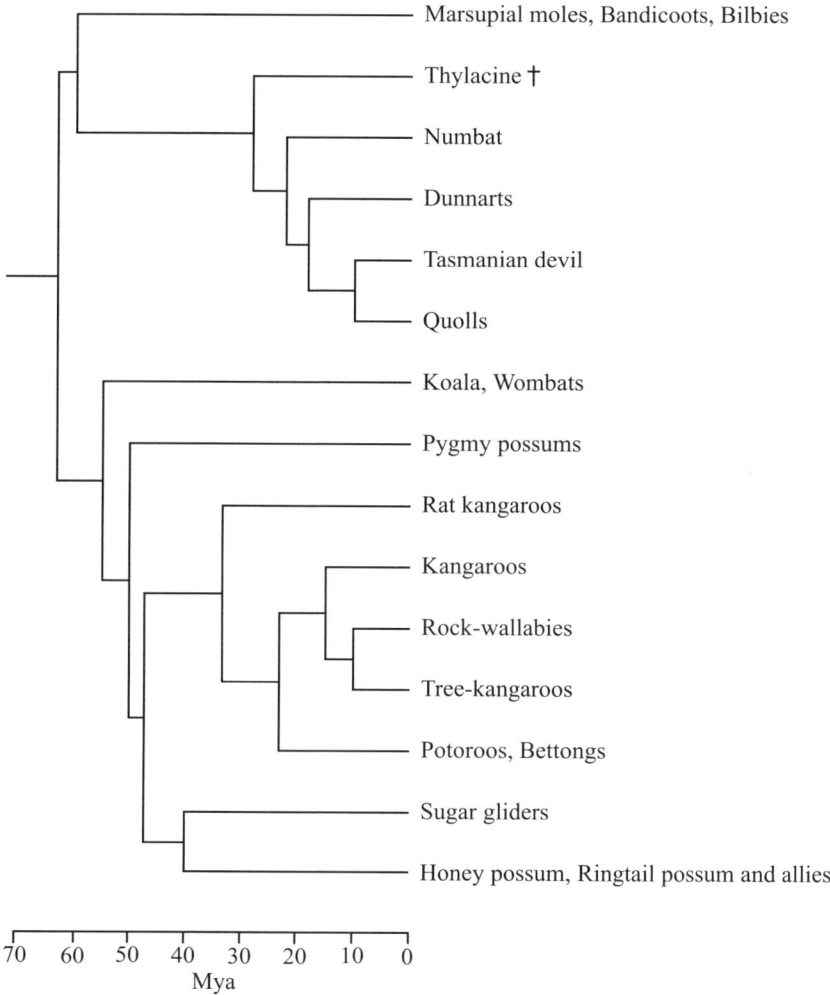

Figure 3.1 Phylogeny of Australian marsupials based on molecular analyses. †, extinct; Mya, million years ago. Modified from Duchêne et al. (2018).[2]

families.[2] While not the last word on the subject (plans are under way to sequence the complete genome of every Australian marsupial), it is the phylogeny followed in this book (Figure 3.1).

Before we meet the oldest lineage, which gave rise to the enigmatic marsupial moles, it is worth exploring incomplete lineage sorting further. It is a phenomenon that has complicated phylogenetic inference in many lineages, not just the monito del monte and the Australian marsupials, but also several eutherian groups, including marine mammals and hominids.[3] Furthermore, ILS can provide insights into speciation times, ancestral population sizes, and patterns of natural selection. Despite its complexity, ILS is worth understanding, as we will revisit the topic in subsequent chapters.

Incomplete lineage sorting

Incomplete lineage sorting (ILS) describes a feature of population genetics where variants or alleles of a specific ancestral gene generate a gene tree that is incongruent with the species tree. In other words, a tree constructed from DNA sequences for a given gene or genes does not agree with the tree representing the correct or 'true' evolutionary pathway of the species in question.

Let us discuss how this situation might arise by envisioning a set of related populations which have all descended from a single ancestral population. Like most founder populations, our ancestral population will have a degree of genetic diversity and contain multiple alleles for many of its genes. In a true or species tree, one would expect these alleles to 'sort' into the various descendent populations, such that population A has only one allele, population B another, and so on – a situation known as 'complete sorting'. However, suppose that each new population contains a random or stochastic mixture of alleles, a more likely scenario if speciation has been rapid or recent. In this case, the resultant 'incomplete sorting' of alleles will result in a gene tree that is 'discordant' with the species tree. Some variants will inevitably assort into a discordant pattern for a population with thousands of genes with multiple alleles. Additional complexities include new alleles that may arise by mutation or others that disappear due to negative selection or genetic drift. Nevertheless, the greater the number of genes sequenced, the less problematic ILS becomes for phylogeneticists.

In 2022, Shaohong Feng and colleagues clarified the situation in marsupials by comparing the whole genome sequences from various species.[4] They found that more than half of the genes did not follow the evolutionary pattern of the species tree, indicating that ILS has been a dominant force in marsupial evolution. As in previous studies, the analysis of a limited set of genes falsely clustered the monito del monte deep within the Australian clade, while the use of complete genome sequences confirmed the species to be a sister taxon.

Feng's group then explored the consequences of ILS in a scientific *tour de force*. They identified several genes affected by ILS and determined the phenotypic effects of the different alleles in transgenic mice. For example, the gene *WFIKKN1*, which affects the shape of vertebral bones, is present in several allelic forms. The team noted that the monito del monte and the koala possess the same allele (glutamine at position 76), whereas the Tasmanian devil carries a different variant (arginine instead of glutamine). Interestingly, the length of the vertebral spinous projections mirrored the allele type, with the monito del monte and koala having a long spinous process. In contrast, the Tasmanian devil has a shorter bony projection despite being more closely related to the koala than the monito del monte is. When the two alleles were introduced into transgenic mice, they resulted in vertebrae with the corresponding phenotype – long spinous projections for the glutamine allele and shorter spinous projections for the arginine allele. Similar findings were obtained for another ILS gene, *PAPSS2*, that influences the structure of the humerus bone.

The discovery that ILS can result in similar phenotypes in distantly related species is relevant to our understanding of evolution. In the past, morphological similarities in phylogenetically distant species were explained by convergent evolution. However, when ILS is involved, the scenario changes and becomes more complex. In the latter situation, the phenotypic similarities may relate to alleles present in the ancestral species. In the case of marsupials, the long vertebral spines did not arise multiple times as independent evolutionary events. Instead, they originated once, with the relevant gene inherited by distantly related species. In other words, the inheritance of the same ancient gene must be considered before any phenotypic similarities in distantly related taxa can be attributed to convergent evolution.

Despite such caveats, the Australian marsupials provide biologists with a dramatic cast of 'alternative mammals' with phenotypes and behaviours that mirror placental mammals on other continents – each an example of true convergence rather than the effects of ILS. In the remainder of this chapter, we will further explore this striking gallery of austral doppelgängers and the evolutionary lessons they provide.

Magnificent substitutes

The marsupial moles are the least diverse but most extraordinarily distinct of the four orders of living Australian marsupials. The two species, the northern and southern marsupial moles, are confined to deserts and spend most of their lives tunnelling beneath the vast, soft red sands that straddle the greater part of arid South Australia, Western Australia and the Northern Territory. Although both species have been known to science for over a century and to the indigenous peoples for millennia, they remain among the continent's least-known creatures. Indeed, marsupial moles are hardly ever encountered, and with only 5–10 sightings a year, they are either very rare or exceptionally elusive.[5]

To survive such an unforgiving environment and their extreme subterranean or fossorial lifestyles, the 'moles' have evolved a raft of specialised and bizarre traits that have created bulldozer-like digging machines. They include a conical skull, tubular body shape, fused cervical vertebrae, absent external ears, reduced tail, and dense, short fur. They have also evolved bony armour to protect their sensitive snout, excavator-like claws on their forelimbs, massive limb musculature and, in common with most burrowing marsupials, pouches that open posteriorly to prevent the incursion of sand. Vision is no longer an asset, and their delicate eyelids are fused shut, covering eyes that lack pupils and lenses. To compensate, they have evolved an acute sense of smell, and their hearing is tuned to low-frequency wavelengths to find their way around and locate their insect prey in total darkness. Indeed, most individuals spend their entire lives underground, as, on the surface, they are sluggish, clumsy, and vulnerable to predation.

Molecular-clock studies show that Australia's moles first evolved approximately 50 million years ago, during the Eocene or Palaeocene.[6] This date is problematic, as desert environments didn't appear in Australia until 1 million years ago, which is not enough time for the moles to have evolved their many unique traits. However, in 2011, Michael Archer and his colleagues from New South Wales resolved this apparent impasse after describing 20-million-year-old fossils from the Riversleigh World Heritage fossil site in north Queensland. Although only a small collection of teeth, partial jaws and limb bones, the fossils provided enough clues to identify the owner – an early 'transitional' marsupial mole that lived in the continent's rainforests. The morphology of the animal's humerus and ulna showed that the species had already evolved to burrow in the wet, soft soils of the forest floor. At the same time, its teeth suggested a diet of soft-bodied subterranean invertebrates.[7] In other words, the earliest moles were serendipitously preadapted to cope with the drier environments that developed later in central Australia, as the rainforests retreated to coastal areas and the first sandy deserts arose in the Pleistocene.

The phenotype of marsupial moles superficially mirrors that of the placental moles – the talpid moles of Eurasia and North America and, most notably, the golden moles of Africa. However, there are subtle differences – tail lengths, tunnelling techniques, presence or absence of a pouch, and the number of fingers evolution has used to fashion their spade-like front paws. Nevertheless, their raft of similar anatomical features, all adaptive solutions to overcome similar environmental challenges, evolved independently in lineages that diverged over 125 million years ago from a non-digging ancestor – a paradigm of evolutionary convergence. Indeed, the placental moles are more closely related to humans than to their austral counterparts.

The dental material from Riversleigh held another secret, proof of evolutionary convergence rather than parallelism. The 'transitional' morphology of the molars proved that their cusp structure – a characteristic L-shaped arrangement that scientists arcanely term zalambdodonty – evolved from the tribosphenic teeth of their ancestors. However, the same L-shaped molar cusps of the placental moles evolved differently, although they look the same morphologically. These observations confirm that highly specialised anatomical patterns, such as teeth structure, can be acquired via alternative evolutionary pathways in distantly related species. Thus, in the case of marsupial and placental moles, zalambdodonty has arisen by convergence and not by the same evolutionary path or parallelism.

Bandicoots (order Peramelemorphia) appeared very early in the evolution of Australasian marsupials, approximately 30–40 million years ago during the late Palaeogene.[8] They are like rabbits with well-developed hindlimbs, reflecting their hopping mode of locomotion, and long ears, emphasising the importance of hearing. Two species of pig-footed bandicoot evolved front toes that resembled tiny hooves like those of placental ungulates, although sadly they were extinct by the 1950s.

The omnivorous bandicoots are the only marsupials that develop a highly invasive chorio-allantoic placenta, like the eutherians. Instead of the yolk-sac placenta of other marsupials, bandicoots use the allantois, the membrane that amniotes evolved to enable gaseous and waste product exchange. However, this anatomical structure only replaces the yolk-sac placenta during the last few days of their short pregnancy. In fact, bandicoots have one of the shortest pregnancies of any mammal – just 12 and a half days for the northern brown and long-nosed bandicoots. Furthermore, the neonate remains attached for a short time by an umbilical cord that runs like a taut string from the placenta to the pouch. As a result, they have the most rapid postnatal growth of any marsupial, suggesting that their placenta is a more efficient exchange organ than most. Intriguingly, the bandicoot's placenta could represent the survival of an early stage in the evolution of mammalian placentae, and it suggests that the role of the allantois became progressively less important during marsupial evolution.[9] As we have discussed, this may reflect marsupial adaptation to Australia's harsh environment, providing a selective advantage for the mother who can abort her offspring in times of hardship.

The next order to evolve, the Dasyuromorphia, gave rise to most of Australasia's marsupial carnivores. The 80 or so species are mainly small, shrew-like taxa, collectively known as marsupial mice, while others are cat-sized, such as the quolls and Tasmanian devil. However, the two best-known species – the ant-eating numbat and the extinct thylacine, or Tasmanian tiger – have a different story to tell and will be the subject of the next chapter.

The 15 species of marsupial mice (genus *Antechinus*) live in the fast lane, dying from exhaustion after an orgy of winter sex, often reported in the popular press as 'big bang reproduction'. Males live for exactly 47 weeks and succumb to stress-induced immune breakdown two weeks after mating. Before their breeding season, males secrete large quantities of testosterone and corticosteroids. Although the hormones encourage promiscuous sexual activity, they lead to stress and eventual failure of the animal's internal organs, with death usually by internal bleeding or infection. Females live longer, up to half raising two litters, while the males die before seeing their offspring. Each local population breeds simultaneously, with most females giving birth on the same day to coincide with the summer insect explosion. Death after a single reproductive event, known as semelparity, is well recognised in insects, including mayflies, and some species of spiders and molluscs. However, it is rare in vertebrates, otherwise occurring only in lampreys, freshwater eels, salmon, some trout, and a single lizard species.

So why would natural selection favour programmed death after a single reproductive episode in these carnivorous marsupials, given that evolution tends to maximise total lifetime reproductive output? One clue is that species where sex is suicidal typically produce more offspring in their single reproductive episode than related species. When animals do not have to withhold valuable resources to ensure future survival and reproduction, they can utilise all available food for a single, massive breeding event. In other words, the increased summer fecundity of marsupial mice more than compensates for the loss of additional reproductive episodes at different times of the year. It seems that the females manipulate male sexual behaviour to increase their own reproductive success. According to Diane Fisher from the University of New Mexico, males increase their sex drive at the expense of survival because females ultimately profit from sperm competition. Simply put, the brief season of plenty allows females to impose severe sexual selection pressure on males by shortening the breeding period and mating with extreme promiscuity.[10] Semelparity must be an advantageous strategy, for the little red kaluta, a nocturnal species which lies on a separate evolutionary branch to other dasyurids, also exhibits suicidal reproduction. Whether semelparity evolved by independent lineage-specific molecular changes or shared molecular events in the two marsupial branches is unknown. However, the recent publication of the chromosome-level genome of an antechinus could pave the way to resolving this issue.[11]

Marsupial cats belong, ironically, to the same order as marsupial mice, although the former are larger adaptations to the predatory way of life. The quolls, or tiger cats, are aggressive species that can take on larger animals, such as placental cats and dogs. Indeed, the recently described bronze quoll from New Guinea was given the scientific name *Dasyurus spartacus* after the Thracian gladiator who led a bloody revolt against the Romans. Like cats, they are primarily solitary, with brown or black fur, and hunt by stalking their prey, including small mammals, birds and reptiles, before despatching them with a powerful bite. Each species lives in a distinct geographical area of New Guinea and Australia. For example, the tiger quoll and eastern quoll inhabit moister regions, while the western quoll has adapted to drier habitats. The quoll lineage evolved around 15 million years ago in the Miocene, and the six extant species diverged around 4 million years ago.

However, the most spectacular marsupial cat is the Tasmanian devil, a close relative of the quolls, now confined to Tasmania after becoming extinct on the mainland. A fierce-looking animal the size of a dog, it is the largest living carnivorous marsupial in the world and is Tasmania's apex predator. Its large head and neck allow it to generate the most forceful bite per unit body weight of any living predatory land mammal. The devil lineage split from that of quolls in the Miocene, between 10 and 15 million years ago. At this time, Australia suffered a climate upheaval which changed the environment from a warm and moist one to an arid, dry ice age that led to mass extinctions. Only a few predators survived, including the ancestors of the quolls and devils, as most of their prey died of the cold. The devils likely arose to fill a niche in the ecosystem as scavengers.[12]

Since 1996, the ferocious devils have been dying from a rare contagious cancer – devil facial tumour disease (DFTD). This affliction is spread by tumour cell inoculation following bites that form part of the species' social behaviour. Despite an 80 per cent loss, some populations have stabilised and developed resistance. To find out why, scientists have compared the genomes of survivors and diseased individuals, and have found that resistance results from rapid evolution in at least seven genes, five of which underpin immune function or cancer risk in other mammals, including humans.[13] The surprise, however, is the speed with which the devil's genome has been modified – in just over 16 years or approximately eight generations.

The devil's story emphasises the importance of gene-pool size and heterogeneity for a species' success. Genetic variation is crucial for long-term survival, allowing for rapid evolution when adverse environmental changes occur. Conversely, small, isolated populations with restricted genetic repertoires are less resilient. Worryingly, DFTD has caused the devil population to crash and fragment with subsequent inbreeding and genetic weakening.[14] Indeed, the reproductive decline may herald the species' demise, as according to some scientists, the devils are on the brink of an 'extinction vortex'. Tasmania's ecosystem is also suffering from a trophic cascade, as the loss of its apex predator is causing dramatic changes in its structure and nutrient recycling. The hope is that conservation efforts can augment isolated populations with genetic material from resistant individuals.

The koala, the marsupial equivalent of the placental sloth, is the lone representative of its family (Plate 6). It is an iconic Australian species, instantly recognisable by its round, humanoid face and distinctive body shape. Fossil evidence identifies as many as 15–20 species following their divergence from wombats 30–40 million years ago, although only the koala survives today. It is a specialist arboreal folivore and feeds almost exclusively on leaves from a few species of *Eucalyptus*. However, *Eucalyptus* leaves are thick and loaded with fibres and pungent oils, which lower the nutritional value and irritate the intestine. To circumvent the problem, koalas have evolved an enlarged appendix-like structure, the caecum, which enables the slow digestion of their food. At the same, mucus-like secretions protect the lining of the digestive tract.

Adult koalas also carefully select the best leaves, often sniffing before tasting them, to target the nutrients and to avoid toxic substances, including terpenes and phenols. To do so, they have expanded the number of *TAS2R* genes that code proteins that detect bitter tastes. These genes have undergone multiple duplications, so koalas now possess 24 copies, more than any other mammal. Furthermore, they have expanded their nasal receptor genes, so-called *V1R* genes, that help identify non-volatile odorants. Koalas have six *V1R* genes, compared to only one in the opossums and none in birds, monotremes, mice, dogs and humans.[15] Finally, the koala's hindgut is home to a variety of specialist microbes, including the Synergistaceae, that aid the fermentation and breakdown of cellulose, lignin and tannins, but especially the toxins of *Eucalyptus*.[16]

Convergent evolution also accounts for one of the most unusual features of koalas. In the 1990s, while working with the animals at a safari park in Adelaide, Maciej Henneberg noticed something extraordinary: koalas possess human-like fingerprints. As a professor of anthropological and comparative anatomy interested in forensic science, he knew that fingerprints were only associated with primates. Furthermore, even under the microscope, he could not distinguish the koala whorls and loops from those of human prints, and suggested that they were good enough to fool detectives at a crime scene.[17]

But why would fingerprints be advantageous from an evolutionary standpoint? According to a recent study, fingerprints may act in conjunction with sweat glands to help maintain grip. A research team from Seoul, South Korea, working with primates, found that fingers and toes release moisture when in contact with hard surfaces, softening the skin to build up friction.[18] The newly pliant skin presses up against the surface and blocks off the sweat pores, allowing evaporation to catch up and maintain friction and grip. This dual mechanism is advantageous in both dry and wet conditions, enhancing hand manipulation and locomotion. The remarkable point is that koala prints have evolved independently, as primate and koala lineages diverged at least 125 million years ago, and no other species possesses them.

Wombats, the koala's sister species, are often considered the marsupial equivalent of the North American groundhog. At around 35 kilograms, they are the world's largest burrowing herbivores and possess powerful short legs equipped with long claws for digging. As a result of their abrasive diet of tough grasses, all wombat teeth lack roots and are ever-growing, like the incisors of rats. Three species exist today, of which the largest and most widespread is the common wombat, but as recently as 40,000 years ago, there were many more, some as heavy as 360 kilograms.[19] Wombats have an extraordinary metabolism, taking around 8–14 days to complete digestion, and, like the koala, possess a large bowel populated by plant-digesting microbes. However, unlike their sister species, their intestines do not host the rare bacteria required to detoxify *Eucalyptus* leaves. In addition, wombat intestines contract in sections for several days, squeezing the faeces to extract the maximum amount of nutrients and water. This extreme adaptation to an arid environment results in the excretion of firm, cube-shaped faeces, the only species known to do this. American and Australian scientists published the explanation for this phenomenon in the appropriately named Royal Society of Chemistry's journal, *Soft Matter*. Not surprisingly, they were awarded the prestigious Ig Nobel Prize in 2019, a satiric prize for 'research that first makes you laugh and then makes you think.'[20]

Kangaroos and allies

The most iconic of Australia's terrestrial animals are the 'large-footed' marsupials (family Macropodidae), which include kangaroos, wallabies, rock-wallabies and tree-kangaroos. While their distant ancestors were arboreal, the Macropodidae descended from the trees around 8 million years ago and diversified to produce nearly 70 species that inhabit Australia and New Guinea today.

The 'kangaroo' lineage consists of four extant species (red, eastern grey, western grey and antilopine) that evolved from a common ancestor that lived around 9 million years ago. Their evolution and diversification resulted from the emergence of the extensive grasslands that arose during the Pliocene when Australia was near its present position.[21] Kangaroos evolved specialised dentition to cope with their diet of grass. Teeth became longer with high crowns to help resist the increased dental abrasion from the diet's silica and dust content. Their large feet and powerful hind legs enable them to bound along at more than 56 kilometres an hour and to leap more than 9 metres in a single hop. While hopping is efficient for fast travel, it is not ideal for strolling since, unlike humans, kangaroos cannot easily move their hind legs independently. To compensate, they have evolved a unique form of walking called pentapedal locomotion, which involves forming a tripod between hands and tail and then moving both hind legs simultaneously.

Hopping, however, did not first evolve within the kangaroo lineage but was a feature of the Macropodidae throughout their 20-million-year history. As evidence, a recent Swedish study by Wendy Den Boer and colleagues examined the fossilised ankle bones of the kangaroos' ancient arboreal relatives. The team concluded they possessed various ways of moving amongst the trees, including bounding, climbing and hopping.[22] Notably, the study quashed the long-held view that hopping first appeared after the continent's climate became drier and grasslands formed. While this trait may have evolved initially to aid escape from predators and free up their forelimbs for better food handling, it proved a very efficient way for kangaroos to move around. Indeed, hopping provided a tremendous survival advantage for those living in arid environments where food and water are in short supply. Interestingly, endurance hopping was only one of a range of gait types employed by the first terrestrial Macropodidae but was the only locomotory mode to survive.[23]

Rock-wallabies are a genus of mainly nocturnal marsupials, which, as their name suggests, inhabit fortress-like rocky outcrops that offer protection from predators and the environment. They fill similar niches to mountain goats in other parts of the world. All 17 species are highly agile and can climb and descend steep, craggy cliffs with ease and precision and climb trees to locate their favourite leaves.

The rock-wallabies intrigue evolutionary biologists, as their chromosomes provide unexpected insights into how new species might form. The Australian scientist Sally Potter and her colleagues from the Australian Museum Research Institute studied six closely related species from northeast Queensland.[24] They noted that despite their remarkable similarity in nearly every respect, each species differs in the number and shape of its chromosomes. Such genetic differences should make gene flow between the species impossible, since hybrid offspring would either be non-viable or have reduced fertility. However, contrary to expectations, they found no correlation between the degree of chromosomal variation and the extent of gene flow between species. The mechanisms driving speciation seem more complex than mere incompatibilities caused by gross chromosomal alterations. However, no explanation exists for how such disparate chromosomal complements could pair up during fertilisation and form viable embryos. One possibility is that a chromosomal chunk containing genes essential for an individual's unique adaptation, hence speciation, becomes fixed and unable to move. In contrast, the rest of the chromosomal content is free to mix. If these different, non-transmissible gene segments exist, it could help explain how the separate rock-wallabies have arisen and aid our understanding of speciation in general.

Tree-kangaroos diverged from rock-wallabies between 5 and 7 million years ago and returned to the arboreal lifestyle of their distant cousins.[25] Although drivers for this evolutionary reversal remain unclear, suggestions include better protection from predators, exploitation of untapped food resources, or escaping faecal-borne infections. Irrespective of the cause, the early tree-kangaroos underwent significant and rapid morphological adaptations. Their feet became short and broad, while their front limbs became muscular and robust, enabling gripping and climbing. The most interesting adaptation, however, was the evolution of flexible ankle joints.[26] Most macropods possess rigid ankles to facilitate hopping, a trait that evolved from the flexible joints of their earliest arboreal ancestors. In other words, tree-kangaroos have re-evolved the ancestral ankle joint, an adaptation which seems to defy Dollo's law of irreversibility, a principle proposed by the nineteenth-century Belgium palaeontologist Louis Dollo.[27]

Potoroo genomes and past habitats

Approximately 25 million years ago, potoroos and their relatives, the bettongs, separated from the kangaroo lineage (Figure 3.1). These small, rabbit-sized creatures primarily feed on underground fungal fruiting bodies or truffles. Their feeding habits make them key environmental engineers, crucial to maintaining the health of forest ecosystems. Unfortunately, all species are now threatened, mainly due to the impacts of human colonisation, including forest fires, land clearing, agriculture, and the introduction of predators such as foxes and cats. Indeed, the critically endangered Gilbert's potoroo has the dubious distinction of being the world's rarest mammal, with an estimated population of less than 100. Intensive conservation efforts are under way, including the creation of offshore safe havens, the control of feral predators, and the use of assisted reproduction.

Genetic studies of the long-nosed potoroo, one of three extant species, have enabled biologists to determine the continent's past environmental conditions. Potoroos like rainforests with thick ground cover and avoid open or dry habitats. During the Pleistocene, when climatic oscillations caused rainforests to contract and expand, populations of potoroo became isolated. Such fragmentations led to reduced gene flow and genetic differentiation, the signatures of which can still be detected in their genomes today. For example, a study conducted by Greta Frankham from the Australian Centre for Wildlife Genomics found that the potoroos of Tasmania were genetically different from those on the mainland.[28] This finding was surprising since, during the Pleistocene, land bridges should have allowed population dispersal and genetic mixing. However, dating the genetic differences revealed that the last gene flow between the island and mainland potoroos was around 2 million years ago. This finding indicates that the subsequent land bridges, the latest approximately 14,000 years ago, must have been too arid, lacking the required forest habitat for potoroo dispersals.

Even more unexpectedly, Frankham's team found a deep genetic split between potoroo populations across the Sydney basin, an area of Australia with no known past biogeographic barriers. Since the genomic split occurred around 6 million years ago during the Miocene, there must have been a significant break in the forest habitat preventing potoroo gene flow. Indeed, the genetic divergence is so pronounced that the long-nosed potoroo populations north and south of Sydney could represent two separate species.

The honey possum's adaptations

The tiny, mouse-like honey possum (also known by its indigenous name, noolbenger) of Western Australia is one of the most unusual marsupials, with no close relatives (Figure 3.1). Since diverging from its ancestors 38 million years ago, the honey possum evolved a suite of bizarre traits to cope with a diet of nectar and pollen. Excluding bats, it is the only entirely nectarivorous and palynivorous (pollen-eating) mammal. Due to its unique suite of morphological and physiological adaptations, scientists arcanely regard the species as a 'paragon of autapomorphic specialisation within the Diprotodontia'.[29]

Western Australia is the only place in the world where the species could have evolved, because of its all-year-round requirement for nectar and pollen. Its adaptive features include a pointed snout with a narrow mouth forming a straw-like structure for extracting nectar. Furthermore, its long, cylindrical tongue is covered in bristles and can reach deep inside tubular flowers. With no need to chew, its dentition has degenerated and is associated with ill-formed maxillary and mandibular bones. Most of its teeth have disappeared or become stump-like pegs, except for the two at the front that help guide the tongue to its target while flicking in and out three times a second. It has a modified gut with a high passage rate, low nitrogen requirements, and a permanently polyuric kidney to cope with the high water content of its diet. Honey possums can produce more urine than their body weight daily.

The species is also an important pollinator and conveys microspores on its head and body as it travels between its nectar sources. Many flowers, including several *Banksia* species, with their fruiting cones and flower heads, depend on the marsupial for pollination, and both partners have evolved to maximise their mutualistic interactions. It will come as no surprise, therefore, that the honey possum is a keystone species in the ecology of the coastal sands of Western Australia.

Gliders and molecular convergence

Australasia is home to approximately 70 species of possum, a number that is likely to increase with further genetic analyses. Three lineages have independently evolved the ability to glide during the last 15 million years, a mode of locomotion that differs from the 'true' or flapping flight of placental bats. During the Miocene, rainforests covered most of Australia, and arboreal marsupials leapt from branch to branch amongst the dense canopies. However, as the continent drifted northwards, trees became less abundant, with fewer branches, so marsupials had to leap greater distances. As a result, some species evolved the ability to glide using a thin membrane of skin called a patagium that stretches between their limbs and body.

Today, six of Australia's mammals can glide between trees, some up to 100 metres. They range from the 1.7-kilogram southern greater glider to the feathertail glider or pygmy gliding possum, the world's smallest gliding mammal at only 10–15 grams. There could be even more glider species in the future, as some possums, including the lemuroid ringtail, are partway through evolving their own patagia. In addition to flaps of skin, gliders have evolved large, bulgy eyes set far apart to allow for more accurate triangulation of flight paths and landing spots. Some gliders can alter their direction in midair by as much as 90 degrees by changing the curvature or angle of the patagium. The feathertail glider has even evolved ridges and sweat glands on its feet to create surface tension so that its footpad acts like suction caps that can adhere to glass.

The fact that flight membranes repeatedly evolved in both marsupials and placental mammals (bats, flying squirrels, colugos) suggested that such convergence may reflect shared developmental programs that pre-date flight. In 2023, a Princeton-led team addressed this question by studying two mammals, the sugar glider (marsupial) and the Seba's short-tailed bat (eutherian).[30] In both species, the fully formed patagia contained several tissues, including hair, a rich nerve supply, connective tissue, and thin sheets of muscle. However, in the earliest developmental phase, the membranes consisted mainly of two layers of skin – the inner dermis and the outer epidermis. The researchers noted that the initiation of these early features correlated with up-regulation of a gene called *Wnt5a*, which codes for a signalling protein involved in cell fate and patterning in embryogenesis. A series of experiments involving cultured skin tissue and genetically engineered mice showed that the expression of extra *Wnt5a* produced early patagium formation in both species. Perhaps even more remarkably, the same pattern of *Wnt5a* expression occurs during the outer ear or pinna development in laboratory mice, a species with no flying ancestry.

These results suggest something profound. The role of the signalling molecule required for patagium formation must have evolved long before mammals took to the air. Indeed, *Wnt5a* originally had nothing to do with flight but contributed to the development of various unrelated traits, including the growth of limbs, brain, skeleton, outer ear and adipose tissue. The researchers went on to find additional evidence of molecular convergence. During early patagium development, many other genes were found to have been redeployed or co-opted, including a raft of genes involved in limb development. Amazingly, the Princeton study reveals that deeply conserved genetic toolkits underpin the convergence of flight in both gliders and bats despite at least a 130-million-year separation of lineages. As we have discussed, evolution is economical and uses whatever is available. The repurposing of *Wnt5a* for the development of patagia is yet another example of exaptation, a recurring theme in our story of mammalian evolution.

New Guinea's marsupials

Many descendants of Australia's early marsupials populate New Guinea today: bandicoots, quolls, tree-kangaroos, wallabies and gliders. For example, of the 14 recognised species of tree-kangaroo, 12 are endemic to New Guinea, while the remaining two taxa inhabit the rainforests of north Queensland. Like most macropodid lineages, tree-kangaroos are thought to have originated in Australia but appear to have subsequently undergone a series of secondary radiations within New Guinea. The first radiation, during the Miocene/Early Pliocene, was associated with the formation of the island's mountains, while another was related to climate change during the Pleistocene.[31] The distribution of Australia's two species, Lumholtz's and Bennett's tree-kangaroos, is relictual, as fossil evidence reveals that their lineage was once widespread throughout the continent's eastern and southern forests. All of New Guinea's tree-kangaroos are scarce and difficult to locate today. I have only ever seen one: a rescued Goodfellow's tree-kangaroo at a locale in the central highlands of Papua New Guinea.

Overland migration, however, was not solely one-directional. Before the Pleistocene, at least one marsupial migration occurred from New Guinea back to Australia, resulting in the evolution of the continent's sugar, squirrel and mahogany gliders.[32] Around 8,000 years ago, rising sea levels created the Torres Strait, which prevented further migrations.

There is another family that has thrived and diversified in New Guinea, one we haven't yet covered – the cuscuses. These marsupials, about the size of a cat, are the largest members of the possum family. With their rounded heads, small hidden ears, thick fur and prehensile tails, which help them climb, they resemble marsupial lemurs owing to their convergent features. The ancestors of cuscuses likely dispersed overland from Australia to New Guinea at the beginning of the Late Miocene. Their arrival at the same time as other marsupial lineages, including bandicoots and quolls, between 11 and 9 million years ago, favours a land route rather than multiple long-range sweepstake dispersals over water.[33] This view is consistent with the origins of modern New Guinea, which formed from colliding tectonic plates around 12 million years ago.[34] For millions of years, New Guinea and Australia shared very similar ecosystems: ancient forests consisting of eucalypts, casuarina trees, tall beech trees and a dense undergrowth of giant ferns. The island's central mountains would not have been as high as today, but they were colder, by as much as 6 degrees Celsius, with marked rainfall and more dense and shorter vegetation. Unsurprisingly, many ancestral marsupials settled and thrived in New Guinea, especially as competition was limited.

The ancestral cuscuses originated on the Sahul Shelf's northern edge, a landmass comprising Australia, Tasmania, New Guinea and the Aru islands. They then dispersed across New Guinea and reached many offshore and distant islands. Such colonisations and subsequent isolation events led to the emergence of new species and account for the endemics found today on the Indonesian islands of Waigeo, Gebe, Sulawesi and Ternate. How the animals made the water crossings remains unclear. Sweepstake dispersals are the most likely explanation, aided by the closing of waterways and the shifting of currents that started around 10–15 million years ago.[35] However, tectonic activity remains a possibility. Between 17 and 15 million years ago, a section of New Guinea's Vogelkop peninsula detached and drifted eastwards to fuse with Sulawesi around 5 million years later. Indeed, Denise Raterman and her colleagues from the University of California believe that the Sulawesi bear cuscus and pygmy cuscus evolved from ancestors transported to the island in this way.[36]

Readers can now appreciate why I came to regret not giving that Waigeo cuscus more attention. At the time of my first encounter, I was ignorant of the species' remarkable, around-the-world, 140-million-year odyssey. Science has now shown that from an early population of therians in Australia, their ancestors spread to Asia, split from the eutherians in North America, travelled to South America via the Caribbean, and crossed Antarctica to reach Australia for a second time. Then, after the sea levels fell, they crossed overland to New Guinea before rafting to Waigeo on a mat of floating vegetation. Their ancestral journey was global, involving five continents and terminating a short distance north of their starting point. Now that's what I call an ascent!

Unlike New Guinea, no mammals made it to New Zealand except fur seals, sea lions and three bat species. New Zealand split away from eastern Gondwana around 85 million years ago, and by the time the monotremes, marsupials and placentals had reached South America and Antarctica, the ark had long gone, and alternative passages were unavailable. Instead, New Zealand's birds came to occupy the vacant ecological niches (discussed below).

Cuscuses and human prehistory

Shimona Kealy, an archaeologist and palaeobiologist from the Australian National University, constructed a molecular phylogeny of cuscuses and gained an unexpected insight into human prehistory.[37] Her study confirmed what was previously suspected – humans must have introduced the northern common cuscus to the Indonesian island of Timor.

Although the northern common cuscus is widely distributed throughout New Guinea, it also occurs on islands that have never been connected to the mainland, including New Britain, New Ireland, the Solomon Islands, and many remote Indonesian islands. Some are volcanic, with regular eruptions that would have repeatedly eradicated all life. Scientists suspected that human involvement was likely, but it was not until the 1980s that evidence was forthcoming. In New Ireland, archaeologists discovered cuscus bones in a cave dating from 10,000–20,000 years ago, coinciding with human occupation.[38] Cuscus introduction also occurred in the Solomon Islands, although slightly later. It is in this context that Kealy's study is so interesting, for it extends the role of human transportation of the northern common cuscus as far as Timor in the west.

These findings imply that the early Melanesians must have managed cuscus populations thousands of years before they planted crops and became farmers. Indeed, cuscus transportation may be the oldest evidence for animal husbandry, even preceding the Agricultural Revolution and livestock cultivation in the northern hemisphere. Such conclusions are at odds with the anthropologist's view of a supposed boundary between prehistoric hunter-gatherers and modern farmers. For the early Melanesians, however, cuscuses provided an essential source of protein and fat, as they were easy to keep and transport in their canoes. Even today, islanders regard the cuscus as a vital food source, and education programmes are being implemented to help people understand how to manage and sustainably harvest the species safely.

Let us now turn to the role of constraint in marsupial evolution, as this phenomenon may have blocked specific phenotypes from ever occurring.

Constraint hypotheses

The dramatic speciation of Australia's marsupials was helped by the continent's wealth of habitats, from tropical rainforests to snow-capped mountains, deserts and grasslands. Evolving in 'splendid isolation' as the continent travelled northwards, and with no mammalian competitors except the fading, ancestral monotremes and the now-extinct multituberculates, the marsupials could adopt almost any conceivable lifestyle.

Why, therefore, are marsupial species fewer and less diverse than placental mammals? Despite both lineages evolving at approximately the same time, marsupials and eutherians have experienced very different levels of success, with marsupials comprising only 6 per cent of modern mammalian species. Although there are multiple examples of convergent evolution of shape and ecology, including 'moles', 'mice', 'squirrels', 'cats', 'tigers' and 'anteaters', marsupials lack specific traits, such as flippers, wings and hooves, that have contributed to some of the most successful eutherian clades.

One possibility is that marsupials are limited to specific shapes, owing to the constraints imposed by their reproductive need to crawl from the vaginas (yes, there are two) to the teat-containing pouch. As discussed, their offspring are altricial, born after only an extremely short gestation, a few weeks at most, and have poorly developed bodies. Neonatal survival relies on well-developed forelimbs and shoulder girdles to help reach the pouch and relatively mature face and jaw bones to enable non-stop suckling.[39] The evolutionary biologist Professor Anjali Goswami expresses it well: 'marsupial newborns look like little jellybeans with hands – everything else is basically cartilage and goo – they don't even have much of a brain.' Because of this reproductive constraint, marsupials appear unable to later remodel their jaws into different shapes for alternative feeding strategies as eutherians can – a fact that may explain, for example, why marsupials have never developed the baleen jaw of placental whales.

An unanswered question is why newborn marsupials must climb at all. Altricial newborns are far from rare among placental mammals, and there is no reason to suppose that most marsupial mothers cannot manipulate their neonates. Bandicoots are the only marsupials to approach a eutherian-like reproduction, with longer placental attachment times and the ability of neonates to wriggle downwards to a posterior-facing pouch.

As a result, bandicoots have some of the most derived marsupial forelimbs, including a degree of digit loss. Marsupial reproduction has also limited the evolution of aquatic species. An exception is the water opossum, a species that inhabits rivers and lakes from Mexico to Argentina. These small semi-aquatic marsupials have evolved a strong ring of muscles to make their pouch watertight, so that the young remain dry even when the mother immerses in water. As a further example of convergence, they have evolved short, dense, water-repellent fur like the otters, their eutherian counterparts. Interestingly, given the constraint hypothesis, only the hindfeet are webbed and used for propulsion, while the forefeet remain free to feel for and grab prey while swimming. Anatomical constraints may also account for the failure of marsupial forelimbs to evolve into flippers or wings and, hence, the absence of marsupial pinniped and bat doppelgängers.

Geography may have also acted as a constraint. Eutherians are more taxonomically diverse across the northern hemisphere, where land masses have been in frequent contact during the Cenozoic era. Indeed, as we will discuss, there have been many episodes of eutherian dispersal between North America, Asia and Europe. In contrast, Australia and South America (until the formation of the Panamanian isthmus) have been almost entirely isolated since the break-up of Gondwana and the formation of the Drake Passage and the Southern Ocean. If competition and faunal exchange drive evolution and speciation, then marsupial biogeography may have also contributed to their limited diversity.

Predictability versus contingency

The existence of convergent evolution led the Cambridge palaeontologist Simon Conway Morris to argue that rewinding and replaying the tape of life would produce near-identical results, even leading to the emergence of human-like, self-conscious intelligence.[40] Stephen Jay Gould, who coined the 'tape of life' metaphor in his book *Wonderful Life*, proposed an alternative view, that contingency would force evolution down novel and unpredictable paths.[41]

As with many ideas in science, there is no straightforward answer to resolve the differences of opinion.[42] However, Conway Morris's key evidence, convergent evolution, does not necessarily contradict the Gouldian stance, since similar phenotypes may result from entrenched developmental constraints driving iterated evolution within different clades. In other words, when natural selection acts on related populations with similar genotypes in similar environments, both groups will likely evolve along parallel trajectories, subject as they are to the same genetic restrictions and choices. However, it should be noted that although Gould stressed the role of unpredictability and chance in evolution, he never promulgated the case for 'hard contingency', or that evolution is utterly random and senseless.

Nevertheless, for Conway Morris' acolytes, there is a problem: convergent evolution can be regarded as biased evidence collected after the event. Indeed, his proffered case studies, of which there are many, are all examples of evolution that has already repeated itself. But in how many instances has evolution failed to do so?

With its lack of terrestrial mammals, New Zealand provides an alternative viewpoint of evolution, a natural experiment in which only avifauna occupied the vacant ecological niches. The result was a bizarre and unfamiliar world where moas and flightless geese became the dominant herbivores. The giant parrot *Heracles*, with a massive beak that could crack open anything, and the oversized coot-like adzebills filled the carnivorous and omnivorous niches, respectively.[43] Furthermore, the Haast's eagle became South Island's apex predator, the only known raptor to do so in a complex ecosystem, while the kiwi lineage evolved to become insectivores. Squat, plump and almost blind, these nocturnal birds are the only avian species with nostrils at the end of their beaks, which they use to locate their insect prey. They nest in underground burrows, and their feathers are more like fur. Indeed, Gould dubbed the kiwis 'honourable mammals', since their phenotypes are more akin to hedgehogs than any bird. The islands' avian-dominated evolution differs significantly from the mammalian-rich communities found elsewhere. However, New Zealand's founder populations were unique, and the 80-million-year 'natural experiment' only confirms that life evolves to fill available niches, rather than helping us resolve the roles of predictability and contingency in evolution.

Recently, protein analysis has added to our understanding of contingency. In 2021, Victoria Xie and colleagues from the University of Chicago developed an experimental method that enabled reconstructed ancestral proteins to evolve.[44] Using synthesised BCL-2 proteins, the team replayed evolution multiple times from various historical starting points under conditions comparable to those that existed long ago. The scientists then compared the final product of each evolutionary trajectory to determine how predictable evolution is. The findings were conclusive: protein evolution is entirely unpredictable. For example, trajectories using the same ancestral protein produced molecules with different structures and functions.

Furthermore, the results from launching proteins from various ancestral starting points were even more dramatically different. When the team introduced the same amino acid changes into other ancestral proteins, they did not produce the expected functional change. Xie's research suggests that early chance events can influence a protein's evolution in unpredictable ways by opening and closing avenues available to it in the future. In other words, protein structure and functional variations are idiosyncratic products of a particular and unpredictable course of historical events. Maybe Gould was right at the protein level when he stressed that 'any replay of the tape would lead evolution down a pathway radically different from the road taken'.

CHAPTER 4

The Tasmanian Tiger's Story

DE-EXTINCTION

The biggest misconception about de-extinction is that it's possible.
Beth Shapiro

Australia accounts for one-third of all the world's contemporary mammal extinctions. Since European colonisation, at least 39 species have been lost, and many more are listed as critically endangered and at a high risk of extinction. One of the most notable of these demises occurred on 7 September 1936 in Tasmania. That evening, the last known Tasmanian tiger or thylacine (derived from the Greek *thylakos*, pouch, and *ine*, referring to), the world's largest carnivorous marsupial, died from exposure, having been locked out of its enclosure at Beaumaris Zoo in Hobart. However, the Tasmanian tiger may have survived in the wild for several more decades, and the species was not officially declared extinct by the International Union for Conservation of Nature (IUCN) until 1982.

Historically, thylacines were widespread throughout New Guinea, mainland Australia and Tasmania. However, around 2,000 years ago they disappeared, except for a small population in Tasmania that had become isolated 14,000 years ago by rising sea levels. Their decline was not a natural occurrence but a result of human activities. Competition from dingoes, which arrived in Australia around 5,000–10,000 years ago, and hunting by indigenous peoples, played a significant role. When British settlers arrived in Tasmania with their sheep in 1803, there were approximately 5,000 thylacines left on the island. Within a few decades, sheep had increased to such an extent that farming dominated the landscape, the economy and the political culture. Sheep ranchers, viewing thylacines as a threat to their livelihood, successfully lobbied the government to offer a bounty of 1 pound for every adult thylacine killed. Between 1888 and 1909, the programme cost the state over 2,000 pounds. It was this government-sponsored eradication effort, combined with habitat loss, feral dogs and disease, that drove the thylacine population to the brink of extinction.

Sandy-brown to grey, with 15–20 distinct dark stripes across its back from shoulders to tail, the Tasmanian tiger was a slender, prognathous meat-eater. Unusually, both sexes had pouches, similar to water opossums. Females used their pouches for rearing young, while males' pouches served as protective sheaths, covering their reproductive organs. With a stiff tail and wide-opening jaw, the thylacine was comparable in size and shape to a coyote or medium-sized dog. Indeed, when placed alongside a dingo, the phenotypic similarities are so striking that it is hard to believe that their lineages diverged at least 125 million years ago (Plate 7).

The extreme convergence of the thylacine and canids (dogs, wolves and foxes), especially concerning cranial shape and biomechanics, relates to their similar feeding ecologies. However, the genetic basis of this convergence remained a mystery until 2018. That year, Professor Andrew Pask and Thomas Feigin from the University of Melbourne sequenced the genome of a 108-year-old, ethanol-preserved, juvenile thylacine specimen from Museums Victoria, Australia. The result provided an unexpected insight into the mechanisms of convergence.[1] Although the Tasmanian tiger's skull was more like that of canids than any other marsupial, there were no similar genomic changes in their protein-coding sequences. This apparent lack of genetic difference challenged the conventional understanding that natural selection primarily acts on genes and their functional differences. However, by undertaking comparative genomics of many different mammals, Pask and Feigin identified hundreds of non-coding regions restricted to the thylacine and grey wolf genomes. These elements, dubbed TWARs (thylacine–wolf accelerated regions), revealed that natural selection acted on these regions rather than the protein-coding sequences.

Previously, DNA that failed to code for proteins was dismissed as useless and termed 'junk DNA'. However, it is now recognised that these chromosomal sections contain critical regulatory elements, including enhancers and repressors, that modify gene expression, especially during early development. Furthermore, TWARs reside near genes that influence the development of the animal's facial muscles, cartilage and bone. In other words, natural selection has worked similarly in the thylacine and wolf by modifying the same developmental processes to create their convergent facial phenotypes.

Many genes have pleiotropic functions, meaning they code for proteins that influence multiple phenotypic traits. As a result, only a limited number of amino acid changes can occur without disrupting their function. In other words, mutations in coding regions that offer a survival advantage for one trait can negatively impact others. Consequently, the role of gene mutations in evolutionary adaptation is restricted. However, as previously mentioned, this evolutionary bottleneck can be overcome by gene duplications, which create redundancy and allow for more experimentation. In contrast, regulatory elements controlling a gene's activity, especially when expressed in a limited range of tissues, are more tolerant of change than the genes they control. This characteristic gives regulatory elements greater evolutionary flexibility, increasing the likelihood of acquiring beneficial mutations without harmful side effects.

The Australian team also discovered evidence of convergence in the regulatory elements that control genes involved in brain development.[2] This finding was unexpected, given the significant differences between marsupial and placental brains. According to Feigin, although we know little about the Tasmanian tiger's behaviour, 'the signatures of convergent brain evolution suggest the tantalising possibility that the thylacine and wolf may have shared more than just their looks.'[3] Currently, it's not possible to investigate this intriguing association further. However, the fat-tailed dunnart, a close living relative of the thylacine, has been developed as a model species for studying brain development and genomics (Plate 8). It may soon be possible to genetically manipulate the dunnart's forebrain to explore the effects of different control elements on marsupial social and hunting behaviour.

Alterations in regulatory elements, rather than the genes themselves, are the primary driver of adaptive convergent evolution in the thylacine and the wolf, but it remains to be seen if these findings apply to all instances of convergent evolution. Nonetheless, Feigin's research contributes significantly to the ongoing evo-devo debate about the relative importance of protein-coding genes versus non-coding elements in development and evolution.

Pask and Feigin's genome analysis also confirmed that the thylacine's closest relatives are the fat-tailed dunnart and the numbat (Figure 3.1). The numbat, also known as the banded anteater, emerged around 40 million years ago and is the sole member of its family. It closely resembles New World anteaters and feeds entirely on termites. Over millions of years, numbats have evolved a tube-shaped mouth and a long, sticky tongue to probe decayed wood for prey. Although numbats have more teeth than any other marsupial, the teeth are degenerate and play a minor role in feeding. The specialised diet has led to a loss of functional sweet taste receptor genes and a reduction of bitter taste receptor genes compared to other marsupials.[4] Numbats also lack a pouch; instead, newborns must crawl to the nipples through coarse hair that provides the only protection from the outside world. Historically, numbats inhabited Australia's arid and semi-arid regions, but their population has declined by more than 99 per cent due to habitat loss and predation. Currently, only around 1,000 individuals remain in Western Australia.

The dunnart lineage emerged around 15 million years ago, with its only surviving species, the fat-tailed dunnart, being one of Australia's smallest carnivorous marsupials. This nocturnal creature feeds on insects, spiders and amphibians. True to its name, it has a long, carrot-shaped tail that stores fat for times of scarcity. The fat-tailed dunnart has recently become a valuable model species for marsupial research, similar to the laboratory mouse due to its ease of care and experimental use. Importantly, because almost all its development occurs after birth in the pouch, researchers can conduct real-time observations without invasive surgical procedures.[5]

Apex predator loss

The iconic thylacine held a distinct position as Australia's sole marsupial apex predator. Its presence played a vital role in balancing the continent's ecosystems by occupying a crucial niche in the food chain. Recent three-dimensional scans of thylacine jaws, generated by advanced computer technology, revealed that the animal's hunting ability was better suited for smaller prey such as bandicoots, possums, birds and reptiles, rather than adult sheep.[6] The disappearance of this apex predator has led to a surge in the population of small macropods, including the red-necked wallaby and the Tasmanian pademelon, on the island. This overabundance has resulted in overgrazing, causing ecological damage to Tasmania's vegetation and threatening other species.

Moreover, the absence of thylacine may have contributed to the proliferation of diseases among its prey, such as the devil facial tumour disease that is currently devastating the Tasmanian devil population. The thylacine's role in removing sick and injured individuals helped control the spread of such diseases.

The decline of Tasmanian devils has unexpectedly influenced the food chain further, leading to genetic modifications in smaller predators like the spotted-tailed quoll. Recent research by Marc Beer and his team at Washington State University reveals that the reduced competition from Tasmanian devils has boosted food availability for quolls and other smaller predators.[7] Consequently, spotted-tailed quolls now travel shorter distances for scavenging and breeding, resulting in decreased genetic exchange among individuals. The loss of genetic diversity involves gene variants that affect muscle development, locomotion and feeding behaviour in the less-mobile quoll populations. The researchers suggest this adaptation is logical, as quolls living in areas with fewer devils need not roam as extensively to compete for food. These findings have important implications. Top predator declines are increasing globally, and Marc Beer's study could serve as a model for future investigations into the evolutionary impacts of altered ecological interactions.

Since its extinction, the thylacine has become one of Australia's most beloved and missed species. It has come to symbolise Tasmania, even appearing on the island's official coat of arms. In 1996, exactly 60 years since its disappearance, 7 September was designated as Australia's National Threatened Species Day, a time to reflect on the past and consider the possibility that many of the continent's other native plants and animals could face similar fates unless urgent action is taken. Given the thylacine's significant place in the Australian psyche, it is unsurprising that some people cannot accept the marsupial's loss. A few optimists still believe that a population may exist in the island's remote outback and that they will eventually be found. While accepting the species' demise, others hope that science will one day reverse its extinction.

Indeed, the possibility that de-extinction might be possible has refocused the careers of several biologists. According to Andrew Pask, the thylacine is a perfect candidate for de-extinction since it died out relatively recently, and comparatively good-quality DNA is available. Additionally, compelling ecological reasons exist to bring it back, as its prey and most of its natural habitat still exist.

Genomic reconstruction and enhancement

One method for reviving extinct species is cloning, a process akin to the technique used to create Dolly the sheep through somatic cell nuclear transfer.[8] In this approach, fertilisation and fusion are bypassed. Instead, the nucleus of the extinct animal is extracted and placed into an egg cell from a closely related species after the egg's original nucleus has been removed. Unfortunately, this method is not viable for thylacines because the available tissue samples are too degraded to provide intact cells. Nonetheless, cloning has shown some success in de-extinction efforts. For example, the last Pyrenean ibex, Celia, died in 2000 when a falling tree crushed her. Scientists attempted to clone her, and after multiple failures they succeeded in 2003. However, the cloned ibex survived only a few minutes due to a lung defect. Despite this setback, the Pyrenean ibex remains the only species to have been brought back from extinction, earning the distinction of having gone extinct twice.[9]

Undeterred, scientists are now investigating an alternative method known as genomic reconstruction. This technique involves editing the DNA of a closely related species' DNA to replicate the extinct species' genome. Andrew Pask has already established the TIGRR Lab (Thylacine Integrated Genetic Restoration Research) at the University of Melbourne to tackle this ambitious and controversial project. Collaborating with Colossal Biosciences, a technology company based in Texas and supported by a 5-million-dollar philanthropic donation, Pask is addressing the project's primary challenges. These include gaining a deeper understanding of the thylacine's genome, utilising marsupial stem cells to create an embryo, and transferring the embryo into a surrogate host's uterus.

The first challenge involves obtaining a complete and intact genetic sequence of the thylacine. Although Pask's team has sequenced most of the animal's DNA, the final few per cent present technical difficulties. The extracted DNA is highly fragmented, and after sequencing thousands of individual pieces, they must be assembled to create an accurate chromosomal complement. Unfortunately, the remaining pieces are short, similar in appearance, and have repetitive sequences, making reconstruction difficult. Pask compares this process to solving a jigsaw puzzle made entirely of baked beans or blue sky: 'Every bit looks the same, and we're trying to figure out how it goes together.'

Fortunately, an unusual aspect of nature has allowed some considerable progress.[10] The thylacine family (Dasyuridae) displays exceptional chromosome conservation with regard to gene number and arrangement. All species lack the extensive genetic reshuffling that characterises other marsupial families, especially the rock-wallabies.[11] As a result, Pask's team can align the thylacine's DNA sequences against the chromosomes of several closely related species with remarkable efficiency and accuracy. Eventually, this 'reference-guiding' technique should enable the thylacine's many small fragments to be spliced into full-length chromosomes.

Next, the TIGRR researchers will compare the completed thylacine genome to the fat-tailed dunnart, a closely related species with a short gestation period that facilitates experimentation. Using the Nobel prize-winning CRISPR gene-editing technology – comprising a set of enzymes that allow the targeting of exact sequences of DNA – the team will modify the dunnart's genome to match that of the thylacine's more closely.[12] Even so, this state-of-the-art approach would take over 100 years to produce a perfect copy. In practice, the project will prioritise which DNA sequences to target to yield a thylacine-like genome. The initial iteration might be only 90 per cent thylacine, but through selective breeding and further genetic manipulation, the result could be nearer 99.9 per cent. In case readers wonder, genetic imprinting, while crucial for normal fetal development, will not pose a problem for such closely related species. At the same time, any critical epigenetic changes are likely to be re-established in the induced embryo.

Most proponents of de-extinction believe that creating exact copies of extinct species should not be the primary aim. Beth Shapiro, Professor of Ecology and Evolutionary Biology at the University of California, writes that in most ongoing projects, 'the goal is to create functional equivalents of species that once existed: ecological proxies that are capable of filling the extinct species' ecological niche.'[13]

Another challenge will be creating new marsupial-specific assisted reproduction techniques to encourage genetically engineered stem cells to form embryos. These embryos would then be transferred to either an artificial womb or a dunnart surrogate for development. However, neither of these two cloning hurdles has yet been mastered. Despite this, Pask is optimistic that 'thylacine-like' joeys could be born within the next 10 years. Then, after many years of careful monitoring, the team plans to release a population of around 100 viable, genetically diverse individuals into the wild.

Early de-extinction efforts show considerable promise for protecting Australia's threatened marsupials, particularly those impacted by recent bushfires. The development of marsupial stem cell biobanks and advancements in genetic enhancement provide new hope. For example, the northern quoll, severely threatened by its predation on the deadly invasive cane toad, could benefit from breakthroughs in the thylacine de-extinction program. These insights might pave the way for genetically engineering toxin-resistant quolls, a potentially transformative step toward ensuring their survival.[14]

De-extinction is not just a scientific concept but a profoundly complex and ethically charged issue. It sparks intense debates, with some seeing it as a moral obligation to correct human-induced wrongs, while others argue that it's a flawed endeavour. Concerns about animal welfare, human hubris, and the allocation of finite conservation resources all come into play.[15] However, it is important to remember that de-extinction is not the start of human manipulation of nature. We've been altering the genomes of flora and fauna for our benefit for millennia, from the first domestication of grey wolves around 30,000 years ago to the selective breeding that has shaped the cereals and meat we consume today.

Finally, before we turn to the story of eutherians, let me conclude with a quote from Sherwin Carlquist. The American biologist wrote in 1965 that 'the Australian marsupials viewed as a whole are a monument to unlikelihood. They are perhaps not so striking individually, but in their entirety, they are a summation of many paradoxes, a unique evolutionary story produced by a unique land.'[16]

I do not doubt that Carlquist would have been amazed at how far modern genetics has advanced our knowledge of this enigmatic group of doppelgängers.

PART THREE

Eutherians

CHAPTER 5

The Aardvark's Story

EVOLUTIONARY DISTINCTIVENESS

Afrotheria represents a triumph of molecular phylogenetics.
Mark S. Springer

The last common ancestor of today's placental mammals was an insectivorous, shrew-like creature with a long snout that inhabited Pangaea. Only later did these early mammals evolve to become more varied in size and shape, although when and how is hotly debated. While the role of plate tectonics is now widely accepted, competing hypotheses for the timing and speed of their early evolution, especially concerning the Cretaceous–Palaeogene (K–Pg) mass extinction event 66 million years ago, remain to be resolved (Chapter 9).

During the Triassic, approximately 250 million years ago, heat from the Earth's core and the resultant convection currents of molten magma began to pull Pangaea apart. Two large supercontinents formed, Laurasia in the north and Gondwana in the south, and their isolated mammalian populations evolved along very different trajectories. The Laurasian species gave rise to the magnorder Boreoeutheria ('beasts from the north') and spawned a host of familiar species, including whales, giraffes, mice, cats, bats and humans. The Gondwanan clade, or Atlantogenata ('originating around the Atlantic'), gave rise to the Afrotheria of Africa and Madagascar and the Xenarthra of South America (Figure 5.1).

The Earth's convection currents are relentless, and approximately 100 million years ago, Africa was wrenched from the grips of Gondwana to remain an insular landmass for the next 60 million years. By the time Africa docked with Arabia, its isolated fauna had evolved to fill a variety of ecological niches. The result was a diverse menagerie of bizarre mammalian taxa – the aardvark, elephant shrews (also known as sengis), golden moles, tenrecs, hyraxes, sea cows and elephants – an assemblage known today as Afrotheria ('African beasts'). Their shared story, which lay hidden from view until the late 1990s, was unexpected and represents a triumph of

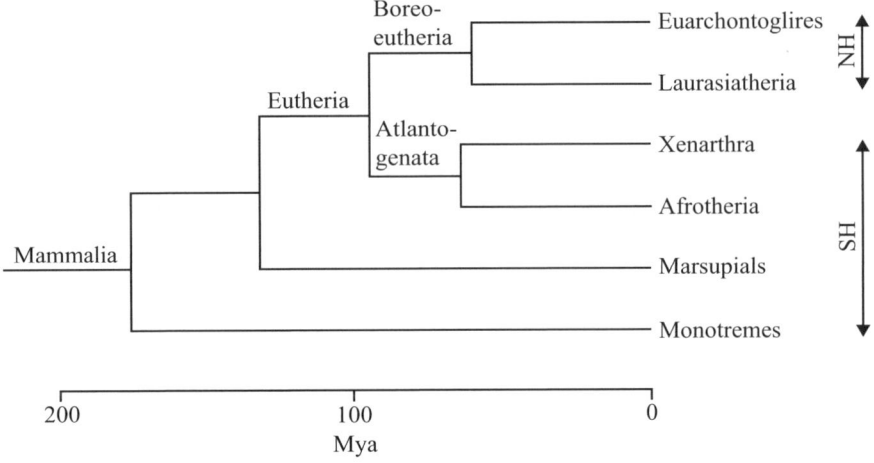

Figure 5.1 Phylogeny of Mammalia. Eutheria comprise the Boreoeutheria in the north (Euarchontoglires and Laurasiatheria) and the Atlantogenata in the south (Afrotheria and Xenarthra). NH, northern hemisphere; SH, southern hemisphere; Mya, million years ago.

molecular phylogenetics over the traditional methods of palaeontology and morphology.[1] For, during most of the twentieth century, phylogeneticists were influenced by the ineluctable views of George Gaylord Simpson.

The scientific clout of Simpson, one of the founders of the 'modern synthesis' of evolution, should not be underestimated. During his time in museums and the field, Simpson recognised that a hierarchical organisation of organisms based on evolutionary relationships was required to advance the study of evolution and biogeography. He published his wide-ranging ideas of phylogenetics in several landmark monographs, including an influential 350-page tome entitled *The Principles of Classification and a Classification of Mammals* (1945).[2] Simpson's views and proposals for mammalian phylogenetics were accepted for decades, conclusions regarded as gospel and doubted by few. Indeed, a younger colleague, Stephen Jay Gould, regarded him as 'unquestionably the greatest vertebrate palaeontologist of the twentieth century, perhaps the greatest of all time ... a brilliant theorist who brought a conceptually backward field of traditional palaeontology into synthesis with the neo-Darwinian consensus.'[3]

Simpson grouped species with similar phenotypic traits not shared by any other taxa together as clades. Such groupings shared anatomical features, or synapomorphies, in the arcane jargon of cladistics, thought to be inherited from a common ancestor. For example, he lumped elephants with other hoofed taxa – odd-toed ungulates, including rhinos and horses, and even-toed ungulates, such as cattle, pigs and deer. He also believed that pangolins, anteaters and sloths had a common ancestor; bats were close relatives of primates; and shrews, tenrecs, hedgehogs and moles belonged to the classical order Insectivora.

There was only one problem with Simpson's apparent logic: convergent evolution. As discussed previously, different lineages can evolve similar phenotypes when faced with the same ecological or environmental pressures. Such adaptations are not inherited from a common ancestor but have developed independently and, as a result, can lead to errors if relied upon as the sole means of determining phylogenetic relationships. Although Simpson was fully aware of convergence, he failed to recognise its pervasiveness amongst Africa's mammals. Throughout the Cenozoic, the isolated afrotherians had adapted and evolved phenotypes to occupy niches similar to mammals in other continents. For example, the hyraxes occupied the niche of rodents; the aardvarks filled the anteater niche; the sengis, golden moles and tenrecs took the place of insectivores; while elephants became the equivalent of Laurasia's large herbivores such as hippos and rhinos.

In Simpson's defence, he lacked the scientific means to untangle convergence from common ancestry. Indeed, he realised the limitations of his system and stated that 'complete genetic analysis would provide the most priceless data for mapping this stream' (i.e. mammalian phylogeny). Sadly, he died before molecular phylogenies became available.

In 1997, a team of biologists led by Mark Springer at the University of California dispensed with the traditional approach to phylogeny. Instead, they used the latest genetic techniques to construct family trees by comparing species' gene sequences. By today's standards, this initial foray was basic, involving only five genes, three mitochondrial and two nuclear; nevertheless, the results were unambiguous.[4] Insectivores are not monophyletic, and golden moles belong to the same clade as the aardvark, elephant shrews, tenrecs, hyraxes, manatees and dugongs, and elephants. This finding led Springer to postulate that a single common ancestor had undergone extensive radiation to fill multiple niches during the aeons of Africa's insularity. Although greeted with deep scepticism by many, if not outright disbelief, several teams embarked on a search for confirmatory evidence. It did not take long to find.

In 2001, van Dijk and colleagues studied proteins rather than DNA and found identical amino acid substitutions in proteins from afrotherians that were absent from other mammals.[5] In other words, despite showing little or no superficial resemblance, afrotherians were truly monophyletic. Two years later, a Japanese team produced similar results after examining a family of retroposons, or 'jumping genes', called SINEs – short interspersed nuclear elements.[6]

These non-coding sequences are widely distributed within eukaryote genomes and have crucial roles in genome organisation and evolution, and modulation of gene expression. Once integrated into DNA, SINEs are inherited unaltered over evolutionary timescales. Crucially, the Japanese team found a novel family of these 'molecular fossils', designated AfroSINEs, that characterised afrotherians but were absent from all other mammals. Furthermore, a subfamily of AfroSINEs, with a unique central deletion, grouped hyraxes, elephants and sirenians (dugongs and manatees) within the clade Paenungulata, while the aardvark, sengis, golden moles, otter shrews and tenrecs belong to a second clade, Afroinsectiphilia (Figure 5.2). Lastly, a study examining the structural organisation of whole chromosomes using fluorescent probes, a technique known as chromosomal painting, further supported the afrotherian concept.[7]

Afrotheria remains one of the most remarkable hypotheses in mammalian evolution, one based on the DNA sequences of living species. Springer's scientific bombshell has profound implications. One-third of all placental mammal orders form an ancient grouping that evolved on continental Africa after it broke away from Gondwana, and plate tectonics played a more significant role in the evolution of early placental mammals than had been previously realised. The discovery of Afrotheria, however, is only one part of a global effort to construct a genealogical tree for all biota, with the ultimate goal being a single tree of life.

Afrotherians have since been shown to share several morphological features, or synapomorphies, that support Springer's findings. For example, all clade members have a similar ankle bone structure, a modified placental membrane or allantois,[8] an increased number of thoracolumbar vertebrae,[9]

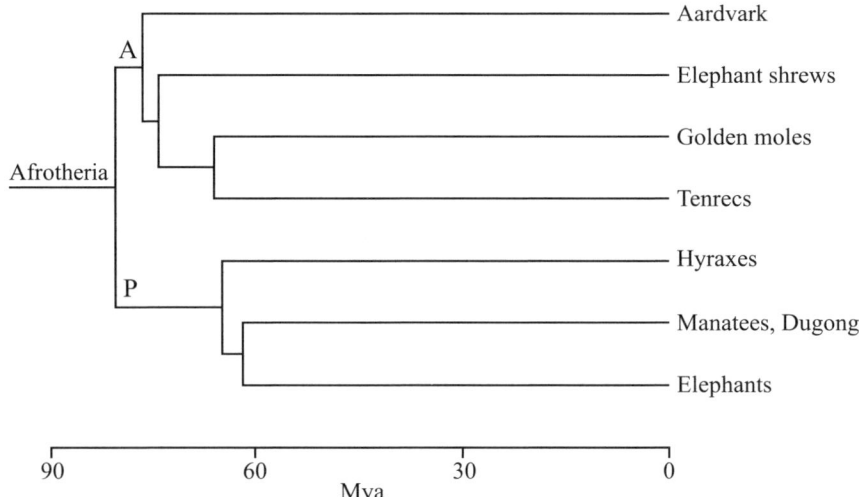

Figure 5.2 Phylogeny of Afrotheria. A, Afroinsectiphilia ('African insectivores'), P, Paenungulata ('almost having hoofs'); Mya, million years ago. Modified from Springer (2022).[10]

and a delayed eruption of permanent teeth. The last feature is associated with the lack of a gene that codes for a protein (Klk4, kallikrein-related peptidase) that shortens the time taken for enamel maturation, and its absence accounts for the delayed dental eruption of all afrotherians.[11]

Palaeontology was not to be outdone, and the lack of informative fossils was about to change. To the west of the Atlas Mountains, near Khouribga, lies Morocco's largest phosphate deposit, the Oulad Abdoun Basin. Round the clock, massive drag-line excavators gauge out the mineral-containing strata for use as fertilisers for farmers worldwide. The whole area has been a metaphorical gold mine for palaeontologists, as many vertebrate fossils have been recovered from the exposed strata and spoil heaps, dating from the Late Cretaceous to the Early Eocene. They include fossils of fishes and turtles, birds, and rare examples of early mammals, often found by amateur collectors living nearby. Crucial to our story is that the site has provided the oldest known fossils of an afrotherian. In 2014, a combined French and Moroccan team led by Emmanuel Gheerbrant described a partial skull, upper jaw, and teeth from *Ocepeia*, a 3.5-kilogram dog-like herbivore.[12] This 60-million-year-old afrotherian, with both herbivorous and insectivorous features, lived along the margins of the ancient Tethys Ocean shortly after the dinosaurs' demise (see Figure 2.1). Gheerbrant now believes that *Ocepeia* is a 'transitional fossil' between the insectivore-like and ungulate-like afrotherians, and that it could help characterise the clade's ancestral phenotype.

Springer's DNA-based phylogeny has superseded Simpson's morphologically derived tree as the accepted standard. Indeed, the emergence of Afrotheria remains one of the most remarkable discoveries in the field of mammal evolution, a grouping that comprises a cornucopia of extraordinary and bewilderingly diverse taxa. Their stories warrant telling, highlighting important evolutionary concepts we have yet to encounter, including relaxed selection or neutral evolution, sweepstake dispersals, adaptive radiations, admixtures and retrogenes.

Evolutionary distinctiveness

Tubulidentata is the oldest and least diverse of the afrotherian orders, with only one living member, the aardvark (Plate 9). The mammal's name is derived from the Afrikaans for 'earth pig', a likeness particularly evident in its nose and face. Aardvarks have long protruding tongues that exhibit convergence with those of pangolins and New World anteaters, none of which are closely related. Molecular-clock estimates suggest that aardvarks diverged from their nearest living relatives, the elephant shrews, golden moles and tenrecs, in the Late Cretaceous, around 78 million years ago. Not only are aardvarks the sole species in their genus, but they also sit in their own family and order. This unusual taxonomic situation arises because all other family members became extinct before the Late Pleistocene despite being distributed in Africa, southern Europe and South Asia. As a result, the aardvark is more isolated and distinct than any other placental mammal in terms of its genetic makeup. Indeed, according to the EDGE scientific method (see below), the aardvark has been assessed as the mammal with the highest score for evolutionary distinctiveness (ED).

The ED score was initially devised by Nick Isaac and his colleagues at the Institute of Zoology in London as an aid, in conjunction with other data, to help identify mammals at the most significant risk of extinction.[13] ED scores are calculated by allocating each branch of a phylogeny a value equal to its length in millions of years (MY), divided by the total number of species that derive from it. A score for a given species is obtained by adding the values for all the branches from which the species is descended, from its terminal branch to the tree's root. For example, Figure 5.2 shows an ED score for the aardvark of 78 (78 million years of evolution divided by only one species). In comparison, some rodents and bat taxa with many close relatives have scores less than 1.0. The aardvark's high score means that if it were to go extinct, no similar species would remain on the planet, and a disproportionate amount of unique evolutionary history would be lost forever.

Despite its high ED score, the aardvark does not feature in the top 100 list for EDGE scores (evolutionarily distinct and globally endangered – a modified ED score taking into account a species' vulnerability to extinction; see also Chapter 9), as its conservation status is only Least Concern. It is the mountain pygmy possum, a critically endangered mouse-like species from the alpine areas of Victoria and New South Wales in Australia, that holds the 'dubious' honour of topping the latest league table (Table 5.1).

Table 5.1 Top five mammal species for evolutionary distinctiveness (ED) and evolutionary distinctiveness and globally endangered (EDGE) scores. Adapted from Gumbs *et al.* 2023.[14]

Top ED scores	Top EDGE scores
1. Aardvark	1. Mountain pygmy possum
2. Pen-tailed treeshrew	2. Aye-aye
3. Platypus[15]	3. Leadbeater's possum
4. Maned rat	4. Cuban solenodon
5. Silky anteater	5. Numbat

Aside from outliving all other family members, aardvarks are intrinsically strange animals. Their anatomy has evolved to support a diet of only ants and termites, a condition biologists call myrmecophagous. Their heavily clawed feet enable the adults to excavate prodigiously, tunnelling at rates approaching a metre in 20 seconds. Such digging prowess allows them to break into large anthills and termite mounds and create extensive underground burrows. Indeed, if pursued into its subterranean home, an aardvark will rapidly seal the tunnel entrance and dig further into the earth in the opposite direction. Aardvarks have also evolved a long thin tongue to penetrate the narrow ant and termite passageways, while the organ's sticky saliva adheres to its prey. Indeed, one individual can target 200 anthills in a single night and consume around 50,000 insects.

Aardvarks are the only myrmecophage on Earth with a functional set of back teeth (they lose their front set on maturity). This evolutionary quirk relates to the aardvark's penchant for a single taxon of fruit-producing plant that, in turn, relies on the mammal for its survival. The *Cucumis humifructus*, or aardvark cucumber, flowers above ground but then bends and pushes back into the soil to produce an underground, orange-sized fruit. Aardvarks seek out this structure for their much-needed water content and, in return, propagate and fertilise its seeds through their faeces. Without this rather unusual symbiotic relationship, *Cucumis* could not exist.

Convergence and relaxed selection

The order Macroscelidea first appeared in the Late Cretaceous, around 75 million years ago, and today comprises 19 species of bizarre but cute animals known as elephant shrews or jumping shrews. They are found widely across southern Africa and, although not common anywhere, can be found in almost any habitat, from desert to forest. While they may look like European shrews, they are not closely related, and their resemblance is yet another striking example of convergence. In fact, as we have seen, they are closer cousins to elephants than to shrews. As a result, many prefer the name sengi (plural sengis), as proposed by the biologist Jonathan Kingdon.[16] Sengis are characterised by a long pointed head and a very long, mobile, trunk-like nose. They have relatively long legs for their size and move in a hopping fashion like jerboas. A gland on the underside of the tail produces a strong, musky scent used to mark territories, which also serves as a deterrent against many predators. Sengis are one of the very few mammals that are truly monogamous. Once paired, they remain faithful to one another through good and bad times, mating for life. However, they rarely spend time together but crisscross their territories with runways and leave messages of their whereabouts by depositing urine and scent. If the animal is disturbed, the pathways provide an obstacle-free escape route.

Golden moles (family Chrysochloridae, from the Greek for 'green gold') diverged from the tenrec lineage around the time of the K–Pg mass extinction event (Figure 5.2). Although they were known to Linnaeus 250 years ago, our knowledge of these blind, subterranean mammals remains limited. They inhabit sub-Saharan Africa, where the 21 species occupy a wide range of altitudinal, climatic and vegetational zones. Despite variation in size and shape, they are all streamlined for ease of tunnelling, with large hands and strong arms for digging like the moles (Talpidae) of the northern hemisphere. The golden moles' specialised silky-smooth hair evolved to allow sand and soil to slip past, enabling an easy passage through the earth. Interestingly, the unique structure of their hair – small size and flattening of scales to produce smoothness – also makes their fur iridescent. While coloured structures typically evolve as sexual ornaments, the blindness of golden moles suggests that their iridescence is an anatomical epiphenomenon and lacks a survival advantage.[17]

The eyes of golden moles, like those of the convergent marsupial moles, are the most degenerate of any mammalian taxa. Both families have vestigial eyes covered by skin and fur and lack optical nerve connections with the brain. Recently, Springer and colleagues have shown that given the absence of selection pressure for a functional eye – termed relaxed selection or neutral evolution – both mole families accrued many inactivating mutations in the genes controlling retinal photoreceptor function and the development of the cornea and lens. The researchers even calculated that these changes occurred sequentially between 40 and 20 million years ago.[18] Unlike golden and marsupial moles, if these random mutations occurred in species that depend on sight for survival, they would decrease the carrier's fitness and be eliminated from the population through a process known as purifying or negative selection. In other words, the 'knock-out' mutations that led to the loss of eye function in golden and marsupial moles were tolerated because vision was unnecessary for a life underground. We will further explore the evolutionary importance of relaxed selection and gene loss when we discuss the whale's story (Chapter 11).

Most golden moles are solitary and live in complex, semi-permanent tunnel systems. An exception is the small Grant's golden mole from Namibia, which instead swims through the desert sands just under or on the surface in search of termites, its favourite prey. When foraging, this golf-ball-sized 'shark of the dunes' frequently dips its head into the sand to detect the extremely low-frequency vibrations of termite mounds. The physiologist Peter Narins showed that the animals do this with the help of the wind, which blows the dune grass and sets the mounds in resonance, producing sand vibrations that the moles can hear.[19] To do so, the species has evolved a phenomenally sensitive and massively hypertrophied malleus, the hammer-shaped bone in the middle ear. The result is that the ear ossicle accounts for 0.1 per cent of the animal's total body mass. To put that in perspective, the figure for a similarly sized mouse is 0.001 per cent, and for humans, 0.00008 per cent. Indeed, the malleus is so big that the golden mole's centre of gravity is shifted, making it more susceptible to imbalance. Furthermore, Narins believes that the animal's middle ear could hold the secret for designing ultrasensitive vibration detectors to help identify earthquake precursors. These subtle sounds can occur hours and even days before any significant slip in the Earth's crust.

The last of the Afroinsectiphilia to evolve were the otter shrews and tenrecs. The confusingly named otter shrews (they are neither otters nor shrews) remained in Africa, while the tenrecs, which diverged around 50 million years ago, emigrated to Madagascar.[20] The three African species of otter shrew are rare and elusive and inhabit the central and western tropical river systems, feeding on fish, crabs and frogs. Although closely resembling true otters, they are unrelated, and their phenotypes – webbed feet, streamlined bodies, dense, water-repellent fur, closable nostrils, small eyes and flattened tails – are further examples of convergent evolution.

The problem of tenrec biogeography

Madagascar, the fourth-largest island in the world, is separated from Africa's eastern shore by the Mozambique Channel, 420 kilometres at its narrowest. This remote island, often regarded as the world's eighth continent, has more unique animal species than any other place except Australia, which is many times larger. This biogeographical observation poses a problem for biologists. How did all the island's species get there, given that Madagascar has been an island for at least 120 million years and that its fauna arrived during the last 60 million years? Important clues are that most species are small and belong to a limited number of groups – tenrecs, lemurs, rodents and carnivores. In contrast, no large mammals from mainland Africa made the journey – no elephants, zebras, rhinos, buffalo, antelopes, camels or giraffes.

To explain the marked imbalance in the island's biological diversity, G. G. Simpson proposed the 'sweepstake dispersal' hypothesis in 1940 in a renowned and much-cited paper.[21] In it, he postulated that the ancestors of Madagascar's present-day mammals must have rafted across on natural mats of vegetation blown out to sea from Africa. Furthermore, he also stated that the island's animals appear to have arrived in occasional bursts of immigration, probably by single species at any one time (or even a single pregnant female) rather than in a continuous, mixed migration. While Madagascar's tenrecs and lemurs each evolved from a single ancestor that once inhabited Africa, the absence of small-bodied canines and felines, monkeys, shrews and most rodents was down to chance: they were 'unlucky' and missed the raft.

The idea of transoceanic dispersal, however, was not new. Alfred Russel Wallace had raised the possibility as early as 1880 in his book *Island Life*:

> *With the smaller, and especially with the arboreal mammalia, there is a much more effectual way of passing over the sea, by means of floating trees, or those floating islands which are often formed at the mouths of great rivers. Sir Charles Lyell describes such floating islands which were encountered among the Moluccas, on which trees and shrubs were growing on a stratum of soil which even formed a white beach round the margin of each raft. Among the Philippine Islands similar rafts with trees growing on them have been seen after hurricanes; and it is easy to understand how, if the sea were tolerably calm, such a raft might be carried along by a current, aided by the wind acting on the trees, till after a passage of several weeks it might arrive safely on the shores of some land hundreds of miles away from its starting-point.*[22]

Proponents of sweepstake dispersal emphasise that however improbable and unbelievable such events are, they will occur given enough time. Indeed, a field observation in the summer of 1995 proved the point. A party of at least 15 green iguanas survived Hurricane Marilyn by clinging to a collection of uprooted trees after being swept out to sea off Guadeloupe. The castaways drifted across the Caribbean for over three weeks before successfully colonising Anguilla to become a potential new species. Locals reported that the vegetative mat was so big that it took days to pile onto the beach.[23]

But Simpson's neat solution to the Madagascan problem faces several challenges. The Mozambique Channel is an extraordinarily ancient and formidable biogeographic barrier, and the prevailing winds and the currents swirling off eastern Africa and the surrounding Indian Ocean would make it virtually impossible for a raft to reach Madagascan shores.[24] Indeed, because of the Earth's rotation, the currents are forced southwards, not eastwards, and follow the outline of the mainland and the continental shelf.

Secondly, at least three species of pygmy hippopotamuses once inhabited the island's forests, one of which may have survived until at least until 1,000 years ago. About the size of cows, the Malagasy hippos were far smaller than their 3-tonne cousin, the common hippopotamus. Even so, they were among the largest animals in Madagascar, along with the Nile crocodile and the giant ostrich-like elephant birds. Professor Brooke Crowley from the University of Cincinnati believes that the island's hippos, the only ungulates to reach Madagascar, likely resembled the secretive and endangered pygmy hippos found today in the forests of West Africa.[25] However, it is possible that their ancestors were full-sized hippos who shrank through insular dwarfism. Either way, sceptics of the sweepstake dispersal hypothesis stress that Madagascar's hippos, large or small, could not have swum or survived a raft journey of several weeks to reach the island, despite their semi-aquatic lifestyle.

As a result, alternative explanations for the arrival of mammals on Madagascar have been proposed, such as crossings via transient marine corridors or island hopping. An international team led by Judith Masters thinks it more likely that tenrecs, along with many other vertebrates, reached the island with the help of short-lived land connections. They believe that at least three such corridors emerged between 66 and 12 million years ago, during episodes of regional uplift, climate change and low sea levels.[26]

Rescue for Simpson's hypothesis, however, came in 2010. Two scientists, Jason Ali from the University of Hong Kong and Matthew Huber from Purdue University in Indiana, adopted the computer modelling techniques used in modern climate studies to predict what the Earth was like in the past.[27] Approximately 50 million years ago, continental drift had moved Africa and Madagascar to about 1,500 kilometres south of their present positions. By plugging information about the ocean and atmosphere of ancient Earth into a supercomputer and analysing over 100 terabytes of data, Ali and Huber found that currents around the two land masses have changed and that they once flowed in the opposite direction, that is, eastwards toward Madagascar. It transpires that the northward drift of Africa and Madagascar subsequently disrupted a major surface ocean current flowing across the tropical Indian Ocean and led to the currents present today. Ali and Huber believe that such favourable currents prevailed between 60 million and 20 million years ago when the ancestors of present-day animals arrived in Madagascar. Furthermore, the currents were strong enough to transport a large raft of vegetation to the island without its mammalian passengers dying of thirst. In other words, Ali and Huber's model supports Simpson's idea that small animals could have rafted to Madagascar, especially if they further reduced their naturally low metabolic rates by adopting a state of torpor or hibernating.

Whether by rafting or walking, the single ancestral tenrec colonising Madagascar found little competition on the island and rapidly adapted and speciated to fill the many vacant ecological niches. These species colonised most of the island, including the eastern rainforests, the drier forests of the central plateau and the southern thorny areas. This extraordinary example of adaptive island radiation has given rise to the approximately 31 species of tenrec alive today.[28] However, tenrecs never colonised islands further east, and those individuals found on the Comoros, Mauritius, Réunion and the Seychelles were introduced by humans around the end of the seventeenth century.

The smallest tenrecs are shrew-size, weighing just 5 grams, while the largest, the tailless or common tenrec, is the size of a small dog and weighs up to 2 kilograms. Although most tenrecs favour insects, the two species of streaked tenrec (lowland and highland) eat earthworms, while the aquatic web-footed tenrec dines on frogs and crustaceans. Indeed, the latter species is thought to have returned to the water independently of its distant otter shrew relatives that remained in Africa.

Although the tenrec species don't look much alike, they bear striking similarities to other mammals elsewhere. For instance, as their names suggest, the greater and lesser hedgehog tenrecs look unmistakably like hedgehogs, with their backs covered in short, sharp spines. They even roll into a ball when threatened to protect their soft underbellies. Comparative genomics suggests that hedgehogs and tenrecs have evolved separately for over 100 million years and that their phenotypic similarities result from comparable environmental pressures.[29] As tenrecs ventured from forest environments to more open habitats, they became bulkier and evolved spines to counter the increased predator risk. However, the growth and transportation of spines is energy-demanding, and to compensate, tenrecs evolved smaller-sized brains. It seems that they 'paid for' the increased cost of spines by reducing the amount of high-energy-consuming brain tissue, mainly since selection pressures for advanced predator detection and risk assessment would have lessened.[30]

There are also approximately 20 species of shrew tenrec that, for most people, are indistinguishable from true shrews. They tend to be small animals found scuttling amongst the leaf litter or climbing high into the canopy. Other tenrecs resemble moles, burrowing into the forest floor, or look rather like rats.

Tenrecs can appear to be relatively primitive mammals. They have bad eyesight and give birth to deaf and blind offspring. They also produce more babies than any other mammal, with up to 30 infants at a time that suckle on a similar number of teats. Given their low body temperature, they do not require a scrotum to cool their sperm like most mammals and retain their testes in the abdomen (Chapter 7). They also possess a cloaca like all reptiles, birds, amphibians and monotremes, an anatomical feature hardly ever found in other placental mammals (the exceptions being golden moles and a few species of shrew).

Having such a large family increases the likelihood of some youngsters getting separated and becoming lost. One species, the lowland streaked tenrec (Plate 10), has evolved a unique solution to this problem. In addition to their barbed quills, used for defence, they also have specialised 'stridulating' quills on their backs, which they rub together to produce a high-pitched grating noise. Such sounds easily cut through the din of the forest and allow family members to stay in touch as they move about in the tangled undergrowth. Amazingly, tenrecs are the only mammals known to use stridulation for communication, a behaviour typically restricted to snakes and insects.

Before we return to South America's 'splendid isolation' and meet the Xenarthra – anteaters, sloths and armadillos – we should discuss a second clade of Afrotheria that gave rise to the hyraxes, sirenians and elephants.

CHAPTER 6

The Hyrax's Story

AQUATIC ORIGINS

The coneys are but a feeble folk, yet make their houses in the rocks.
Proverbs 30:26

As highlighted in Chapter 5, molecular genetics revealed the unexpectedly close relationship of the hyraxes (Hyracoidae) to two other afrotherian orders – the manatees and dugongs (Sirenia) and the elephants (Proboscidea). All three orders constitute the clade Paenungulata (from the Latin *paene*, almost, and *ungulatus*, having hoofs), and all diverged within the early Cenozoic – the hyraxes first, quickly followed by the elephants and sirenians (see Figure 5.2).[1] While the hyraxes remained within Africa and the adjacent Arabian Peninsula, the aquatic sirenians and the terrestrial proboscideans dispersed more widely. The morphology and physiology of all three orders support the genetic findings and are compatible with the existence of a last common ancestor during the Cretaceous. Furthermore, biochemical analyses of muscle proteins and earlier palaeontological evidence hint that their ancient ancestor may have been aquatic rather than terrestrial.

Sex, song and sea

The hyrax, also known as coney or rock badger, first appeared in the fossil record around 37 million years ago, but many species existed by the Middle to Late Eocene. Some were tiny, mouse-sized, while others became as large as tapirs. Later, during the Miocene, competition from the newly arrived bovids from the north, including wildebeest, antelopes, warthogs, rhinos and buffalo, displaced the hyraxes to marginal habitats, given the interlopers were more efficient grazers. Nevertheless, hyraxes remained widespread and diversified until 2 million years ago, when the five extant species emerged. Over millions of years, climatic and geographical events can isolate populations, leading to speciation by vicariance. As the African forests expanded and contracted, new species would have evolved in isolated forest fragments known as refugia, or become restricted to interfluvial areas following the emergence of new river systems. Indeed, there is increasing evidence that a species of West African tree hyrax, the Benin tree hyrax, arose by vicariance during the Plio-Pleistocene because of the insurmountable barriers posed by the Niger and Volta rivers.[2]

Hyraxes may appear like guinea pigs or rabbits with short ears, but on closer inspection their many unusual characteristics betray their sister relationship to sirenians and elephants (Plate 11). One striking similarity is that males in all three orders lack scrotums, their testes remaining undescended inside the abdomen close to the kidneys. Female hyraxes and elephants also share comparable placental structures and fetal membrane arrangements.[3] Both have long pregnancies, 7–8 months for hyraxes and 21–22 months for elephants. While the hyrax's pregnancy is shorter, it is exceptionally long for a small mammal – especially compared to cats and dogs, which typically gestate for only 6–8 weeks. Another shared trait is the unusual distribution of mammary glands, with a pair located close to the front legs.

Hyraxes and proboscideans have similar foot structures, thick padded soles used as cushions and flattened nails on the ends of their digits, unlike the curved, elongated claws seen on other mammals. In the case of the hyrax, the pads are moistened by sweat, with a hollow in the sole's centre formed by a unique muscle arrangement to create a suction-cup-like effect, helping them maintain grip on rocky surfaces. Hyraxes use their molars to tear apart leaves and grasses rather than using their incisors. Instead, their two upper incisors are large and tusk-like and grow continuously through life, like those of elephants. In contrast, all other mammals – narwhal, muntjacs, pigs, hippopotamuses and walrus – have tusks derived from canine teeth.

Rock hyraxes also possess a sensory specialisation shared only with the sirenians – body vibrissae or tactile hairs. These cutaneous filaments cover the entire body and appear as long black structures interspersed among the shorter fur. In contrast, sirenians only have vibrissae. The hyraxes' expanded sensory system allows for a greater spatial resolution, facilitating navigation in the rocky crevices and small caves and enhancing survival where visual clues are minimal. In Hebrew, rock hyraxes are known as *shaphan* ('the hidden one'), and hiding may be their best defence, aided by tactile feedback from their vibrissae. Elephants also have vibrissae, but only on the tips of their trunks.[4] These observations suggest that total body vibrissae – face and body – represent the ancestral condition of the Paenungulata.[5]

Hyraxes and elephants are highly intelligent mammals known for their exceptional long-term memory. They lead social lives and rely on complex vocal communications in order to interact within their groups.[6] As with bird communication, hyraxes produce structured and organised vocalisations referred to as syntax. Their repertoire includes around 20 distinct sounds – trills, yips, chucks, wails, snorts, squeaks and tweets – whose patterns and arrangements vary by region, resembling human accents and dialects. The evolution of such complex language relates to living in large colonies, where communication offers a survival advantage. A follow-up study also found a connection between male reproductive success and the ability to maintain rhythm during courtship songs.[7] Males tend to sing in crescendo and reach peak complexity towards the end of their songs, likely to sustain the interest of the females. Songs with precise frequency and rhythm signal a male's health and suitability as a mate. Such advanced communication skills may be attributed to the fact that hyraxes, like elephants, have a relatively large hippocampus, a complex brain structure embedded deeply within the temporal lobe that undertakes significant roles in learning, forming new memories, and spatial awareness.[8]

Since the time of Charles Darwin, scientists have often pondered whether human music evolved from animals' courtship songs. Rhythmic structure is a fundamental feature of many musical styles, and maintaining a regular rhythm is critical for good musical performance. Only a few mammals sing – whales, gibbons and hyraxes – but the styles and rhythmic signatures of their songs support the idea of common purpose. Or at least, as suggested by the leading researcher in the field, Vlad Demartsev, there is a common neural preconditioning for rhythmic patterns in animals and humans.[9]

Whether modern Paenungulata evolved from an aquatic or an amphibious ancestor has long been debated.[10] In 2013, convincing evidence came from an unlikely investigation, the analysis of the muscle protein myoglobin.[11] Breath-holding and diving require a high muscle release of oxygen, and nature achieves this feat using the oxygen-binding protein myoglobin. In diving mammals, the protein is present in such high concentrations that their muscles appear almost black rather than the typical red colour of meat. Since proteins tend to aggregate and be less effective at high concentrations, it was a mystery how such high levels could exist, and the protein still function.

Researchers from Liverpool University, led by Scott Mirceta, hypothesised that an increase in the protein's net surface charge might cause the molecules to repel each other electrically and prevent damage and loss of function. And the team confirmed their prediction: all diving-mammal myoglobin possesses the necessary electrical alteration. Intriguingly, this adaptive change was present in hyraxes, proboscideans and sirenians, where it resulted from amino acid substitutions similar to those of cetaceans and pinnipeds. These findings suggest that the common ancestor of all paenungulates was amphibious, and that the clade was the earliest eutherian radiation to adapt to an aquatic lifestyle. In other words, an early hyracoid branch must have taken to the water around 64 million years ago, soon after the asteroid strike, and spawned the sirenian and proboscidean lineages.

The sirens of myth

Mermaids are, without doubt, creations of human imagination. But in a world steeped in folklore, from the nefarious sirens of Greek mythology to the seductive creatures of more recent literature, sightings continue to occur to this day. Historical accounts, such as those written by Christopher Columbus during his explorations of the Caribbean in 1492, are now thought to reflect descriptions of sea cows or manatees. While mistaking a ponderous, blubbery sirenian for a scantily clad, fish-tailed maiden may seem hard to believe, their true story is no less strange: evolving first in water, followed by a terrestrial phase, and then returning to water again.

The order Sirenia, also known as sea cows, includes four living species: three manatees and one dugong. These remarkable creatures are the only fully aquatic mammals that feed exclusively on vegetation, primarily grasses, in warm, shallow coastal waters. Sirenians are believed to have diverged from the elephant lineage around 58 million years ago, likely in the marshy wetlands of Africa. Fossil evidence, dating back to the Eocene and discovered in regions such as Tunisia and Jamaica, sheds light on their evolutionary journey from land-dwelling ancestors to fully aquatic animals.[12] One notable fossil, *Pezosiren* from Jamaica, provides key insights. This pig-sized creature had short, functional limbs and dense bones, characteristics that helped counteract the buoyancy of its large lungs, much like modern sirenians. Its anatomy suggests it spent much of its time in shallow water, similar to modern hippos. However, *Pezosiren* could swim when necessary, using spinal flexing, pelvic paddling, and its short limbs for propulsion.

Later, the direct ancestors of modern sirenians ventured into the shallow coastal waters off North Africa, paddling in the proto-Mediterranean or Tethys Sea. Gradually, they became fully aquatic, modifying their forelimbs as flippers and replacing their hindlimbs with a spade-like tail. Then, around 34 million years ago, they spread far and wide, first to northern European waters before undertaking a key trans-Atlantic dispersal, reaching southern North America and the Caribbean. They also spread eastwards to the Indian Ocean. However, the Earth's climate deteriorated around this time, with rapidly falling temperatures and plummeting sea levels, causing widespread extinctions of both terrestrial and maritime species. Those sirenians to the east suffered and eventually disappeared. But in the west, they survived and, near the beginning of the Oligocene, diverged to give rise to the manatee and dugong lineages.

In 2022, Stephen Heritage and Erik Seiffert from the Duke Lemur Center Museum of Natural History at Duke University assembled the largest dataset of living and fossilised sirenians, in order to understand their evolution and historical dispersal routes. The team also collected the species' genetic, anatomical and biogeographical profiles.[13] The generated models suggest that the early manatees evolved within continental South America after swimming upstream from the rivers that flowed northwards into the Caribbean. During the last 20 million years, most of western Amazonia was a million-square-kilometre wetland – the Pebas mega-wetland – consisting of shallow lakes and swamps that drained into the Caribbean. As South America drifted away from Africa, it crossed a subduction zone, where slabs of the Earth's crust sink into a softer mantle. The tectonic upheavals resulted in an intense uplift along the continent's western half. At the same time, the continent's northeastern region subsided by as much as 400 metres. The overall result was that South America tilted like a giant seesaw, and the resultant 'slide' effect created the Amazon river system, with the mega-wetland's waters redirected over 6,400 kilometres into the South Atlantic Ocean.[14] This geographical upheaval allowed the manatees to disperse out of South America during the Late Pliocene. Then, during the Pleistocene, around 1.3 million years ago, they moved into the Caribbean, along the coasts of North America, and back across the Atlantic to West Africa.

The dugong lineage, in contrast, diversified near Florida, producing a wealth of species that included the ancestor of today's dugong. Then, before the closure of the Central American Seaway, they moved into the Pacific, crossed the South China Sea, and reached the Indian Ocean, aided by wind-driven surface currents. Such long-distance ocean crossings would have lacked rooted seagrasses, so how dugongs could have sustained the energy levels required for such extensive journeys is a mystery. Nevertheless, such dispersals occurred several times during sirenian evolution, including a relatively recent reverse dispersal of manatees back to Africa that gave rise to the African manatee.

A second dugong diaspora ventured northwards 15 million years ago, probably driven by global cooling and the decimation of seagrasses off California. They reached the Bering Sea area and adapted to a novel diet of kelp. They lost their teeth and developed an array of bristles on their upper lips and cornified pads on their palates and jaws. The Steller's sea cow was the result: an enormous creature – adults weighing up to 4,000 kilograms – that evolved many adaptations to survive the cold subarctic waters. Sadly, their extensive layers of blubber, thick bark-like skin and long, broad lungs meant they could not fully submerge, and they became easy targets for the early explorers. Sailors hunted this largest of known sirenians to extinction a mere 27 years after its discovery by Bering's ill-fated crew in 1741.

The three species of manatee, the Amazonian, West Indian and African, rarely wander far from fresh water. While their vision is poor, they have excellent hearing and tactile senses, enabling them to navigate murky coastal waters. Propulsion is achieved by undulating their bodies and flapping their paddle-shaped tails and flippers. The latter are modified five-fingered forelimbs with vestigial nails, evolutionary remnants of their terrestrial past. All living sirenians have a low metabolic rate and lack a thick insulating layer of fat, which restricts their colonisation to the warm waters of the tropics and subtropics. They feed on submerged and shoreline seagrasses, using their pronounced, prehensile upper lip to grasp their food, much as their close relatives, the elephants, use their trunks.

Sirenian evolution mirrored the rise of seagrass meadows, which flourished during the Late Oligocene and Early Miocene when plate tectonics produced vast areas of shallow-water habitat. Seagrass, however, is not nutritious, and the manatees evolved immensely long intestines with hind-gut fermentation to digest the plant cellulose and extract its essential nutrients. Dietary sand and grit wear away their teeth, which get replaced in a conveyor-belt-like fashion: new teeth erupt at the back of the jaw and inch forward to replace those that fall out in the front.

Dugongs reside in the coastal waters of almost 40 countries across the Indo-Pacific region, stretching from East Africa and the Red Sea to Australia. While they resemble manatees, dugongs have smaller bodies, fluke-like tails, and distinctive skull and tooth structures. Additionally, their longer, downward-turned noses are specifically adapted to aid in bottom-feeding.

Aquatic adaptations

Comparative genomic approaches reveal that sirenians evolved specialised proteins that protect against the high iodine content of the aquatic vegetation they feed on.[15] Iodine is required to synthesise thyroid hormones essential for energy metabolism, thermoregulation and the integrity of many tissues. However, deficiencies and excesses of the element can lead to thyroid dysfunction. Despite high intakes, blood hormone levels match those of carnivorous dolphins, a fact related to mutations in many of the genes involved in the thyroid hormone pathway, especially those involved in the transportation of iodine across membranes.

Polypeptides controlling circadian rhythms have also mutated to allow activity heavily reliant on lunar tidal currents and water temperature fluctuations. Furthermore, mutations in the genes responsible for coding haemoglobins, the primary oxygen-carrying proteins in vertebrates, have produced molecules with enhanced oxygen binding capacities, allowing for extended dive durations.[16] Since the manatees' and dugong's common ancestor lived approximately 30 million years ago, these evolutionary protein modifications were likely critical for their transition from a terrestrial to an aquatic lifestyle.

Unlike the living sirenians, Steller's sea cows underwent convergent evolution with cetaceans during their adaptation to the cold subarctic waters of the North Pacific. They developed a tough, bark-like skin due to inactivation of lipoxygenase genes, mutations which cause congenital ichthyosis in humans, a disease characterised by a thick, dry outer layer of skin.[17] Steller's sea cow haemoglobin also possessed a unique amino acid substitution not found in any other mammal. The resultant structural change in their blood protein led to a reduced oxygen affinity and a higher solubility and blood level. Replacing just one of haemoglobin's many amino acids produced the opposite physiological effect to that seen in the blood of living sirenians. Indeed, the novel protein contributed to the evolution of the Steller's sea cow's higher metabolic rate and tolerance to cold.[18]

As mentioned earlier in this chapter, all paenungulates have vibrissae to a greater or lesser extent. Manatees have over 5,000 scattered over the body, 3,300 on the post-facial area and 2,000 on the face.[19] The animal's visage has a strange appearance, consisting of a muscular and prehensile area between the upper lip and nostrils. This prominent feature, the oral disc, gives manatees a browbeaten expression that makes them so captivating. However, the area is essential as it acts like an elephant's trunk, with unique muscles allowing each side to move independently. By flexing and flaring the oral disc, manatees can investigate novel objects and potential food items with the dexterity and sensitivity of fingers (a process called oripulation). Indeed, the tactile acuity of the manatee's oral disc is likened to that of the Asian elephant's trunk, with the wealth of environmental information entering the brain, where large areas are devoted to processing the sensory input.[20]

The perioral vibrissae act as a tactile sensory organ, like whiskers in other mammals, and as an aid to grasping plants during feeding. Such a combination of motor and sensory usages is unique to sirenians.[21] In contrast, the post-facial vibrissae are not actively involved in tactile exploration. Instead, they may function to detect water movements, thus functioning like a mammalian version of the lateral line system found in fishes and amphibians.

CHAPTER 7

The Elephant's Story

ADMIXTURES, RATCHETS AND RETROGENES

If elephants didn't exist, you couldn't invent one. They belong to a small group of living things so unlikely they challenge credulity and common sense.
Lyall Watson (1939–2008)

In northern Ghana lies Mole National Park, a magical wilderness seemingly untouched by the modern world. It may be a challenging place to reach, but its stunning setting, unrivalled wildlife and astonishing beauty make the effort worthwhile. Extending in a giant crescent from southeast to southwest runs an escarpment, a sculptured swathe of red-coloured sandstone that dominates the pristine landscape. Thirty metres below, among the wooded savanna, lie waterholes where crocodiles lurk, antelopes drink, and monkeys grunt in the nearby trees. However, the most sought-after inhabitants are the African savanna elephants, the largest terrestrial mammals (Plate 12). Unlike elsewhere in Africa, Mole's majestic creatures appear non-hostile and tolerant of humans, offering a unique opportunity for a close encounter. Some years ago, I was privileged to observe a family gathering at close quarters as they loitered and squelched in the surrounding mud. With calves close by, the matriarchs appeared no more concerned at my presence than they were by the cattle egrets foraging between their legs. Their trunks never rested, however: swinging, coiling, scanning, twisting, touching and sensing. Lyall Watson was right: the massive body, huge ears and tusks, and bizarre proboscis 'challenge credulity and common sense.' But how did such an odd, keystone behemoth evolve? It's a question science is beginning to answer.

Elephants belong to the taxonomic order Proboscidea, which consists of one living family, Elephantidae, and several extinct families. As previously discussed, they are the best-known afrotherians and belong to the clade Paenungulata, along with hyraxes and sirenians (see Figure 5.2). The first elephants appeared over 60 million years ago in Africa and, as one might expect, were much smaller than the three species alive today. The oldest and most primitive, *Eritherium*, was only a 5-kilogram, fox-sized animal that lived in swampy vegetation in present-day Morocco. It lacked a trunk but had an enlarged first incisor that may have represented a primitive tusk.[1] As the Palaeocene transitioned to the Eocene, the forest-dwelling proboscideans grew steadily larger, reaching 200 kilograms within 5 million years, although still bearing little resemblance to modern elephants.

The early proboscideans then evolved a precursor trunk, like those of tapirs, which, millions of years later, transformed into a snorkel-like structure, allowing more extended periods underwater. Over time, this remarkable fusion of the nose and upper lip became longer, stronger and more flexible, with nearly 40,000 muscles (in comparison, the entire human body has only 600–700) and no bones.[2] However, its immense size and weight considerably increased the stress on the facial bones. Coupled with the need to provide space for the attachment of the trunk musculature, this led to a marked enlargement of the elephant skull, with a much higher forehead. With an almost infinite freedom to move, the trunk became multifunctional, used for breathing, drinking, feeding, grabbing, smelling and trumpeting. Its extraordinary flexibility, aided by the folds and wrinkles in the overlying skin, allowed the animals to undertake various tasks, from grabbing delicate vegetation to ripping apart tree trunks. Indeed, the evolution of the elephant's proboscis was a game-changer and provided access to vegetation that was out of reach for other species. It also allowed elephants to drink and feed without constantly having to move – an energy-saving innovation that unleashed the evolution of even bigger and heavier species.

Like all Paenungulata, elephants possess vibrissae to enhance tactile awareness.[3] Dense arrays of whiskers are located at the end of their trunk tips, although not in the area used for pinching objects. Hundreds of neurones innervate each vibrissa, and as a result the infraorbital nerve that supplies the trunk evolved to become thicker than the elephant's optic nerve.[4]

The prominent tusks of elephants are modified second incisors of the upper jaw. Early proboscideans possessed teeth entirely capped with enamel. Over aeons, the overlying enamel became restricted to a thin longitudinal band on one side that allowed the teeth to grow continuously, like the incisors of modern rodents. Eventually, the enamel was lost completely, resulting in the dentine-only tusks of modern elephants.[5] During the Oligocene and Miocene, evolution experimented with many shapes and sizes of tusk: upturned, downturned, straight, shovel-shaped, and matching upper and lower sets. They served a variety of functions: digging, gathering food, lifting objects, stripping bark from trees, defence – and allowed the exploitation of different habitats. The rate at which different tusk phenotypes can evolve has recently been revealed in a surprising way. After the Mozambican Civil War

(1977–1992), when ivory poaching went unchecked, a sizeable percentage of female African savanna elephants was born tuskless. Genetic studies showed that this trait was sex-linked (located on the X chromosome) with recessive lethality and related to at least two genes controlling tooth development. Before the war, both genes were rare in the population, but within only 20 years, the selective pressure of poaching had favoured their inheritance and the birth of tuskless females.[6] In contrast, affected males, who lack a normal X chromosome, died *in utero*.

Once the Afro-Arabian tectonic plate had collided and docked with the vast Eurasian landmass, an important migratory corridor opened, allowing the early proboscideans to explore new habitats in Eurasia. Eventually, they reached North America via a land bridge that intermittently connected Siberia to Alaska, now submerged under the Bering Sea. The Eurasian proboscideans were exposed to many new habitats and climates and evolved faster than their cousins left behind in Africa. As Juan Cantalapiedra of the University of Alcalá in Spain noted, this led to a remarkable diversity of coexisting species – a richness of mega-herbivores unparalleled in modern ecosystems.[7]

Habitat fluctuations are continuous due to the Earth's ever-changing climate, and while many novel species evolved, adaptation was crucial for survival. By 3 million years ago, a harsh global cooling prevailed, and only the Elephantidae could withstand the relentless selection pressures. They included the ancestors of the three extant species – the African savanna, the African forest and the Asian elephants – and their extinct relatives, the straight-tusked elephant and the mammoths. Straight-tusked elephants, the largest known terrestrial mammal, once roamed Europe and western Asia (reaching as far north as Britain) during the Middle and Late Pleistocene. Like their descendants, they lived in herds and were likely exploited by early humans, including Neanderthals. In contrast, mammoths comprised several species, of which the once circumpolar woolly mammoth survived in small populations on Arctic islands well into the Holocene around 4,000 years ago. However, the North American Columbian mammoth, a more temperate species, disappeared by the end of the last ice age approximately 11,000 years ago.

Admixture and isolation

Understanding how extinct and modern elephants are phylogenetically related has been a formidable scientific challenge. In the past, palaeontologists deduced such relationships by comparing incomplete bone fragments. Even when whole skeletons are available, the task remains formidable. Genome sequencing has revolutionised the field, and the results were unexpected, highlighting the role of hybridisation in speciation.

Although not an exceptionally stable molecule, DNA can occasionally be extracted from fossilised bones. The chances are better for younger specimens or those preserved in cold environments, such as permafrost. Technological advances, including extraction techniques, new-generation sequencing (NGS) and powerful supercomputers, have also increased the success rate. Indeed, the oldest DNA ever sequenced comes from a million-year-old mammoth tooth recovered from the Siberian steppe – a feat thought impossible only a few years ago.[8] Even so, the DNA had degraded into tiny fragments, and, according to the research team, its assembly was like solving a billion-piece jigsaw puzzle. To simplify the task, they compared the sequenced sections to a near-complete genome of the African elephant, which provided positional clues similar to the illustrated lid of a jigsaw box. It was worth the effort, since the tooth belonged to an unknown species of mammoth that probably colonised North America.

In 2018, Eleftheria Palkopoulou and colleagues generated high-quality genomes for the three living elephant species and less complete sequences for two species of mammoth and the straight-tusked elephant.[9] Contrary to expectations, sequence comparison showed the straight-tusked elephant was not closely related to the Asian elephant, as suggested by palaeontology and mitochondrial DNA studies. Instead, straight-tusked elephants evolved from a genetic mixture of three populations related to the ancestors of African elephants, woolly mammoths, and present-day forest elephants. In other words, multiple episodes of interbreeding must have occurred between different elephant species, involving ancestors of both living and extinct taxa. In the case of mammoths, interbreeding also happened between the Columbian mammoths of the temperate south and the American woolly mammoths of the north.

The Palkopoulou study highlights critical insights into evolution and speciation. Random, spontaneous mutations do not solely drive these processes; new species can emerge through gene flow, or 'admixture', between distinct taxa long after their divergence. The traditional view of a family tree with discrete branches representing separate species tracing back to a common ancestor may be overly simplistic. In reality, many species' ancestors likely interbred, creating a complex genetic tapestry resembling a network of interconnected branches. Gene flow and hybridisation are, in fact, pervasive evolutionary mechanisms, playing vital roles in the development of numerous species, from bears to humans. The once-dominant perspective of Ernst Mayr – that hybridisation disrupts and impedes evolutionary divergence – is increasingly untenable.[10] Instead, admixture is now considered a recurring and integral aspect of mammalian evolution.[11]

Another conclusion of the Palkopoulou study is the lack of genetic admixture between the African savanna and African forest elephants. This finding was surprising, given that hybridisation is a well-recognised phenomenon in areas where the species meet, producing fertile offspring. Using complex statistical methods, Palkopoulou found that their genomes show no evidence of admixture despite their lineages diverging approximately 500,000 years ago. In other words, even though savanna and forest elephants hybridise in the contact zone, this admixture does not affect the overall genetics of either population. One species prefers to remain in the depths of the tropical West and Central African jungle, while the other likes to roam the open savannas. Isolation, as well as hybridisation, has shaped elephant evolution in unexpected ways.

Ratcheted evolution and Liem's paradox

Modern African elephants have evolved specialised teeth for grazing, a result of their ancestors adapting to new diets millions of years ago. Adrian Lister from the Natural History Museum in London explains that East Africa's landscape shifted from woodlands to drier grasslands around 10 million years ago. During this transition, elephants' teeth developed taller crowns (hypsodonty) and additional enamel ridges, allowing them to grind and break down grasses with high dust content. To explore the connection, Lister studied soil samples and fossils spanning 20 million years. His findings revealed that while elephants and their relatives switched from browsing to grazing around 8 million years ago, their dental adaptations took an additional 3 million years to fully develop. This observation suggests that by experimenting with novel feeding strategies, elephants placed themselves in a position where natural selection favoured individuals with better-adapted teeth. By challenging the more passive view of natural selection, whereby an environmental change selects elephants with stronger teeth, Lister believes that the elephants' trial with grazing (a form of phenotypic plasticity that can presage evolutionary change) may have helped shape their evolutionary destiny.[12]

A follow-up study by Saarinen and Lister incorporated palaeoclimatic data from sediment cores from deep-sea drilling in the Arabian Sea.[13] The dates of dusty, grass-rich periods were compared to the age of the fossilised elephant teeth (accurately determined, as they were found alongside well-dated early hominid fossils). As Africa became more arid, elephant teeth became more resistant to wear, enabling their survival. However, the teeth didn't become less resistant to abrasion when the climate reverted to wetter conditions. They remained unaltered until drier, grass-dominated conditions returned, driving the next burst of evolution. In other words, the dental crowns of elephants became higher and higher but never reduced – a phenomenon the researchers likened to an evolutionary ratchet. Although such evolutionary changes have not previously been observed, Lister and his colleague think ratcheted change could be an important and relatively common evolutionary mechanism. The shortage of fossils and limited climate data over such long periods may have hidden the process from a scientific view. While the ratchet mechanism allowed elephants to survive dry conditions and the spread of grasslands, it may have led to overspecialisation and contributed to the extinction of extreme grazers that couldn't adapt to changing conditions. In contrast, modern elephants are dietary generalists, which may have helped them survive while their relatives fell by the wayside.

The observation of elephants' stepwise dental evolution correlating with peaks in selection pressure supports the hypothesis that traits, including hypsodonty, are adaptations to extreme rather than average conditions. This fact explains Liem's paradox, in which mammalian species at any given time often have more specialised dentition than their observed diets would predict. The paradox resulted from work by Karel Liem of Harvard University on a Mexican fish called the Minckley's cichlid (pronounced 'sick-lid') – a species with specialised pebble-like teeth that evolved to eat hard-shelled snails common in its environment.[14] Liem discovered that the fish would often ignore snails if softer foods were available. In other words, they avoided food their bodies had evolved to eat. Proboscidean hypsodonty, in which the teeth extend far above the gums, increases dental longevity and allows the consumption of abrasive foods. Although the hypsodont state has evolved convergently in many mammalian lineages, it has never reversed to produce less prominent crowns or a brachydont state. This observation implies that animals have evolved to survive hard times but will bypass their unique adaptations to eat preferred, more accessible food in times of plenty.

Why are elephants not bigger?

Since their first appearance, rapid and repeated morphological change towards larger body size has shaped mammalian diversity, a phenomenon most likely due to selection pressure.[15] So why has the elephant lineage plateaued in size and not become as massive as the largest dinosaurs or whales? The answer is dependent, in part, on several biological constraints faced by terrestrial mammals.

The first relates to the so-called cube-square law. This mathematical principle, first described by Galileo Galilei, states that the ratio of two volumes is greater than that of their surfaces. In other words, as elephants grew, their volume increased faster than their surface area, so they required much stronger limbs to support their weight. Mathematics reveals that if the largest known terrestrial mammal, the straight-tusked elephant, had doubled in size, it would have collapsed. While the cross-sectional areas of its leg bones would have been four times greater, its weight would have increased eight-fold, with the result that the stress on its leg bones would have doubled, leading to fractures. To overcome this limitation, elephants had to evolve even bulkier and more robust limbs. Eventually, a limit was reached when the legs became so sturdy and cumbersome that mobility, hence survival, would be compromised. However, the buoyancy of water counteracts gravity's effects. Consequently, the blue whale has been able to evolve without the same musculoskeletal constraints experienced by similarly sized terrestrial animals.

A further limitation on mammalian bulk is the requirement for heat loss. Unlike the reptilian dinosaurs, mammals are endothermic and must maintain a constant body temperature, regardless of the ambient conditions. Because warm-blooded mammals have a faster metabolism, they must consume approximately 10 times the amount of food required by reptiles of the same weight. This fact alone explains why elephants and all other large terrestrial mammals are herbivores. Such animals need plenty of available resources and graze for most of the day, foraging over large areas. Carnivores, in contrast, require speed and agility to catch their scarcely distributed prey, a constraint that prevents lions and cheetahs from becoming the size of elephants. Large herbivores, however, risk generating too much heat. To enable the digestion of cellulose-rich plants, elephants have co-opted the help of gut bacteria to break down the tough fibrous components by fermentation. This adaptation is a double-edged sword: elephants obtain the energy they need, while fermentation generates unwanted heat. The larger an animal becomes, the bigger its volume-to-surface-area ratio and the less efficient its heat loss.

Phenotypic adaptations, however, have helped to ease the overheating problem. African elephants, for example, have evolved huge ears or pinnae that constitute 20 per cent of their surface area. These organs contain many surface blood vessels, which, when dilated, transform the pinnae into radiators that efficiently dump heat into the air. Elephants also flap their ears back and forth, creating local airflows to increase heat dissipation.[16] The animals' low hair density also aids heat loss. Although thick hair evolved to keep mammals warm, low-density hairs provide a larger surface area and act as little heat fins that offset any increase in insulation. In effect, elephant hair significantly increases the animal's effective heat transfer coefficient and provides a thermoregulatory heat sink. This finding dispels the idea that body hair functions only as an insulator, and it may help resolve the paradox of why hair evolved in a much warmer world than today.[17] Despite these evolutionary innovations, though, it is likely that the straight-tusked elephant approached the maximum size possible for terrestrial mammals.

Peto's paradox and retrogenes

Increased cancer risk is also a potential constraint to evolving larger body masses. Theoretically, bigger animals should be at greater risk of cancer if every dividing cell has the same chance of becoming malignant. Indeed, this relationship holds within a species. Cancer incidence increases with height for numerous human tumours, while bigger dog breeds have more cancers than smaller breeds. Similarly, if an animal has an extended lifespan, its cells have more time to accumulate the many mutations required to divide uncontrollably and become cancerous. Since larger species tend to have longer lifespans and hence more cell divisions – elephants live up to 70 years – this should compound the problem.

However, in 1977, Professor Sir Richard Peto, working at the Institute of Cancer Research in London, highlighted a simple but puzzling fact: given the size difference between mice and ourselves, humans should have a greater cancer rate than mice – but we don't.[18] Paradoxically, cancer risk appears not to be increased in the largest mammals. For example, cancer rates in elephants are nearly five times less than in humans despite the fact that they possess 100 times more cells and have approximately the same lifespan. If elephants had the same cancer risk per cell division as humans, most would not survive to reach sexual maturity, preventing the evolution of the world's largest land mammals. This phenomenon, known as Peto's paradox, implies that larger animals must have developed mechanisms to offset the heightened cancer risk from numerous cell divisions. This conclusion implies that the largest and longest-living species will likely possess the most effective defences against cancer. Indeed, if scientists can uncover the genetic basis for Peto's paradox, doctors may be able to use nature's solution to enhance cancer therapies. This intriguing possibility prompted several research groups to search for clues in the elephant genome, and they all came up with the same answer – the *TP53* gene.[19]

The *TP53* gene encodes for the tumour suppressor protein p53, which protects cells from cancer in several ways. The protein prevents precancerous cells from dividing, so that their defects can be corrected, or if such cells are beyond repair, it induces the cell's suicide programme, or apoptosis, to eliminate them. Simply put, p53 acts as a 'guardian of the genome' according to its co-discoverer David Lane. Indeed, humans who inherit a faulty or missing *TP53* gene have a high risk of developing a wide range of cancers.[21]

In 2016, a team from the University of Chicago studied the genomes of 61 animals of different sizes. While most possessed one copy of *TP53*, several species had extra copies, known as *TP53* retrogenes (discussed below). Intriguingly, African elephants had more than any other mammal, with 20 copies – one ancestral *TP53* gene and 19 retrogenes. To investigate why this should be so, the researchers analysed DNA from Asian elephants and several other closely related but extinct species, including the American mastodon and woolly mammoth. The results indicated that the proboscideans gained additional *TP53* retrogenes as they evolved larger body sizes. Indeed, the first retrogene insertion occurred in the paenungulates' common ancestor around 64 million years ago, followed by an expansion during the last 40 million years, coincident with the evolution of larger elephant body masses.[20]

But what are retrogenes, and how did they drive elephant evolution?

Gene duplications can arise in two ways: copies derived directly from DNA with genetic features similar to their parental genes (Chapter 1) and retrogenes. The latter are RNA-based copies that lack the genes' non-coding sequences, both introns and control elements. The journey from gene to protein is complex and involves initial transcription to form mRNA that includes the gene's non-coding introns. A protein-and-RNA complex called the spliceosome cuts and removes these interspersed sections to prevent the synthesis of a non-functional or incorrect protein. Once edited, the modified mRNA makes its way to the ribosome to direct the synthesis of its respective protein. However, the 'stripped-down' mRNA can sometimes be reverse-transcribed by a specific enzyme to recreate a copy of the original gene without introns. These novel sections of DNA become retrogenes if inserted into the genome in different locations. Because retrogenes lack the regulatory elements of their parents and can have slightly different base sequences, they may evolve new expression patterns and acquire novel functions.[22]

A recent study by researchers across Europe has helped explain the role of the extra *TP53* retrogenes.[23] There appear to be subtle but significant structural differences between each protein. As a result, the various p53 molecules interact differently with target proteins in the cell, making it harder for a cancer to disable enough of them to escape detection. It's the same principle as the use of combination chemotherapy by oncologists to outwit cancer's defences and improve clinical outcomes. Proboscideans also gained extra copies of *LIF6*, leukaemia inhibitory factor 6, another tumour suppressor gene whose protein works alongside p53. While additional copies of *LIF6* are present in manatees and hyraxes, they are pseudogenes and lack an 'on' switch. However, around 25–30 million years ago, as the elephant's hare-sized ancestors grew in stature, they hotwired a second *LIF6* gene. In the presence of damaged DNA, elephant p53 can bring the 'zombie' copy of *LIF6* back to life to help destroy cancerous cells.[24] The revived LIF6 protein then responds by puncturing mitochondrial membranes and removing the cell's power source, leading to rapid cell death.

However, while *TP53* retrogenes enabled the elephant lineage to increase in size, they don't explain why the duplications should have occurred initially in a distant and relatively small ancestor. One theory proposed by the evolutionary biologist Professor Fritz Vallrath from Oxford University relates to the location of the afrotherian testes.[25]

Unlike most placental mammals, the testes of Afrotheria, except those of aardvarks, fail to descend into a scrotum and remain deep inside the abdomen, a state known as testicondy. Afrotherian abdominal testes result from knock-out mutations in two genes that guide testicular descent in other placental mammals. Interestingly, this finding implies that the common ancestor of all afrotherians must have had functioning copies of both genes, as pseudogenes would not have appeared by chance. Furthermore, because genes accumulate mutations at a predictable rate after a loss of function, researchers have been able to estimate that testicondy arose independently at least four times, ranging from about 25 million years ago in golden moles to about 80 million years ago in elephant shrews. In other words, testicondy evolved after afrotherians split from other placental mammals around 100 million years ago, implying that the common ancestor of all mammals' must have possessed scrotal testes.[26]

The scrotum is an evolutionary adaptation that provides a cooler environment for the production of sperm (spermatogenesis), a necessity that emerged when mammals became warm-blooded. Sperm are highly sensitive to heat; even typical body temperatures can impair their quality and increase rates of spontaneous mutations. To ensure optimal sperm production, a temperature several degrees below the body's core temperature is required. For instance, while a mouse's core body temperature is 36.6 degrees Celsius, its scrotal testicles are maintained at approximately 34 degrees. Sperm produced at 37 degrees are less viable, showing significant DNA damage and high mutation levels, while exposure to higher temperatures triggers sperm apoptosis. Unlike most mammals, elephants have undescended testicles, subjecting their sperm to potentially harmful temperatures. According to Vallrath's hypothesis, *TP53* retrogenes evolved primarily to promote the production of healthy sperm under warmer conditions, with any cancer-protective activity being a serendipitous pleiotropic effect.

A warning from the grave

Analysis of mammoth DNA has also revealed a likely cause for their demise, a post-mortem finding that provides a cautionary tale for conservation enthusiasts. Two researchers from the University of California, Rebekah Rogers and Montgomery Slatkin, realised that mammoth DNA offered a rare chance to explore how genomes react to pre-extinction pressures, and their results were both enlightening and worrying.[27]

Woolly mammoths were among the most widespread large herbivores, inhabiting North America, Siberia and Beringia during the Pleistocene and early Holocene. While global warming and human hunting pressures led to their extinction in mainland Asia and America around 10,000 years ago, populations on isolated Arctic islands, such as Wrangel Island, existed until 3,700 years ago. But why did these isolated, non-predated populations fail to survive?

Rogers and Slatkin provided the answer. They identified many structural abnormalities in the DNA of a Wrangel Island mammoth, including deletions, retrogenes, substitutions and loss-of-function mutations. This plethora of detrimental changes is consistent with genomic meltdown and results from the mammoth's reduced population size and lack of opportunity for acquiring novel alleles from genetically different populations. Indeed, the finding supports the nearly neutral theory of genome evolution, which predicts that selection is less able to weed out harmful DNA changes in small, isolated populations.[28] In other words, DNA mutations accumulated quickly in the small Wrangel Island population, hastening the mammoths' extinction, especially in the face of coexisting climate changes.

Like the Tasmanian devil, the mammoth's story emphasises the importance of gene pool size and heterogeneity for a species' success. Genetic variation is crucial for long-term survival, allowing for rapid evolution when adverse environmental changes occur. Conversely, small, isolated populations with limited gene pools are less resilient. These findings are not academic. Sadly, the populations of several extant mammal species, including the Iberian lynx and the mountain gorilla, have reached sizes similar to those of the pre-extinction woolly mammoth. Assuming that the nearly neutral dynamics of genome evolution affect most mammals – and there is no reason to suppose otherwise – the survival of these iconic animals remains precarious even with current conservation measures.

Could de-extinction let us see the woolly mammoth again?

An American company, co-founded by Harvard University geneticist George Church and entrepreneur Ben Lamm, is working with the University of Melbourne's Andrew Pask (whom we met when discussing the de-extinction of the thylacine) to resurrect the woolly mammoth. Colossal Biosciences' approach is to insert key mammoth genes into the DNA of the Asian elephant, the mammoth's closest living relative. The goal is to create an embryo with these modified genes that could develop in an African elephant surrogate or artificial uterus. Any fit offspring would be released into a Siberian or Alaskan park. However, these animals would not be 'true' mammoths but rather elephants engineered for Arctic conditions, featuring traits like small ears, a longer, thicker coat, an increase in fat storage, and a haemoglobin molecule with higher efficiency in carrying oxygen at low temperatures.[29]

Readers might wonder why scientists cannot change every base in the elephant's DNA to recreate a complete mammoth genome. Such an approach, however, would involve making a mind-blowing number of changes. Suppose we assume that the woolly mammoth and Asian elephant diverged from their common ancestor approximately 4 million years ago, and that their subsequent DNA substitution rates are like those of other mammals. In that case, we can expect about 70 million genetic differences between the two species, the same number of base changes that differentiate the genomes of humans and chimpanzees. While less than 2 per cent of elephant DNA would require editing, the 'mammoth' task is beyond the scope of current methodologies.[30]

The release of mammoth-like creatures would not be without controversy, and would also be subject to political and economic considerations. Already, Colossal Biosciences has stated that Russia's invasion of Ukraine has disrupted its Siberian plan and that locations in Alaska are being considered instead. Furthermore, as we have discussed, if mammoths became extinct because of their inability to adapt to the warmer climate that followed the last ice age, then evidence will be required that they could survive the current period of increasing global temperatures. While the project has garnered considerable media attention, the technical and political obstacles are formidable. It seems unlikely that any of us will witness herds of mammoths roaming the snowy tundra in the near future.

CHAPTER 8

The Sloth's Story

REGRESSIVE EVOLUTION AND PSEUDOGENES

I would come back as a sloth. Hanging from a tree, chewing leaves sounds great.
Sir David Attenborough

It is time to return to South America and reacquaint ourselves with the continent's early, tribosphenic, shrew-like eutherians. Once Africa, with its unique cargo of Afrotheria, had wrenched itself free and commenced its long drift towards Europe, South America underwent a phase of 'splendid isolation'. This southern land mass would remain a gigantic island for over 50 million years, until the Panama isthmus formed around 3 million years ago, linking it to North America. The prolonged separation from the rest of the world coincided with a crucial time for the evolution of mammals elsewhere.

As a result, South America became a giant natural experiment culminating in a unique and bizarre mammal fauna – the Xenarthra (from the Greek *xen*, foreign, and *arthra*, joints), one of placental mammals' four major clades or superorders (see Figure 5.1).[1] The collective name reflects the presence in most species of extra articulations between the lumbar vertebrae and fused pelvic bones that strengthen the spine for digging. Recent biogeographic reconstructions indicate that the cradle of xenarthran evolution lay within the tropical rainforests of Amazonia and the Guiana Shield, with later dispersals occurring into more open and dry habitats.[2] Once the Panamanian land bridge had formed, many xenarthrans migrated to North America during the Pliocene or Early Pleistocene as part of the Great American Biotic Interchange. After surviving for tens of millions of years, most of the large terrestrial and aquatic taxa (ground sloths, glyptodonts and pampatheres) died out during the Quaternary Extinction Event, shortly after the colonisation of the Americas by early humans. As a result, extant xenarthrans comprise a meagre 31 species consisting of anteaters, tree sloths and armadillos.

In 1931, George Gaylord Simpson first suggested that the last common ancestor of the xenarthrans was a burrowing animal, a species that we now know lived around the time of the K–Pg boundary approximately 66 million years ago.[3] Unlike most creatures, its fossorial lifestyle may have helped it survive the devastating Chicxulub impact and the resultant infrared radiation (Chapter 9). Evidence for an earlier subterranean history is reflected in the anatomy of living xenarthrans, whether terrestrial or arboreal – markedly curved claws, modified shoulder blades that allow a forceful retraction of the upper arm and, as already highlighted, altered lumbar joints to stabilise the spine.[4] But perhaps the most convincing evidence for a past life underground lies at the back of their eyes.

Despite being diurnal and living in trees or on the ground, all existing xenarthrans lack retinal cone receptors. Their eyes contain only rods – retinal components that provide low visual resolution and are most effective in low-light conditions. The lack of functioning cones, known as rod monochromacy, leads to reduced visual acuity, colour blindness, and an inability to see clearly in daylight, as rods quickly become saturated in bright light. Most xenarthrans, therefore, rely on vision primarily at night or within their burrows. However, species that inhabit the forest floor may receive enough daylight to perceive shadows and faint movements. Impaired vision also explains why xenarthrans are often involved in fatal vehicle collisions, and understanding this evolutionary limitation should aid in conservation efforts.

Recent molecular genetic analyses indicate that the clade's poor vision was indeed inherited from a distant burrowing ancestor. In 2015, researchers Christopher Emerling and Mark Springer from the University of California examined the genomes of living and extinct species for inactivating mutations in their photoreceptive or opsin genes.[5] Their analysis revealed that two mutations that reduce cone function occurred over 80 million years ago in the common ancestor of xenarthrans. During the next 40 million years, additional mutations knocked out cone function in the common ancestor of sloths and anteaters (pilosans), while a different set of mutations occurred in the early armadillo lineage (cingulates). The existence of multiple non-functional genes, or pseudogenes, affecting cone function implies that the early xenarthrans inhabited low-light conditions. This conclusion is supported by evidence that rod monochromacy is also found in other creatures living in dark environments, such as deep-sea fishes, deep-diving cetaceans and subterranean vertebrates.[6] Additionally, the selective pressures associated with their ancestral burrowing lifestyle – characterised by the loss of cones and specific skeletal adaptations – likely constrained xenarthrans' evolutionary potential. This limitation explains why they never developed alternative modes of locomotion, such as gliding, flying or running, nor evolved into active predators.

The loss or degeneration of a once-useful trait, known as regressive evolution, is not restricted to xenarthrans but is a widespread evolutionary phenomenon. Examples include the absence of legs in snakes, teeth in birds, wings in flightless rails and penguins, legs and hair in cetaceans, tails in higher primates, and hair in humans. All such trait losses followed niche change and resulted from impaired function of the relevant genes due to relaxed selection pressure, genetic drift or direct selection against the redundant trait to reduce energy costs. In other words, the pseudogenes that underpin regressive evolution are inactivated original genes. As such, they differ from the non-functional duplicated genes that litter all mammalian genomes.

Finally, before we discuss the evolution of the three suborders of Xenarthra – anteaters (Vermilingua), sloths (Folivora) and armadillos (Cingulata) – we need to highlight their teeth, or rather a lack of them, for they provide another textbook example of regressive evolution.

In contrast to the edentulous condition of birds, baleen whales and pangolins, xenarthrans underwent different degrees of dental degeneration during their 65-million-year radiation. For example, sloths (Plate 13) and armadillos lack front or premaxillary teeth and only possess a few uniform molars, while anteaters have no teeth. However, even those of sloths and armadillos are degenerate, peg-like structures with open roots and no enamel. Unlike most mammals, they also lack deciduous or milk teeth and have only one set that grows and wears away throughout their lives. Indeed, the morphology of xenarthran teeth has drifted so far from those of their tribosphenic shrew-like ancestors that, until recently, it has been challenging to identify homologies or commonalities with other mammalian teeth.[7]

An international study led by Christopher Emerling reported that inactivating mutations in five enamel protein-associated genes occurred in a common ancestor of sloths and anteaters, resulting in a complete loss of the outer coating of their teeth.[8] While sloths halted further dental regression – an event likely related to their adaptation to a leaf-based diet – anteaters lost additional tooth-related genes, ultimately becoming edentulous. They include a small deletion in a crucial dentine-controlling gene and a single base insertion in a gene involved in tooth retention. Furthermore, the two lineages of armadillo independently lost the function of genes relating to the gum–tooth junction and enamel formation, which led to a similar dental phenotype. One lineage, which includes the fairy, giant, hairy, naked-tailed and three-banded armadillos, lost all traces of enamel. The other lineage, which includes the nine-banded armadillo, produces offspring with thin enamel that wears away with age.

In summary, xenarthran tooth regression occurred gradually, in a stepwise fashion, over tens of millions of years. Furthermore, dental degeneration evolved in parallel in the three xenarthran lineages rather than being inherited from their last common ancestor. Like many recent studies, Emerling's findings highlight the importance of identifying pseudogenes to reconstruct evolutionary history, especially when fossil evidence is sparse or absent.

Ants, tongues and body size

Today, anteaters (Vermilingua, meaning 'worm-tongue') comprise three genera – the small and entirely arboreal silky anteater, the medium-sized and semi-arboreal northern and southern tamanduas, and the terrestrial giant anteater.

Compared to sloths and armadillos, the evolution of anteaters is poorly known, as few fossils have been described. However, molecular-clock studies suggest the clade diverged from that of the sloths around 55 million years ago, between the Palaeocene and Eocene epochs. Subsequently, the silky anteater lineage arose about 40 million years ago, during the Oligocene, followed by the tamandua and giant anteater lineages around 10 million years ago. Following the formation of the Panamanian isthmus, taxa from all three extant lineages invaded Central America, where they can be found as far north today as southern Mexico. According to Frédéric Delsuc from the University of Montpellier, the diversification of anteaters, like other xenarthran taxa, followed periods of significant climate change caused by periods of Andean uplift and associated tectonic processes.[9]

All anteaters evolved powerful digging forearms, sharp claws, tube-like snouts and long tongues to facilitate their specialised insectivorous diet. However, each species has its dietary preference – small species feed on arboreal insects, while large species feed on terrestrial ants and termites. To avoid being bitten or stung, anteaters have evolved a strategy of hoovering up large numbers of prey as quickly as possible. To do so, they evolved a most unusual tongue.

The giant anteater's tongue is the longest of any mammal, reaching 60 centimetres in length, although it is not attached to the throat but fixed directly by a muscle to the sternum. This unique anatomical arrangement allows the organ to extend 45 centimetres from the mouth to penetrate and probe deep labyrinthine tunnels of its prey. In addition, it is covered with backward-curving spines or papillae and coated with copious amounts of sticky saliva produced by large salivary glands.

Daniel Casali, a Brazilian scientist at the University of Minas Gerais, believes that dietary pressure led to the development of increasingly long tongue muscles that can stretch, react quickly, and move in and out up to 160 times a minute.[10] Anteaters also evolved elongated snouts to accommodate their long tongues, and modified palates to prevent the retracted organs from obstructing their airways. However, they cannot open their jaws more than a couple of centimetres and cannot chew their food. Instead, they swallow their prey continuously and rely on a muscular gizzard to pulverise the insects. Remarkably, anteaters do not require hydrochloric acid for digestion, as the formic acid in ants and termites allows their prey to essentially digest itself.

Casali also highlighted that despite possessing tongues that superficially appear like those of the distantly related pangolins, they are, in fact, very different anatomical structures. In other words, the same morpho-functional adaptation of anteater and pangolin tongues arose convergently via distinct evolutionary pathways.

Finally, the diet of anteaters imposed harsh restrictions on their body size so that, unlike sloths and armadillos, no megafauna ever evolved. Social insects are small and have low nutritional value, with almost no fat content, except for their larvae and the winged reproductive females. To compensate, anteaters evolved a low metabolic rate: at 32.7 degrees Celsius, the giant anteater has the lowest body temperature of any eutherian. However, they must eat fast and frequently to avoid their prey's defensive mechanisms. As a result, they spend only a limited time at each nest before moving on to find other food sources, a high-energy-demanding strategy. As body size enlarges, the required foraging area increases exponentially, limiting the evolution of larger body sizes.[11]

Life in the slow lane

The six living species of sloth belong to two genera: the two-toed sloths (*Choloepus*, two species) and three-toed sloths (*Bradypus*, four species), which diverged approximately 28 million years ago.[12] Despite their names, all sloths have three toes on their hindlimbs; however, two-toed sloths have only two fingers on their forelimbs. These species are the last remnants of a once-diverse evolutionary group that included a wide range of taxa with a much broader geographic distribution. Unlike their modern counterparts, ancient sloths were primarily ground-dwelling, with some, such as the giant ground sloth (*Megatherium*), growing as large as modern elephants.

Phylogeneticists once believed that the two genera were closely related sister groups. However, recent morphological and molecular research suggests that both lineages originated from distinct, distantly related clades of extinct ground sloths.[13] This finding means that the two genera independently developed their arboreal lifestyles, including traits such as weight reduction, a folivorous diet, slow movement, low metabolic rate, variable body temperature, and suspensory behaviour. As Timothy Gaudin, a biologist from the University of Tennessee, points out, living sloths represent 'one of the most striking examples of convergent evolution known among mammals'.[14]

But there are differences, other than the number of toes, that betray the lineages' long separation. Three-toed sloths are generally smaller, diurnal and more docile than their two-toed counterparts. They also have rougher, scruffier fur. The most remarkable distinction, however, lies in their neck length and the number of cervical vertebrae. Two-toed sloths have slightly shorter necks with five or six cervical bones, whereas three-toed sloths boast longer necks with up to nine cervical vertebrae.

Amazingly, sloths are the only mammalian taxa, other than manatees, that depart from the 'rule of seven' for cervical vertebrae. All the remaining 5,000-plus eutherians, from solenodons to humans, have seven 'ribless' vertebrae in the neck, even creatures with long necks, like giraffes. Eighteenth-century naturalists were aware of this remarkable fact, and ever since, scientists have debated how and why sloths broke ranks with their evolutionary peers.

Robert Asher from the University of Cambridge explains that the extra cervical vertebrae in three-toed sloths are essentially thoracic vertebrae – typically associated with ribs – that have taken on the appearance of neck bones.[15] This adaptation results from a downward shift in the positions of the sloth's shoulders, ribcage and pelvis along the vertebral column, effectively lengthening the cervical spine compared to other mammals. Interestingly, in two-toed sloths, the reverse evolutionary process occurred. Additional research has linked these shifts to changes in homeotic, or *Hox* genes, which are key regulators of body segment development.[16] In most mammals, such genetic alterations lead to severe consequences, including cancer, stillbirths and neurological disorders.[17] However, sloths seem to avoid these adverse effects, likely due to their low metabolic rate and sedentary lifestyle. This 'life in the slow lane' may have allowed sloths to evolve their unique vertebral structure, though the evolutionary purpose of this adaptation remains a topic of active scientific debate.[18]

Sloths have long been chastened and ridiculed for their apparent inactivity, and are the only animal whose common name derives from one of the seven deadly sins. The influential French naturalist Georges Buffon fuelled the widely held misconception with a misinformed and prejudiced account of the animal in his monumental *Histoire Naturelle* (1749). Despite never having seen a sloth, Buffon wrote the following widely quoted description:

> *Slowness, habitual pain, and stupidity are the results of this strange and bungled conformation ... These sloths are the lowest form of existence ... One more defect would have made their lives impossible.*[19]

Sloths have undergone a recent reappraisal by the public. Instead of regarding them as anachronistic animals about which little is known, they have become, to many, the most endearing and lovable of mammals. At the same time, scientific studies have transformed our understanding of sloths. No longer considered misfits in the modern world, they are now recognised as exquisitely adapted animals honed for over 30 million years to survive in an exceedingly harsh and challenging niche. Buffon's derogatory assessment could not have been more wrong. Sloths are masters of a highly specialised and rare lifestyle – a canopy-dwelling leaf-eater. But why have so few mammals made the canopy their home, and why have even fewer mammals evolved to become strict folivores, given that a third of the Earth is covered by forest?

Leaves are a nutritionally and energetically poor food source. As a result, most herbivorous mammals, such as giraffes, deer and cattle, have large bodies that house the complex and lengthy intestines needed to extract nutrients from their challenging diets. Large mammals, however, cannot be supported by the canopy's branches, which severely constrains the upper weight limit of modern sloths. The challenge for scientists is to understand how sloths evolved to exist on a diet of leaves and yet have remained relatively small. In 2016, a research team at the University of Wisconsin-Madison, led by Jonathan Pauli, set out to answer these questions by studying the energy expenditure of sloths in Costa Rica.[20]

The team found that sloths use very little energy, with three-toed sloths having the lowest energy expenditure of any mammal – only 460 kilojoules per kilogram per day. This amount equates to just 110 kilocalories, roughly the same as provided by one mid-sized banana. To survive, sloths have evolved specialised behaviours and thermoregulatory strategies. They are highly efficient in their lifestyle, maintaining small home ranges and limiting their foraging to just one or two trees, where they can remain for days. Their movements are slow, and they sleep up to 15 hours a day, coming down only once a week to urinate and defecate (it takes them 2–3 weeks to digest a meal). Hanging upside down makes them less visible to predators, and their claws allow them to grip branches without expending energy.

Sloths are heterothermic and can alter their body temperature by adjusting their internal thermostats. Three-toed sloths, for instance, are able to modify their body temperature by as much as 5 degrees Celsius to match the ambient conditions and considerably reduce their energy output. Indeed, their metabolic rate is only 40–45 per cent of that expected for their body weight. Sloths also rely on heat produced by gut fermentation, topped up by solar energy when exposed to sunlight on the tops of trees. However, the metabolism of the two-toed sloths is not as efficient, forcing them to lead a slightly more active lifestyle, feeding over a more extensive area and in a greater variety of trees, occasionally supplementing their diet with small mammals.

The result of all this slothfulness is that a third of the animal's body weight may consist of ingested food, stored urine and faeces. This extraordinary observation raises the question of how sloths can breathe while hanging upside down with a full stomach and the weight of a week's excretory products pressing on their lungs. In 2014, a team from the University of Swansea provided the answer.[21] Sloths have evolved attachments or fibrous adhesions that anchor the internal organs to the rib cage, preventing diaphragmatic pressure. Without these anatomical innovations, it would be energetically very demanding, if not impossible, for sloths to breathe, let alone move, while hanging upside down. These observations explain why more sloths and other arboreal leaf-eaters have yet to evolve, as energy constraints reduce the likelihood of adaptive radiation. As a result, arboreal folivory remains one of the world's rarest dietary regimes.

Given the significant advancements in our understanding of the sloth's physiology, the term 'sloth' should not be viewed as a negative trait but rather as a highly successful evolutionary strategy. Indeed, the sloths' unique adaptations have allowed them to survive, unlike their terrestrial omnivorous relatives, who vanished around 10,000 years ago.[22]

Evolutionary constraints

Armadillos are the only surviving members of the order Cingulata, which once included creatures like glyptodonts and pampatheres. Known for their armoured shells, or carapaces, made of calcified skin tissue, they earned their Spanish name *armadillo*, meaning 'little armoured one'.[23] Armadillos split from sloths and anteaters roughly 65 million years ago, with their most recent common ancestor living in the Eocene around 40 million years ago. Today, most species live in South America's tropical and subtropical regions, particularly around Paraguay. However, the northern naked-tailed armadillo and the nine-banded armadillo inhabit Central America, with the latter's range extending into the southern United States. They are all terrestrial insectivores equipped with strong limbs and sharp claws to help them search for food and dig burrows up to 6 metres long.

Armadillos come in various sizes, ranging from the small, chipmunk-sized pink fairy armadillo to the much larger, pig-sized giant armadillo. They also vary in colour, including shades of black, brown, grey, red, salmon and yellow. All armadillos have a tough, armoured shell covering their back, head, legs and tail, providing protection from predators and rough vegetation.

This unique structure consists of two main layers: an outer dark-brown layer of keratin, the same fibrous protein found in human hair and nails, and a layer beneath it, composed of tightly packed bony tiles or osteoderms. These triangular or hexagonal structures are bonded together with collagen fibres, creating a shell that combines the characteristics of both hard and soft tissues.[24] The protective covering is also flexible, with bands of interspersed pliable tissue that expand and contract as the armadillo moves. Some species also shield their vulnerable underbellies by tucking their legs under their shoulders and hips, while others curl into a defensive ball. However, two biologists, Mariella Superina and James Loughry, stress that the benefits of a carapace have come at a cost – a constraint on the development of several traits.[25]

Unlike other mammals, armadillos cannot breathe deeply if blood oxygen levels fall. Instead, given their rigid thorax, they take more rapid and shallow breaths. Although the carapace means armadillos don't have to outrun predators, their slowness has resulted in a low metabolic rate and a reduced body temperature. Shields have also constrained fur development and hence thermoregulation. Contrary to expectation, most armadillo carapaces have high thermal conductance, which maximises heat loss and helps keep the animals cool. To remain warm, armadillos hunker down in their burrows, only to emerge when the ambient temperatures are tolerable. In effect, the evolution of a shield has restricted the animals' geographical range to tropical and subtropical environments while promoting the evolution of their burrowing behaviour. Some species possess hairs, although their density is less than that of other fur-bearing mammals.

However, the smallest armadillos in the world, the pink and greater fairy armadillos (or pichiciegos) from Argentina, Paraguay and Bolivia, have overcome this constraint. These rarely seen species spend most of their time underground and have evolved a soft carapace that controls heat loss rather than providing protection. The thin structure overlies a wealth of blood vessels that empty or fill to regulate heat loss, allowing the animals to emerge for short spells at night when temperatures are low.[26] Their strict fossorial lifestyle likely evolved in response to the significant aridification during the Oligocene, around 32 million years ago, when their lineage diverged from other armadillos. Molecular phylogenetic studies suggest that the two species of fairy armadillo separated approximately 17 million years ago. This split coincided with marine incursions in the Middle Miocene along the continent's south-central river systems, pointing to a vicariant origin when geographic barriers disrupted their ancestral range.[27]

The presence of a carapace has led to significant adaptations in the armadillos' reproductive system. While a shell might seem like a barrier to reproduction, male armadillos have evolved an exceptionally long penis relative to other mammals – up to 60 per cent of their body length. This adaptation allows males to bypass the female's bony armour and reach the vagina for successful penetration. A fully calcified carapace would also hinder birth, so the shield remains soft and flexible while the fetus develops and only hardens after birth. Since soft carapaces leave neonates at risk of predation, birth occurs underground, and many species' young stay in their burrows until fully weaned. As the fictional mathematician Ian Malcolm from *Jurassic Park* said, 'life finds a way'.

Evolutionary constraints manifest in various forms and are pervasive, significantly shaping and limiting life's morphological and physiological possibilities. For armadillos, their protective shield exemplifies how such constraints can narrow the range of potential variations, influencing the course of new adaptations. Similarly, other notable evolutionary constraints include the reproductive strategy of marsupials, which has impeded the evolution of flying and aquatic species, and the pleiotropic effects of *Hox* genes, which restrict the number of vertebrae in the mammalian spine to seven.

Finally, studies of armadillos, especially the nine-banded armadillo, have provided insights into the role of epigenetics, or the modulation of gene expression as a source of individuality. This species is unique among vertebrates in that it produces a litter of identical quadruplets every pregnancy, a condition termed polyembryony. Once fertilised, the ovum divides into four genetically identical embryos that share a single placenta. While genetic heterogeneity has long been recognised as evolutionarily advantageous – it enhances the likelihood that some individuals will survive in an ever-changing environment – the benefit of cloned offspring remains unclear.[28]

Irrespective of any evolutionary benefit, armadillo polyembryony is a valuable model that helps scientists understand the influence of non-genetic effects on phenotypes. Although the nine-banded armadillo's four offspring are clonal, they often appear different, including variations in size, facial colouration, behaviour, and even the number of cervical vertebrae. How can this occur, given the offspring have inherited identical genes?

Jesse Gillis, from the Cold Spring Harbour Laboratory, suggests that random or stochastic differences in gene expression are key to the puzzle. Such modifications start early, when each armadillo embryo comprises only around 25 cells. The most apparent phenotypic differences result from random X-chromosome inactivation, leading to the expression of only the paternal or maternal X chromosome (Chapter 2). However, Gillis's team found many more chance events affecting alleles on other chromosomes, creating permanent individual variability.[29] The nine-banded armadillo has revealed the importance of epigenetic-induced phenotypic variation as an evolutionary mechanism. In other words, inherited changes in gene expression resulting in novel traits can occur despite identical gene sequences. Such knowledge could help physicians understand the aetiology of various diseases. For instance, nearly two-thirds of identical twins have different risks for type 1 diabetes, schizophrenia and autism: that is, one twin may develop the condition while the other twin does not.

But it is now time for us to leave the armadillos and their evolutionary constraints, and to head northwards, where we will encounter another group of our Cretaceous ancestors: a group of tribosphenic insectivores that thrived in Laurasia, the ancient northern supercontinent.

CHAPTER 9

The Solenodon's Story

EDGE SCORES, VENOM AND THE K–PG EVENT

There are a whole slew of reasons why the solenodon's star should rise.
The Guardian

While afrotherians and xenarthrans were evolving in isolation in the southern hemisphere, the early shrew-like mammals inhabiting Laurasia in the north undertook a different evolutionary trajectory (see Figure 5.1). Around 85 million years ago, this group of placentals, the so-called Boreoeutheria ('true beasts from the north'), diverged to produce two mammalian superclades: Laurasiatheria, or 'mammals from Laurasia', and the awkwardly named Euarchontoglires, derived from the Greek *Euarchonta* (true rulers) and *Glires* (dormice). The Laurasiatheria gave rise to an impressive array of species, including solenodons, moles, hedgehogs, shrews, camels, llamas, pigs, ruminants, hippos, whales, pangolins, bats, cats, dogs, cats, bears, seals, horses, rhinos and tapirs. In contrast, the Euarchontoglires produced only five extant orders: rodents, lagomorphs, treeshrews, colugos and primates.[1] Despite their limited diversity, Euarchontoglires are significant, as they include the highly speciose rodent order and the primates, including humans.[2]

The remaining chapters of this book will focus on the evolution of the Boreoeutheria, a clade encompassing most placental mammals existing today, totalling over 5,000 species. Despite their origins in the north, the relentless jostling of continental plates facilitated their dispersal southwards, where they interacted and competed with their distant cousins, the Afrotheria and Xenarthra. We will commence our account with Laurasiatheria, a superorder defined by shared genetic sequences rather than any distinctive anatomical features. The oldest lineage produced the solenodons (Plate 14), species that warrant highlighting as they likely resembled our earliest placental ancestors (Figure 9.1).

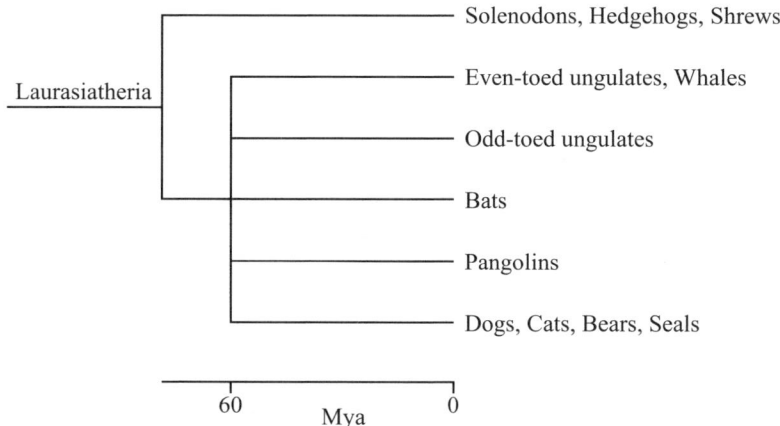

Figure 9.1 Phylogeny of Laurasiatheria. The rapid divergence of the laurasiatherians (polytomy) has prevented an accurate determination of their branching order. Mya, million years ago.

Fossil-calibrated phylogenies

Only two solenodon species survive today, the Hispaniolan and Cuban solenodons. Both are venomous, burrowing, nocturnal insectivores that emerge from rocky clefts, hollow logs or excavated burrows to prey on invertebrates and lizards. Utilising their sensitive snouts, they forage by rooting in the ground and tearing into rotten logs and trees with their foreclaws. The flexibility of their snouts is remarkable, with the Hispaniolan solenodon even possessing a ball-and-socket joint at its base to enhance mobility. This unique feature enables them to explore narrow crevices where potential prey may be hiding.

While the larger Hispaniolan species is relatively widespread, the Cuban solenodon is rare. Declared extinct in 1970, with no sighting reported since 1890, it seemed to have disappeared. However, in 1974, a glimmer of hope emerged when an individual was discovered in Cuba's Alexander Humboldt National Park, one of the Earth's most biologically diverse tropical island ecosystems. This discovery sparked a series of systematic searches conducted over many years by teams of dedicated biologists and local rangers. Their efforts bore fruit, and by 2016, 11 animals had been captured, studied and released. Even more encouraging was the realisation that the species also inhabited the surrounding forests and plantations, indicating that populations of this highly threatened mammal may be more robust than initially feared.

Solenodons, along with their close relatives – moles, hedgehogs and shrews – belong to a group called Eulipotyphla. This term, originating from Greek words meaning 'truly fat' and 'blind', encompasses a group considered the most ancient within the laurasiatherian clade.[3] They are all insect-eating mammals and began diversifying before the K–Pg extinction event that wiped out the dinosaurs.[4] The dating of their Cretaceous origins highlights the importance of palaeontology in constructing phylogenetic trees and elucidating ancient dispersal routes.

In a 2016 molecular study, Sato and colleagues concluded that solenodons diverged from their closest relatives around 58 million years ago, during the Palaeocene.[5] This relatively recent divergence suggests that the ancestors of solenodons may have rafted across the Caribbean Sea to reach Cuba and Hispaniola. Interestingly, the study also found that all the major eulipotyphlan families diverged around the same time, resulting in a lack of any clear relationships among them – a situation known as polytomy. However, most polytomies are likely due to analytical limitations and incomplete data rather than actual evolutionary history. Additionally, Sato's findings conflict with the fossil record; for example, the oldest hedgehog fossil is about 62 million years old, which would make solenodons a 'zombie lineage' since their divergence appears more recent than the fossil evidence suggests.

After reviewing Sato's paper, Mark Springer and his team grew sceptical of its conclusions and opted to reanalyse the molecular data.[6] By incorporating the fossil dates for the earliest hedgehogs, as well as those of shrews, carnivores and rodents, Springer was able to recalibrate the molecular timelines, grounding them more firmly in the fossil record. This adjustment brought the molecular data in line with previous studies, placing the origins of solenodons back in the Late Cretaceous, during the age of dinosaur dominance. Additionally, the Cretaceous timeline supports the plate-tectonic explanation for the presence of both solenodon species in the Antilles. According to this model, a fragment of continental North America broke off around 76 million years ago, transporting ancestral insectivores across the Caribbean. Over time, this drifting landmass evolved into the islands of Cuba and Hispaniola, each home to its distinct solenodon species.

The limitations of molecular clocks have been acknowledged since they were first used. Mutations occur randomly, much like radioactive decay, although they can exhibit a more predictable average rate over many events. Furthermore, different genes mutate at varying rates; mitochondrial genes tend to mutate faster than nuclear genes, while essential housekeeping genes usually have lower mutation rates.[7] In addition, the genomes of smaller species, such as rodents with shorter generation times, mutate more quickly than those of larger species. Even with advanced statistical methods like maximum likelihood and Bayesian phylogenetics, there is still a risk of reaching incorrect conclusions. To minimise such errors, molecular clocks are best calibrated with radiometrically dated fossils, ideally providing a lower limit for the timing of a branch point. In essence, phylogenies anchored with accurately dated fossils offer more reliable dates for the divergence of a particular clade.[8] For this reason, I have opted for a Cretaceous date for the emergence of solenodons, despite a further molecular study by Sato's group reiterating a post-K–Pg timing.[9]

EDGE scores and molecular convergence

Dr Samuel Turvey, a Research Fellow at the Institute of Zoology in London and co-creator of the EDGE score, was part of the team that rediscovered a small population of solenodons in Haiti. As discussed in Chapter 5, EDGE scores are composite metrics that combine a species' conservation status with its evolutionary or genetic distinctiveness (ED). After witnessing the failure to save the Yangtze River dolphin,[10] Turvey was determined to prevent a similar fate for the solenodons. Given the rarity of the Cuban species, with only one closely related taxon remaining, it holds a unenviable position near the top of the mammalian EDGE-score table (see Table 5.1). Additionally, threats from deforestation, habitat degradation and predation by feral cats and dogs have made the Cuban solenodon a conservation priority. Like all species with high EDGE scores, its extinction would result in a significant loss of unique evolutionary history and biodiversity.[11]

Solenodons are often regarded as 'living fossils', implying that they offer a glimpse into the distant past and could shed light on how our early ancestors looked and behaved. However, such a term is misleading and inaccurate. Even species that show little or no outward change continue to evolve with ongoing genetic recombination. Both solenodon species have undergone genetic changes over the past 76 million years as they adapted to shifting environments and lifestyles. For instance, they evolved a unique trait – a venom delivered through grooves in their lower teeth – a feature absent in the earliest boreoeutherians, which likely had venomous spurs on their hind legs instead. Today, solenodons, along with some shrews, hedgehogs and moles, are among the few mammals capable of incapacitating their prey with oral venom. As we will explore, the composition of solenodon venom offers a fascinating example of convergent evolution at the molecular level.

A recent study led by Nicholas Casewell from the Liverpool School of Tropical Medicine explored the origins and evolution of venom in eulipotyphlans by analysing their genomes.[12] The research revealed that solenodon venom contains a variety of toxins, which are produced by multiple duplicated kallikrein-1-like genes. These genes encode serine proteases that break down other proteins, including those regulating blood pressure. When these proteases were injected into mice, they caused a significant drop in blood pressure, confirming that they evolved to incapacitate prey. The study also found that some shrews, which are closely related to solenodons, produce similar toxins, suggesting that their common ancestor may have had this trait. However, if this were the case, then most shrew descendants would have had to have lost the ability to produce venom. A more likely explanation is that solenodons and other eulipotyphlans independently developed venom after diverging from a non-venomous common ancestor. This theory is further supported by different species within the group using distinct sets of proteases.

Casewell's team concluded that the venom system evolved four times in the eulipotyphlans, more than in all other mammals combined – a phenomenon even more frequent than in any other class of vertebrates except bony fish. The underlying reasons for such remarkable molecular convergence remain elusive. However, venom evolved mostly in shrews, species with high metabolic rates and a need for frequent feeding. Hence, venom likely emerged as a beneficial adaptation to facilitate the near-continuous feeding requirements. Moreover, the independent recruitment of similar proteases suggests that mammals are subject to evolutionary constraints with limited options for venom production. Since kallikreins are commonly found in mammalian saliva, small modifications over time seem to have led to the creation of increasingly potent proteins, paving the way for venom evolution.

A 'Goldilocks' species

One spring day, around 66 million years ago, the world abruptly and irreversibly changed.[13] Unaware of the impending catastrophe, the early insectivores would have been safely hunkered down after a long night foraging on the forest floor. Then, without warning, a massive asteroid, 10 kilometres across, streaked steeply across the sky, illuminating their island habitat, before crashing into the shallow seas near Chicxulub on the Yucatán Peninsula.[14] The impact fractured and melted the Earth's crust, sending molten rock high into the sky, where it solidified into glass shards that rained over North America and the Caribbean. This cataclysmic event led to significant temperature spikes and massive tsunamis. Wherever the solenodons were hiding, the strange rabbit-sized mammals must have been uncomfortably close to ground zero, for the Caribbean islands were much closer to the mainland during the Early Palaeocene than today. Yet, while the dinosaurs, plesiosaurs, pterodactyls and up to 75 per cent of all living creatures perished, including many mammals, the solenodons somehow survived. According to Samuel Turvey, 'it's truly remarkable that they survived this direct hit whilst global ecosystems collapsed around them – and we have little idea how they did it.'

The solenodons are undoubtedly champion survivors. Not only did they survive the K–Pg extinction event, but they also tolerated the super-hot greenhouse Earth of the Eocene with its higher temperatures and acidification of oceans.[15] Throughout this period, a profound 'modernisation' of terrestrial mammals resulted in many new lineages, including rodents, bats and whales, that rapidly impacted the communities they entered (discussed in later chapters). Solenodons survived further global ecological transformations, including the Late Quaternary ice age, which saw the extinction of various megafauna species, such as mammoths, giant rhinoceros, cave bears and lions, giant elks and hyenas.[16]

Then, around 6,000 years ago, the fauna of the Caribbean faced an existential threat with the arrival of *Homo sapiens*. Before human colonisation, Hispaniola was home to a rich variety of at least 25 mammal species. However, following human settlement, these species began to disappear, with only solenodons and a unique rodent called the Hispaniolan hutia surviving. To uncover the cause of this decline, Samuel Turvey and his colleagues explored caves on the island once inhabited by early settlers. Beneath layers of sediment and detritus, they unearthed skeletal remains of various species, including giant hutias, spiny rats and pygmy sloths, but found relatively few solenodon bones. Turvey believes that the solenodons were a 'Goldilocks' species – just the right size for survival. When explorers arrived in the fifteenth century, black rats from their ships wiped out many indigenous species. However, the Hispaniolan solenodon, too large to be preyed upon by rats and too small to interest human hunters, managed to escape these threats. In essence, solenodons were perfectly sized for survival.[17]

The Chicxulub impact

I have frequently referred to the 'K–Pg boundary', the most significant event in the story of mammals, without discussing its elucidation. While many readers will be familiar with the serendipitous nature of its discovery, it is nevertheless a story worth recounting. For the K–Pg boundary sets the scene for one of the liveliest debates in mammalian evolution. Why did three-quarters of the planet's plants and animals, including the dinosaurs, die out while mammals survived?

Deciphering the nature of the K–Pg boundary hinged on the discovery of iridium, a very hard, silvery-white transition metal closely related to platinum. Despite its brittleness, iridium's corrosion-resistant properties and ability to withstand high temperatures make it a valuable alloying agent in manufacturing aircraft, spark plugs, and medical devices such as pacemakers. Iridium is a rare element, and its high density and tendency to bind with iron resulted in most of it sinking into the Earth's core when the planet was young and still molten. It is the rarity of this element that allowed the K–Pg boundary to be deciphered.

In the early 1970s, the American geologist Walter Alvarez was conducting fieldwork near Gubbio in Italy, when he stumbled upon a thin layer of reddish clay that separated the Cretaceous and Palaeogene layers of limestone. This layer, devoid of foraminifera, was a stark contrast to the whitish rock beneath it, which was teeming with fossilised foraminifera, microscopic single-celled organisms with external shells. Alvarez's curiosity was piqued by this anomaly, leading to a significant breakthrough in our understanding of the K–Pg boundary.

Alvarez wanted to know how long it had taken for the clay to form, so that he could estimate the length of time that the foraminifera took to recover. After discussing the idea with his Nobel-prize-winning father, the physicist Luis Alvarez, Walter sent clay samples to the University of California for rare-metal analysis. Alvarez senior knew that meteorites are rich in iridium, and reasoned that the quantity of the element present in the samples could help determine the time taken for the layer of clay to form, since the dust from meteorites is deposited on the Earth's crust at a relatively constant rate.

The results were unexpected. The clay samples revealed iridium levels 300 times higher than those in the surrounding layers. Crucially, similar findings were observed in samples obtained from the boundary layer at various locations from around the world. Today, this global iridium layer is regarded as the 'golden spike' marking the end of the Cretaceous period and Mesozoic era.[18] The father-and-son team concluded there could only be one explanation: the high concentrations of iridium, rare in the Earth's crust but abundant in asteroids and meteorites, must have resulted from a large extraterrestrial object colliding with Earth. Given the worldwide presence of this anomaly, they believed they had sufficient evidence to propose a 10-kilometre-wide asteroid as the culprit, basing this deduction on the iridium levels in the clay and the average iridium content of meteorites. Their conclusion was published in 1980 in the prestigious journal *Science*.[19] However, the idea of an asteroid impact was met with scepticism and even ridicule, not least because no evidence for a 200-kilometre-wide crater had been found – Walter's estimate was based on predicted meteor size. Critics preferred alternative explanations, including climate change that dramatically altered the biosphere at the end of the Cretaceous. However, the idea of increased volcanism was the most widely accepted. Indeed, the Deccan Traps in India, the largest continental flood basalt province on Earth, covering over half a million square kilometres, was widely argued as the 'true' culprit for the K–Pg mass extinctions.

However, Walter and Luis Alvarez and their critics were unaware that evidence for a candidate impact crater was already in the public domain. In 1978, two geophysicists working for an oil company had detected anomalies in the Earth's magnetic and gravitational fields in the coastal waters off eastern Mexico. Further investigations suggested the presence of a large circular feature, approximately 180 kilometres wide and 48 kilometres deep, on the Yucatán Peninsula near the small fishing village of Chicxulub (pronounced 'cheek-shoo-loob'). One of the scientists involved was so convinced that his data reflected a cataclysmic event in geologic history that he obtained the oil company's permission to present the findings at an international conference. However, the symposium was poorly attended, as most 'big shots' in the field of impact craters were at another meeting, and his findings received little interest. Understandably, the geophysicist became disillusioned and returned to his day job, prospecting for oil. Then, by chance, Alan Hildebrand, a planetary scientist, was informed of the findings by a journalist from the *Houston Chronicle*. In April 1990, Hildebrand, with the help of the geophysicist, persuaded the oil company to release the drill samples taken in the 1950s from the area of interest. The findings were conclusive. The cores contained shock quartz and small glass spheres, called tektites, that only form in the heat and pressure of a high-yield nuclear detonation or an asteroid impact. After 10 years of searching, the 'smoking gun' for the 'Alvarez hypothesis' had been found.

Conclusive evidence linking Chicxulub and the K–Pg boundary came in 2021 when several international groups reported high levels of iridium within rock cores retrieved from deep inside the impact crater itself.[20] Furthermore, the ground-zero iridium constituted such a thick layer that scientists estimated that its deposition had occurred within two decades of the meteor strike. In other words, the dust forced into the atmosphere remained aloft for up to 20 years before falling back to Earth to coat the planet's surface. This time frame would agree with the length of time required for all the creatures that became extinct to have starved to death. According to the study's lead scientist, Professor Steven Goderis, 'there is no longer any doubt: the asteroid impact and dinosaur extinction are indisputably linked.'

Further analysis of the crater's geochemistry indicates that the impacting object belonged to a class of carbonaceous chondrites, a primitive type of meteor with a high carbon ratio that likely formed in the early stages of the solar system. Until recently, the source of such a dark, primitive impactor remained a mystery. However, scientists now believe they have identified the culprit's origin. A study by Nesvorný and colleagues proposes that the impact was caused by a meteor from the solar system's main asteroid belt, between Jupiter and Mars.[21] This distant zone contains many asteroids of the right type, rocks with a chemical composition that reflects very little light and makes them appear darker than most. Using a computer model, the investigators found that gravitational and thermal forces from surrounding planets can periodically nudge large asteroids out of their stable orbits. Crucially, a 10-kilometre-wide asteroid was calculated to be pushed into a collision course with Earth once every 250 million years – an event five times more common than previously thought. Furthermore, the study concluded that 60 per cent of the larger asteroids that have hit the Earth during its 4.54-billion-year history likely originated from the outer region of the belt: a zone where most rocks are of the dark variety.

Researchers have recently identified another crater on the seabed off the coast of West Africa, smaller in size but potentially connected to the Chicxulub impact.[22] This new feature, called the Nadir Crater, is roughly 8.5 kilometres in diameter and dates to around 66 million years ago, raising questions about a possible link between the two craters. Could Chicxulub and the Nadir Crater have formed from the fragmentation of a larger asteroid, or as part of an impact cluster? This scenario assumes that any asteroid orbiting the Earth would be subject to its gravitational forces, potentially causing it to break apart. In this event, asteroid fragments could have impacted the Earth at separate locations, possibly within days of each other.

Another question is how the Chicxulub impact led to the extinction of three-quarters of all life forms.[23] While scientists may never know the precise details, there are predictable consequences of a 10-kilometre-wide asteroid hitting the Earth. For a rock the size of Manhattan, travelling at 20 kilometres per second, the energy delivered would have been equivalent to more than 4 billion atomic bombings of Hiroshima and Nagasaki.[24]

The initial result was likely a massive plume of 25 trillion tons of molten rock, with temperatures hotter than the sun's surface, shooting through the atmosphere and into space. The majority, however, would have rained down over North America, incinerating everything in a 1,000-kilometre radius. If the deluge of molten rock and superheated gas were hot enough, the impact could have induced the largest mega-tsunamis in the Earth's history, a wave up to 1,500 metres high, devastating the Earth's coastal plains. With an open seaway between North and South America, the tsunami would have raced worldwide, scouring and destroying the sea floor for up to 10,000 kilometres from the impact site.[25]

Massive earthquakes would have been generated at ground zero, with the resulting shockwaves triggering further earthquakes and volcanic activity around the globe.[26] The release of dust and particles rich in sulphur may have covered the entire planet for many years and blocked the sunlight from reaching the Earth. As a result, photosynthesis would have been inhibited in both oceanic and terrestrial ecosystems, effectively shutting down large swathes of the food chain. Volcanic ash ejected into the atmosphere would have added to the impact's dust and increased the planet's albedo, the fraction of solar energy reflected into space, producing a nuclear-winter-like effect. As much as 100 billion tonnes of sulphur and 10 trillion tonnes of carbon are estimated to have been vaporised by the impact, causing the production of vast quantities of carbon dioxide and sulphur dioxide. These and other gases would have reacted with water vapour in the atmosphere and caused acid rain.[27] The acidity of the oceans would have increased, inhibiting the growth of coral reefs and disrupting the marine ecosystem from the bottom up, features confirmed by studies of fossils from Seymour Island off the tip of the Antarctic Peninsula.[28] Lastly, Douglas Robertson, from the University of Colorado, envisages an incandescent Hadean landscape in which re-entry of ejecta produced a widespread infrared pulse sufficient to ignite global firestorms. He calculated that such fires started almost immediately and burned for days or weeks, leading to the complete destruction of terrestrial ecosystems.[29]

The Chicxulub impact undoubtedly triggered a global mass extinction, wiping out roughly 75 per cent of all species. But it was the intense climatic shifts that followed, rather than the impact itself, that caused the most devastation and loss of life. Amazingly, however, despite its global scope, the Chicxulub asteroid was not Earth's most severe extinction event. That honour belongs to the catastrophe at the end of the Permian period, about 250 million years ago. Known as the 'Great Dying', this event, caused by volcanic eruptions and global warming, eliminated 96 per cent of marine species and around 70 per cent of terrestrial species.[30] Despite repeated mass extinctions, life on Earth has shown remarkable resilience, though ecosystems can take up to 2 million years to recover and become fully functional again.[31]

The consequences of the Chicxulub impact raise a fascinating question: How did a small group of mammals – ancestors to creatures like the solenodons – manage to survive while the successful and widespread dinosaurs, along with pterodactyls and plesiosaurs, all perished?

Survival against the odds

It is likely that only 10 per cent of mammalian taxa survived the Chicxulub impact with its transient high temperatures and protracted nuclear winter. Those that did tend to be ground-dwelling or semi-arboreal species with various advantageous traits: subterranean living, omnivorous feeding habits, hibernation capacities, small size, and ecological flexibility.

The ability to dig burrows would have helped the early monotremes and therians to survive, protecting them from the many heat pulses and wildfires that briefly ravaged the planet. In contrast, large species, especially dinosaurs, were left exposed and, with nowhere to shelter, were burnt to death. After several days of scorching heat, the Earth's surface temperature returned to tolerable levels, allowing the emergence of our rat-sized ancestors from their underground sanctuaries. However, they would have encountered a forbidding landscape – a barren world with little available food. With most above-ground vegetation gone, many creatures that survived the asteroid's immediate effects would have fallen victim to starvation. Crucially, the omnivorous diet of the earliest mammals enabled them to survive on what was available: insects, seeds and aquatic plants. The more varied an animal's diet, the better their survival chances until the conditions improved.

Hibernation has long been known to increase winter survival rates and protect mammals from extinction. However, it was widely assumed to be an adaptation restricted to species inhabiting high latitudes and temperate zones.[32] But a discovery by Barry Lovegrove and colleagues in 2014 challenged this notion.[33] The three scientists from the University of KwaZulu-Natal found that the tailless or common tenrec, native to subtropical Madagascar, can undergo continual hibernation for up to 9 months. Furthermore, they don't exhibit periodic interbout arousals, a mechanism observed in other hibernating mammals that prevents brain damage caused by oxygen deprivation from reduced blood supply. According to Lovegrove's team, hibernation reduces predator risk and confers a survival benefit for the common tenrec when it is inactive and not directly engaged in reproduction. Considering their basal phylogenetic position, long-term hibernation may represent a behavioural trait dating back to Cretaceous mammals, a legacy of over 100 million years of ecological constraint by dinosaurs.

The small size of early mammals can be attributed, in part, to the pressures exerted by competition and predation from dinosaurs. A diminutive body size requires fewer daily calories for sustenance, an advantage for species during periods of environmental stress. Also, small mammals tend to have shorter reproductive cycles and faster recovery times, both valuable assets in difficult times. As previously discussed, elephants have gestation periods of approximately 22 months, while mice have a remarkably short gestation of only 20 days. Additionally, the time taken for large species to reach sexual maturity is particularly prolonged. Indeed, despite their rapid growth rate, dinosaurs like the *Tyrannosaurus rex* required up to 20 years to reach sexual maturity.[34]

Even though the earliest members of the superorder Euarchontoglires, including ancestral primates, had already adopted an arboreal lifestyle, many survived the asteroid's resultant deforestation. A likely explanation was the presence of forest refugia – distinct ecological niches that might have been less prone to complete tree loss, such as marshes and swamps. Alternatively, early Euarchontoglires might have had sufficient behavioural flexibility to survive in non-arboreal environments before becoming the first to reoccupy tree habitats as the forests underwent regeneration.[35]

Diversity also played a pivotal role in our ancestors' survival. Mammals were numerous and widespread and evolved many different lifestyles. While dinosaurs and their ecosystems remained relatively stable throughout the last 18 million years of the Cretaceous, mammals explored novel avenues of adaptation, giving rise to taxa that climbed, glided, swam and burrowed. As any gambler knows, the more diverse your bets, the better the chance of success.[36]

However, despite mammals' advantageous behavioural and anatomical traits, it remains a mystery how the early solenodons, inhabiting islands near ground zero, survived. Undoubtedly burrowing played a role, given that modern solenodons den in holes and that animals that spend time underground are more resilient to anthropogenic extinction pressures.[37] Alternatively, given the gaps in our understanding of proto-Caribbean tectonics, the solenodon's ancestors may have colonised the region from the nearby North American mainland after the meteor strike.

While representatives from other animal classes, such as crocodiles, turtles and snakes, also survived, mammals and birds emerged as the primary beneficiaries of the extinction event. Both groups rapidly expanded to reach nearly every corner of the globe. The sudden disappearance of many life forms, especially the non-avian dinosaurs, created an extraordinary evolutionary opportunity for mammals and birds. Once freed from their predators and competitors of the Mesozoic era, both lineages underwent remarkable adaptive radiations, marked by rapid bursts of speciation to fill the newly available ecological niches. The remaining chapters will explore how these transformative events affected mammals, resulting in the 6,500-plus species alive today. But, before doing so, it's worth reflecting on the unlikelihood of our existence.

The rise of mammals and the eventual emergence of humans hinged on a random, catastrophic event 66 million years ago. Had the Chicxulub impactor missed the Earth or skimmed harmlessly off the atmosphere, the world would look very different today. Indeed, the consequences would have been far less severe had the rock struck deeper ocean waters than the Caribbean's shallow seas. Yet, as fate would have it, the asteroid hit the proverbial bullseye, an event that turned out to be an extraordinary stroke of luck for the mammalian lineage, including ourselves.

Mammal radiation: resolving controversies

The debate surrounding the timing and pattern of placental mammal diversification around the K–Pg event remains unresolved. As is often the case in science, uncertainty catalyses new hypotheses, with each research group seemingly wedded to a favoured explanation. Five models are under serious consideration, among which three have emerged as prime contenders: the Explosive model, the Long Fuse model, and the Short Fuse model (Figure 9.2).

The Explosive model assumes that most, if not all, of the placental radiation occurred after the K–Pg mass extinction. It envisages that during the Palaeocene, placental mammals diversified quickly in response to the many niches left vacant by the demise of the non-avian dinosaurs, a narrative favoured in the popular literature. In its most extreme form, only one placental ancestor had to have crossed the K–Pg boundary. However, the model relies on a literal reading of the fossil record, which contains no definitive crown placental fossils pre-dating the extinction event. Readers should note that the earliest fossil of a species will always be younger than its actual time of origin. Typically, fossils from a given lineage reflect when the population

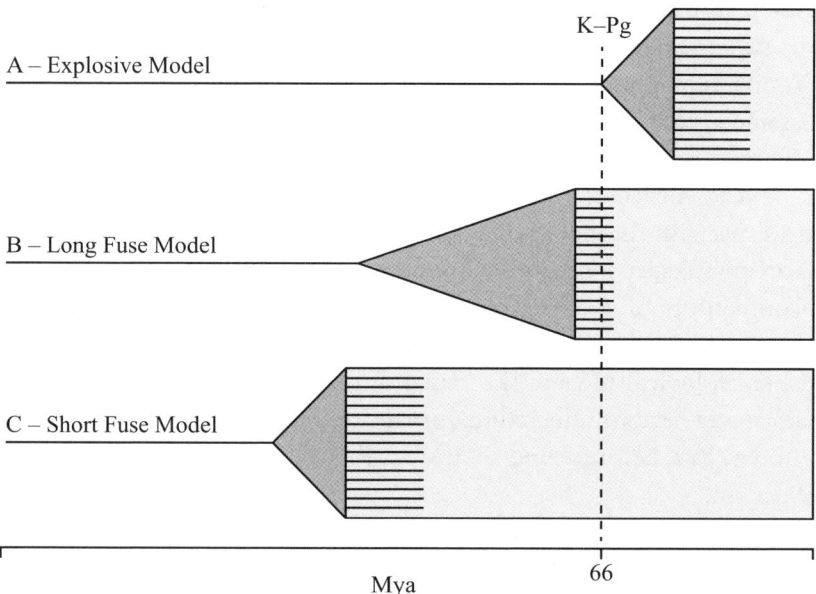

Figure 9.2 Models for the diversification of placental mammals in relation to the K–Pg boundary. (A) The Explosive model proposes that all placental mammals originated and diversified after the K–Pg boundary. (B) The Long Fuse model suggests that placental mammals originated in the Cretaceous, but intraordinal diversification began after the K–Pg boundary. (C) The Short Fuse model sees placental origination and diversification occurring during the Cretaceous. K–Pg, Cretaceous–Palaeogene boundary; Mya, million years ago. Modified from Springer et al. (2019).[38]

became abundant and stable rather than the exact time of their evolution.[39] The absence of early fossils may reflect a slow rate of morphological evolution during the Cretaceous relative to molecular changes. If this is true, early crown species would be difficult to separate from stem members.[40] The Explosive model also suffers another weakness – it remains to be confirmed by rigorous molecular analysis.

The Long Fuse model occupies the middle ground and proposes that the placental orders originated in the Late Cretaceous and that subsequent diversification didn't occur until the Palaeocene. Such an evolutionary pattern may have been driven by the Cretaceous Terrestrial Revolution, also known as the Cretaceous Angiosperm Revolution.[41] Spanning from 125 to 80 million years ago, this transformative period witnessed a remarkable global proliferation of flowering plants and insects, marking one of the most significant radiations in the history of life. The unprecedented change in plant communities led to new and plentiful food sources, including fruit and nectar, providing opportunities for the diversification of placental mammals and modern birds.[42] Furthermore, unlike the Explosive model, this hypothesis is strongly favoured by molecular data. Indeed, placental phylogeneses derived using molecular clocks consistently estimate a pre-Palaeogene origin for placental mammals. However, a word of caution: Many genetic studies likely contain biases from unrecognised DNA rearrangements, including hybridisations, introgressions and incomplete lineage sorting, that can lead to false conclusions. Significantly, however, a recent palaeontological study that addressed the recognised limitations of the fossil evidence provided support for the Long Fuse model.[43]

The third model – the Short Fuse model – proposes that the diversification of placental orders and crown groups occurred during the Late Cretaceous.[44] In this scenario, the critical evolutionary events took place well before the K–Pg boundary, long preceding the extinction of the dinosaurs, with the Chicxulub impact playing a minimal role. Compared to the other two models, however, the Short Fuse hypothesis has limited support from palaeontological or molecular studies.

Overall, the Long Fuse model aligns closely with the estimated 76-million-year history of the solenodon and seems the most plausible. That said, caution is warranted, as the early history of eutherian mammals is fraught with unresolved issues. Despite these challenges, evolutionary biologists remain hopeful that new fossil discoveries, higher-quality genomes and advanced analytical techniques will help refine the timeline of placental radiation. This effort is important for understanding mammalian evolution, as clarifying the events surrounding the K–Pg boundary would shed light on the rise of the placental lineage, one that would lead, some 66 million years later, to *Homo sapiens*.

CHAPTER 10

The Camel's Story

HIGH LATITUDES AND DOMESTICATION

On horseback you feel as if you're moving in time to classical music; a camel seems to progress to the beat of a drum played by a drunk.
Walter Moers

Strathcona Fjord is as remote and barren as it gets, lying on the western flank of Nunavut's Ellesmere Island. Yet, for vertebrate palaeontologist Natalia Rybczynski from the Canadian Museum of Nature, this unforgiving terrain has shaped her professional career. During the brief arctic summer of 2006, while painstakingly sifting through the fjord's sandy-rich fossil beds, Rybczynski noticed what she thought was a piece of fossilised wood and carefully wrapped it in tissue paper. Only after she returned to the expedition's field camp did she realise it was a fossilised bone, a piece of a tibia from a mammal larger than anything previously recovered from the site.

Over subsequent years, further excavations yielded nearly 30 bone fragments, none exceeding 7 centimetres in length. Their survival was fortunate, given that the island's repeated freeze–thaw cycles could easily have ground the bones to dust. Although the identity of the species remained a mystery, the shape of the bones indicated that they came from an artiodactyl, a group of even-toed ungulates encompassing deer, camels, pigs and giraffes. Unable to identify the species, Rybczynski sent the valuable fragments to Mike Buckley at Manchester University for analysis. Without accessible DNA, the team employed a novel approach: protein fingerprinting. Buckley's group extracted trace amounts of type I collagen, the primary structural protein in bone that is less biodegradable than other molecules. Crucially, the protein's amino acid sequence is sufficiently variable between mammal genera to be useful taxonomically. Once it had been analysed, the Manchester team compared the result to a database of collagens from different mammals and, unexpectedly, found the closest match to be the dromedary or Arabian camel.[1]

The Ellesmere Island fossils were found embedded within a sediment layer laid down during the Pliocene, over 3.4 million years ago. Based on bone proportions, the camel was deduced to be a giant, roughly 2.7 metres high and a third larger than modern species. Before winter, it may even have weighed as much as 900 kilograms. Rybczynski's discovery is an important one. It is the first evidence that camels inhabited the High Arctic and hints that specific adaptations of modern camels may owe their origins to a past life in the freezer.

Rainforests to deserts

The 'Age of Mammals' commenced in the Palaeocene during the 10 million years that followed the Chicxulub impact. As noted, placental mammals were previously almost all small, terrestrial and sub-arboreal insectivores, displaying minimal ecological diversity.[2] However, during the Palaeocene, the Laurasiatheria suddenly increased in size and heterogeneity to produce the first large-bodied herbivores, specialised carnivores, and gliding, flying and fully aquatic species. Unfortunately, their definitive phylogeny remains elusive because of rapid divergence, compounded by the associated problems of hybridisation and incomplete lineage sorting (Chapter 3).[3] Rather than obscure the evolving storyline by reviewing the multitude of conflicting studies, I have assumed that all lineages of Laurasiatheria diverged simultaneously, except solenodons, which are widely accepted as the earliest to evolve. Ideally, as new research strategies are developed and more comprehensive datasets become available, clarity will replace uncertainty, and the polytomy evident in the phylogram shown in Figure 9.1 will soon be rendered obsolete.

This chapter and the next three focus on the group of laurasiatherians known collectively as Cetartiodactyla. This is a recent grouping that comprises two orders: Artiodactyla, or even-toed ungulates (camels, pigs, hippopotamuses, antelopes, giraffes, okapi), and Cetacea (whales, dolphins, porpoises), the latter lying embedded within the Artiodactyla (Figure 10.1). Together with their relatives, the Perissodactyla or odd-toed ungulates – horses, tapirs and rhinoceroses – they form the hoofed mammals. Many of these species are notable for their historical significance to humans.[4]

The artiodactyls are thought to have evolved from 'archaic' condylarths, a vaguely defined group of small placental mammals that originally possessed clawed feet. Artiodactyls are the fifth-largest mammalian order, including the most diverse and extensive terrestrial species. Despite typically inhabiting open habitats, even-toed ungulates thrive in many environments and on every continent except Australia and Antarctica. They range in weight from

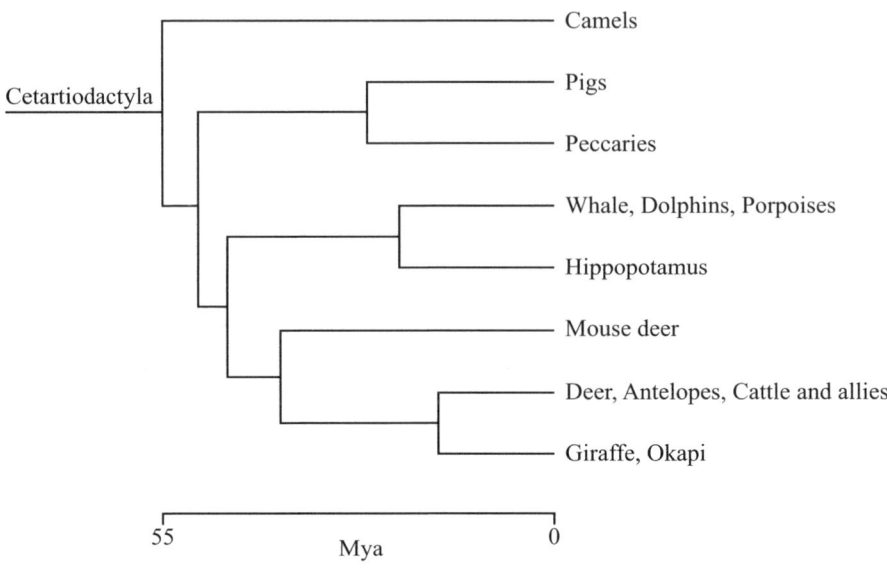

Figure 10.1 Phylogeny of Cetartiodactyla, a group comprising two orders: even-toed ungulates (Artiodactyla) and whales, dolphins and porpoises (Cetacea). Modified from Price et al. (2005).[5]

the massive hippopotamus, which can weigh up to 4,000 kilograms, to the diminutive 2-kilogram Java mouse-deer.

Artiodactyls are paraxonic; that is, the plane of symmetry of each foot passes between the third and fourth digits – the first digit, or pollex, is lost, while the second and fifth digits are usually reduced in size. However, the third and fourth digits remain large and weight-bearing. This arrangement contrasts with the mesaxonic limb support of the odd-toed ungulates, in which the weight-bearing axis passes through the third or central digit (Chapter 14). While perissodactyls are hindgut fermenters, artiodactyls are foregut fermenters and possess multi-chambered stomachs that deal with different stages of digestion, all aided by bacteria. Partially digested food from these chambers is regurgitated into the mouth for further grinding or 'chewing the cud'.

One of the earliest Artiodactyla offshoots were the camelids (family Camelidae), a group which today consists of species living in hostile environments where most large mammals would perish. Surprisingly, their story starts in the rainforests of North America approximately 50 million years ago. The earliest identified species, *Protylopus*, was small compared to modern camels, roughly the size of a large dog, weighing around 25 kilograms and with a length of 80 centimetres. Based on its dentition, it likely fed on succulent leaves of forest plants. Its forelimbs were shorter than its hindlimbs, all of which had four toes. Furthermore, the shape of its toes suggests that it had hooves rather than the foot pads of modern camels, making it the first known ungulate.

However, by the middle Oligocene, 25–30 million years ago, the descendants of *Protylopus* began to look more like modern camels. In 1848, a fur trapper working in the Badlands of Nebraska found a strange collection of fossil animal bones. Not knowing what to make of them, he forwarded them to the palaeontologist Joseph Leidy, who recognised that several bones belonged to a mammal resembling a small camel. The *Poebrotherium* (Greek for 'grass-eating beast') was larger and quicker than *Protylopus*, with a face like a modern guanaco. Despite its name, it was not restricted to one habitat and was a mixed feeder, venturing onto the savannas that formed during North America's drier spells.

Throughout the Oligocene, at least a dozen camel species evolved, each with larger bodies and longer legs and necks that increased their browsing efficiency. Then, around 25 million years ago, the camelids split into two.[6] One tribe (Lamini) dispersed to the south and eventually crossed the isthmus of Panama around 2.7 million years ago during the Great American Biotic Interchange. They are represented today by the South American guanaco and vicuña, and the domesticated llama and alpaca (Figure 10.2).

The second tribe (Camelini) moved northwards into the Arctic and produced species better adapted to harsher climates, including the poorly defined *Paracamelus*. The giant camel identified by Rybczynski on Ellesmere Island was similar to these hardy animals and would have roamed across

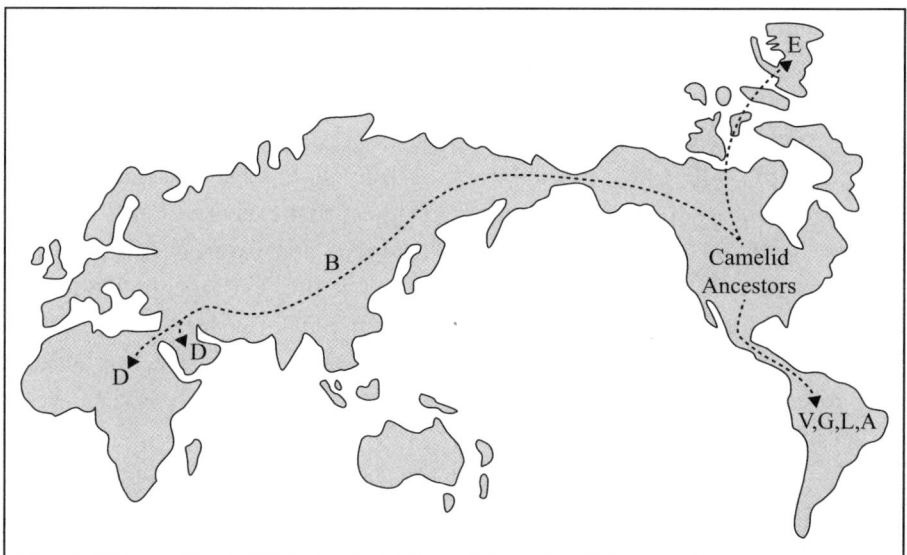

Figure 10.2 Historical dispersal of the camelid family from their origins in North America. E, Ellesmere Island; B, Bactrian; D, Dromedary; V, Vicuña; G, Guanaco; L, Llama; A, Alpaca.

open forests interspersed with peatlands, living alongside bears, rabbits and beavers.[7] Average temperatures were higher than today, although still below freezing for most of the year. While the camelids living further south had teeth adapted for grazing, the dentition of the giant arctic camels was better suited for grinding foliage, like those of modern camels. Indeed, Rybczynski suggests that many of the most distinctive features of modern camels evolved during their time in the high latitudes. Wide flat feet for walking on sand would have been just as helpful for soft snow; large eyes to aid foraging during the winter darkness; and humps that provided fat storage for when food supplies were scarce.

Around 7 million years ago, the camels spread into Asia, becoming the ancestors of today's three species – the two-humped Bactrian, the wild Bactrian and the single-humped dromedary (Plate 15). During the Pleistocene, when ice sheets formed, sea levels fell and exposed a vast intercontinental highway of grassland, Beringia (or the Bering Land Bridge), that extended from the Lena River in Russia to the Mackenzie River in Canada. However, those left behind in northern North America were ultimately doomed. They survived the Pleistocene interglacials when forests and shrublands dominated the area, but the arrival of full glacial conditions around 75,000 years ago led to their extinction. The *Paracamelus*-like species retreated to the temperate regions of America before disappearing alongside mammoths, mastodons and many other megafauna during the Quaternary Extinction Event that coincided with the arrival of humans from Asia around 13,000 years ago.

Unravelling the story of the camels' migration into Asia and their subsequent domestication remains a challenge. For example, in the early 2000s, archaeologists discovered bones from an enormous and previously unknown species at an oasis in Syria, dating from 100,000 years ago. This animal, nicknamed the 'Syrian camel' by its Swiss–Syrian team of discoverers, stood 3 metres tall, almost twice the size of its living relatives. Human remains discovered nearby, including bony tools, suggest that early humans killed the species, most likely around their watering holes. However, despite existing for thousands of years, there is no evidence that the 'Syrian camel' was ever domesticated, and the cause for its demise remains a mystery. Nevertheless, the findings are remarkable, not just because of the animal's size, but because they reveal that a species of dromedary camel existed nearly 90,000 years earlier than thought.

Genome sequencing indicates that the split between the dromedary and the Bactrian camel occurred around 4–5 million years ago.[8] Complicating their subsequent story are the compounding issues of domestication and hybridisation, such that there are now three recognised species – two domesticated species (dromedary and Bactrian) and the wild Bactrian. Molecular studies suggest that the domesticated Bactrian evolved from an unknown species that split from the wild Bactrian 1.1 million years ago.[9] It is estimated that the initial domestication of Bactrians occurred around 5,000 years ago, around Iran and Turkmenistan, and as far back as cattle, horses and dogs. Domesticated animals provided milk, meat, fibres and transport, encouraging regional cultural and economic development. Further west, the domestication of the dromedary took place 2,000 years later, and contributed to the opening of the trans-Saharan and trans-Arabian trade routes.

All camelids possess the same number of chromosomes, 37 pairs, and all inter-species hybrids are fertile.[10] Following domestication, Bactrians were frequently crossbred with dromedaries in areas of sympatry to produce offspring that mature faster, produce more milk and are more robust. They are also larger than both parental species and have been used traditionally as draft animals. Wild Bactrians are now critically endangered, with only small populations surviving in China and Mongolia. In contrast, the wild ancestors of dromedaries became extinct less than one millennium after the domestic form appeared.[11]

Physiological adaptations

Camelids are ungulates, from the Latin *unguis*, nail, despite lacking hoofs. However, the ungulate way of walking evolved independently in the even-toed and odd-toed lineages and is, therefore, a descriptive term rather than a taxonomic one. Nevertheless, the fossil record confirms that camels are true ungulates, as early species walked on their toes (unguligrade) and likely possessed hooves. Sometime in the past, camelids evolved two-toed feet, with the third and fourth digits lying flat on the ground (digitigrade), each having a toenail and soft pad. The surface area on the pads can increase to help with weight distribution, an adaptation that probably evolved during their time in the Arctic's snowy wastes. In addition, the South American camelids can adjust their toe pads for a better grip on steep and rocky terrain. All camelids have a similar gait, in which both legs on the same side move simultaneously, resulting in a swaying motion for any rider, thus reinforcing the aptness of their 'ships of the desert' analogy.

Camels boast a plethora of physiological adaptations.[12] They store fat in their humps, enabling them to survive for extended periods without food or water. They are resilient to sustained periods of high temperature, tolerating daily body fluctuations between 34 and 41 degrees Celsius, partly due to a fast and concentrated release of heat-shock proteins. A quarter of their body weight can be lost without needing food or fluid replacement, while most mammals succumb if 15 per cent of their weight is lost. Camel kidneys have evolved to conserve water efficiently by producing highly concentrated urine. Additionally, intestinal water loss is minimal, as their faeces are almost dry. They also have unusual red blood cells. Unlike most mammals' concave or spherical red cells, camel red cells are elliptical. This distinctive shape allows for continued blood flow even in a dehydrated state and ensures that cells can always cross the walls of small blood vessels. The cell membranes also have an altered distribution of phospholipids, making them very resistant to osmotic lysis. As a result, the red blood cells can increase their volume 2.5 times before bursting, an essential adaptation to the low osmotic pressure that follows after downing a large volume of water. Camels can also tolerate a high dietary salt intake, eight times as much as cattle and sheep, and their glucose levels are twice those of other ruminants. Yet, intriguingly, they do not develop diabetes or hypertension. Eyelashes protect their eyes from the strong sunlight, and they are more resistant to respiratory diseases than most other desert mammals.

Recent genomic investigations have attempted to understand how such adaptations might have evolved. In 2012, The Bactrian Camels Genome Sequencing and Analysis Consortium published the sequence of the animal's entire genome of over 20,000 genes. Crucially, they reported that nearly 3,000 genes showed exceptionally high rates of evolutionary change that had occurred over a relatively short period. Involved genes include those that control carbohydrate and lipid metabolism and enable the efficient storage and release of energy in extreme environments.[13] Bactrians also possess many copies of the cytochrome P450 family of genes. The corresponding proteins control the dilation of blood vessels in the kidney, enhancing water retention, and preventing hypertension when salt concentrations are high.

Solar radiation is another challenge for desert-dwelling mammals. Long-term exposure causes several eye conditions, and camels have avoided such complications by undergoing rapid positive selection for protein-coding genes involved in photoreception and visual protection. Finally, the overexpression of alpha-actin in camel heart muscle suggests an adaptive trait for tolerating alternating haemoconcentration and haemodilution associated with alternating bouts of drought and rehydration.

South American camelids

The Lamini tribe diverged to produce numerous species, but only the vicuña and guanaco survive in the wild today. Their speciation likely resulted from adopting different foraging behaviours, leading to niche partitioning and reduced competition.[14] Vicuñas are restricted to the Andean plateau, while guanacos are more widespread, inhabiting the continent's steppes and grasslands as far south as Patagonia. However, the guanaco's southern expansion was recent, occurring after the Last Glacial Maximum, and probably relates to climate warming and niche availability following the megafaunas' extinction in the Late Pleistocene.

The vicuña, the smallest of the camelids, inhabits an arid world with intense solar radiation and low oxygen levels. To protect against subzero temperatures, they have evolved a fleece of the finest wool, valued and harvested since ancient times. Their bodies have a high volume-to-surface-area ratio, to reduce heat loss further. Unlike most mammals, their red blood cells have evolved to cope with high-altitude hypoxia – elevated counts to enhance oxygen transport, elliptically shaped cells to protect against dehydration-induced damage, and a haemoglobin molecule with a high oxygen affinity.[15] Also, they have evolved enlarged lung capacities and efficient hearts for optimal blood delivery to vital organs, phenotypic adaptations mirroring those of other high-altitude mammals and likely resulting from molecular convergence.[16]

Guanacos are also hardy animals and can even be found in the Atacama Desert, the driest non-polar location in the world. Along the desert's mountainous Pacific coastline, the resilient guanacos survive in what are known as 'mist oases' or lomas. Here, where the cold coastal waters meet the warmer land, a phenomenon is triggered in which the desert air cools, causing fog and water vapour to form. Sea breezes carry the moisture across the desert, where cacti and lichens absorb it. The desert guanacos can only survive by feeding on the plant's flowers and the moisture-laden lichens.

Despite their regional importance – and they are now farmed globally – the history of the two domesticated species, llama and alpaca, remains poorly understood. Both arose around 7,000 years ago after communities across the Andes domesticated the wild species. Until recently, however, their evolutionary history was thought to be straightforward – llamas were derived from guanacos and alpacas from vicuñas. However, genetic studies have shown that this scenario is too simplistic, as the early camelid herders undertook widespread breeding. Repeated inter-species and back-cross breeding has led to a tangled genetic legacy, making it difficult to decipher. There is even evidence that some domesticated animals inherited DNA from an ancient guanaco population that no longer exists.

Like their domesticated cousins in the Old World, the complex genetic history of llamas and alpacas has hampered efforts to elucidate their precise origin, and their evolutionary path remains unresolved.[17]

CHAPTER 11

The Whale's Story

LOSS OF GENE FUNCTION

The cetaceans are on the whole the most peculiar and aberrant of mammals.
George Gaylord Simpson (1902–1984)

Approximately 400 million years ago, the common ancestor of all tetrapods ventured onto land, setting the stage for a wealth of species, including mammals, to evolve. Fast-forward to around 50 million years ago, and some early mammals undertook a *volte-face*. They reversed course and returned to the water, undertaking one of the most dramatic transitions in vertebrate history, accomplished in just 8 million years. Over time, these aquatic species diverged markedly from their land-dwelling relatives, evolving into the cetaceans we know and love today – whales, dolphins and porpoises. The evolution of cetaceans is one of the best-known examples of macroevolution documented in the fossil record. But what is their phylogenetic history? Where should cetaceans be placed in the mammalian tree of life? The answer to this long-standing puzzle emerged in the 1990s, thanks to genetics and new fossil finds.

Early DNA analyses showed that whales are deeply embedded within the artiodactyls, but there was no consensus on the closest lineage. Then, in 1996, John Gatesy from the University of Arizona approached the problem by comparing the milk casein genes of whales to those of artiodactyls and other placental mammals.[1] The results were startling – phylogenetically, whales lie closer to hippos than to any other even-toed ungulate (Plate 16). This unexpected finding challenged existing ideas and was a significant breakthrough in the understanding of cetacean evolution. Despite anticipating scepticism from the scientific community, Gatesy was confident enough to stress in his paper that palaeontologists were mistaken, and that the fossil record was incomplete. And it wasn't long before fossil hunters found the missing evidence confirming the evolutionary link between cetaceans and artiodactyls, and revealing that whales and hippos are each other's closest living relatives (Figure 10.1).

In 2001, fossilised bones from several species of terrestrial whale were recovered from Eocene deposits in Pakistan. Among them, taxa like *Artiocetus* and *Rodhocetus* possessed a unique anatomical feature that was thought to be exclusive to artiodactyls: a 'double-pulley' shaped astragalus located within the ankle.[2] This diagnostic bone, characterised by grooves at both ends, restricts flexion to the vertical plane and prevents dislocation during rapid movement. Despite obviously being whales, these newly discovered taxa had feet resembling those of even-toed ungulates. Essentially, the cetacean astragalus was an inheritance from their artiodactyl ancestors, a trait lost as they evolved a fully aquatic lifestyle.

But Asian fossil beds held a further surprise, a 'missing link' between cetaceans and hippopotamuses. This pivotal species, named *Indohyus*, was unearthed by a team led by Dutch-American palaeontologist Hans Thewissen in 48-million-year-old deposits.[3] Found in Kashmir, *Indohyus* was roughly the size of a raccoon. Thewissen identified key traits linking this semi-aquatic creature to modern whales. One notable feature was the ectotympanic bone, which encloses the middle ear. Unlike the uniformly thick ectotympanic bone found in artiodactyls, *Indohyus* exhibited a thinner external side, a characteristic unique to all cetaceans to enhance underwater hearing.

However, other features of *Indohyus* are more hippo-like. Although its legs are slender, the bones are relatively heavy, with a thick outer cortex, like those of wading mammals, an adaptation to counteract buoyancy so they can remain underwater. Thewissen also found isotopes of oxygen from fresh water and brine in the fossilised teeth, indicating that *Indohyus* drank both and must have inhabited the coast, wallowing in the shallow subtropical waters of bays and estuaries of the ever-shrinking Tethys Ocean. The carbon isotope ratio also implies that *Indohyus* was herbivorous, feeding mainly on land plants, similar to modern hippos. While it may seem strange for an early deer-like artiodactyl to be semi-aquatic, such a lifestyle resembles one modern-day artiodactyl. The diminutive water chevrotain or African mouse-deer lives in tropical forests and feeds mainly on fruits and flowers, although it will take insects, crabs and fish. However, it never strays far from rivers or lakes and escapes predators by jumping into the water, often remaining submerged for prolonged periods.

Thewissen suggests that such predator-avoidance behaviour might have underpinned the cetacean's remarkable return to the sea. An alternative hypothesis proposed by the American palaeontologist Philip Gingerich emphasises the opportunity that water offered for new food sources.[4] Despite being predominantly herbivores, many extant ungulates demonstrate flexibility of diet, occasionally taking meat when available. For instance, white-tailed deer have been observed taking dead river herring or alewives as a convenient source of protein. Instances of deer taking songbird chicks have also been recorded, while similar behaviour may occur in domestic cattle. Furthermore, hippos occasionally feed on impala, wildebeest, and even other hippos, while swine are notorious for being omnivorous. These observations highlight that artiodactyls can be more predatory and carnivorous than previously thought. Knowing this, it is not hard to imagine early artiodactyls progressing evolutionarily from scavenging dead fish to preying on dying or injured fish to actively pursuing healthy ones. In other words, Gingerich believes that the driving force was not competition or avoidance of predation but simply one of opportunity.

The realisation that whales lie deeply embedded within Artiodactyla gave taxonomists a headache. Whales already belonged to the order Cetacea, and since you cannot have an order within an order, Cetacea was no longer a valid term. Instead, a new composite was coined, Cetartiodactyla, which encompasses both whales and even-toed ungulates (see Figure 10.1).

Making the transition

The exceptional cetacean fossil record provides a detailed history of how the early species altered their body plan to become fully aquatic mammals. The rapid change from terrestrial furry quadrupeds to streamlined oceanic leviathans was helped by the relative absence of large predators in the post-Cretaceous oceans. However, delving into the myriad species and their sequential transformations falls beyond the scope of this book.[5] Instead, I will focus on the fundamental shifts and the genetic modifications underpinning them, whether anatomical, physiological or biochemical.

The first 4 million years of whale evolution occurred amid the Tethys Ocean's shallow waters on the northern shores of the Indian subcontinent. At this time, India was still an island, and for a few more million years the quiet backwater to the north acted as a laboratory for evaluating an aquatic lifestyle. The results were encouraging, and around 45 million years ago, four-legged species paddled to Africa, then onwards to eastern North America, to eventually reach Peru 2 million years later.[6]

Then, around 39 million years ago, the early whales, or archaeocetes, gave rise to crown cetaceans, species that began to resemble the whales we recognise today. Their forelimbs became flippers, used to steer and balance, while their back legs became superfluous and disappeared through genetic drift. The pelvis shrank, lost contact with other bones, and became surrounded by muscle. But it never became a useless vestige, remaining linked by muscle to the genitals in males to provide penile control.[7] With the loss of limbs, locomotion was provided by a tail fin or fluke, a novel swimming device supported by wider vertebrae. However, flukes are not vertical like the tails of fish but horizontal, moving up and down to generate thrust, rather than sideways. This crucial difference relates to the whale's terrestrial ancestors, whose vertebral columns didn't bend easily from side to side. Sceptics can verify this by observing their pet dog's spine move up and down as it runs.

One of the most notable adaptations in cetaceans relates to their skull structure, which supports their ability to breathe, feed and navigate underwater. The bony nares, or nostrils, migrated to a more posterior position, evolving into the blowhole that allows for surface breathing when the animal is horizontal. This shift led to a shortening of the skull's posterior section behind the snout or rostrum, accompanied by a unique overlapping or underlapping of facial bones, a process known as cranial telescoping.[8] This term, inspired by the collapsible mariner's telescope, in which wider sections slide over narrower ones, effectively describes how the whale's facial bones have slid back over one another. This intricate evolutionary change marks one of the most dramatic transformations in mammalian skull structure.

During the late Eocene, crown whales diverged to produce two clades that persist today – the toothed whales (Odontoceti) and the baleen whales (Mysticeti), each boasting unique and intriguing adaptations. Toothed whales evolved echolocation as a sophisticated sensory tool for navigating, hunting and communicating underwater, a capability distinguishing them within the realm of marine mammals. In contrast, baleen whales, named for their specialised keratinous baleen plates, developed the remarkable ability to filter-feed, marking a distinctive feeding strategy with an uncertain evolutionary path.

Toothed whales have been remarkably successful and today are the most diverse marine mammals, with nearly 80 living species. They include beaked whales, sperm whales, dolphins and porpoises, which all utilise high-frequency or ultrasonic clicks and whistles emitted from their nasal passages for echolocation. The reflected sounds are then interpreted through fatty structures surrounding their ears and in their lower jaws, which transmit echoes to the middle ear. By processing the echoes, toothed whales construct a detailed image of their surroundings, enabling them to discern prey location, as well as their speed and direction of movement, effectively 'seeing' within their underwater environment.

In 2022, Ellen Coombs from the Natural History Museum in London made a significant discovery regarding toothed whales.[9] She found that during their evolution, the animals' skulls became increasingly asymmetrical or 'wonky', with the bones on one side occupying different positions than their counterparts on the opposite side. A mass of underlying fatty tissue called a melon lies between the blowhole and the tip of the snout, which accounts for the asymmetry. This strange structure focuses the echolocating sounds and ensures that the sound transmission characteristics of the whale's tissue and the surrounding water are similar. The result is a reduction in acoustic energy wastage, maximising echolocation efficiency. However, the other internal structures are squashed into the skull's left side to accommodate the melon, further contributing to the asymmetrical skull shape. After studying skulls from many extinct and living cetaceans, Coombs showed that the skulls of baleen and toothed whales started to evolve differently around 36 million years ago. While the skulls of baleen whales remained symmetrical, those of toothed whales first became lopsided around 30 million years ago and grew increasingly wonky as they evolved.

Coombs also documented that Odontoceti species inhabiting extreme environments, such as narwhals, belugas, river dolphins and deep-diving sperm whales, depend more on echolocation than other whales and, as a result, have even more oddly shaped skulls. Living in challenging environments, whether shallow water, polar ice, murky rivers or deep oceans, requires specialised echolocation systems with a more diverse sound repertoire, leading to increased cranial asymmetry.

Despite including the blue whale, the largest vertebrate to have ever lived, baleen whales began their evolutionary journey as relatively small creatures, around 3 metres in length, with a superficial resemblance to dolphins. Gigantism is a relatively recent phenomenon, emerging during the Pliocene about 4.5 million years ago. Before discussing how such leviathans evolved, however, we should explore how early toothed mysticetes transitioned into filter-feeding specialists.

The earliest whales, equipped with teeth-lined jaws, were skilled at stabbing and tearing their prey, a trait inherited from their terrestrial ancestors. Odontoceti, like the sperm whale, continued with this strategy, preying on octopus, squid and fish. However, Mysticeti embarked on a remarkable evolutionary path as the Earth's climate cooled and ocean currents shifted. They developed a novel feeding method – filter-feeding – a strategy previously unseen in the mammal kingdom. This transition, a testament to cetacean adaptability, remains a fascinating aspect of their evolution.

All mysticetes, such as the grey whale (Plate 17), feed by opening their mouths, retracting their tongues, and gulping large volumes of water. Once the mouth is closed, the water is pushed out through the baleen, which acts as a sieve, capturing their primary diet of shrimp-like krill and copepods. To support their nutritional needs, filter-feeders have evolved enlarged mouths and flexible lower jaw joints to allow wider gapes. Among the clade, one subgroup that includes the blue whale, known as rorquals, evolved a pleated throat that expands during feeding, allowing for an even greater gulp volume.

The baleen sieve comprises rows of keratin plates, the same protein that makes up our hair and fingernails, spaced approximately 1 centimetre apart. Between 200 and 400 plates are arranged transversely in two racks that erupt from the gums on each side of the upper jaw. However, the plates of baleen are not teeth. Although mysticetes grow teeth, they are rudimentary and reabsorbed in the womb before enamel formation. Baleen growth then occurs in the same place as the teeth and is fed by the same nerves and blood vessels.

However, the origin of filter-feeding remains shrouded in mystery. Did baleen and filter-feeding evolve alongside teeth, or did whales become edentulous and rely on suction-feeding before the evolution of baleen? During the last 20 years, a spate of new fossil finds has revealed a complex picture of early whale evolution.

In 2018, Carlos Peredo, a postdoctoral fellow from the National Museum of Natural History in Washington DC, reported a mysticete from Oregon's Oligocene deposits that possessed neither teeth nor baleen.[10] Instead, it relied on suction to vacuum up its prey. But why would evolution favour such a change? What advantages could outweigh the loss of teeth, a crucial feeding tool? According to Peredo, teeth are a costly investment. Growing teeth with strong enamel takes a lot of energy and resources. While hoovering up food limits prey size, it is more economical and allows gathering small food items in large quantities. Furthermore, the climate was in flux, and the altered food availability could have contributed to the phenotypic shift.

A different finding was reported by Jonathan Geisler: a 30-million-year-old toothed whale from the Oligocene of South Carolina that filter-fed with its teeth.[11] Its incisors captured large prey, while the gaps between the lower molars acted as filters. Unlike the archaeocetes before them, their upper and lower teeth didn't interdigitate but widely overlapped, allowing water to pass through the interdental slots. Geisler's find supports the 'dental filtration hypothesis' that proposes a transitional state where teeth acted as filters before the evolution of baleen. Indeed, a similar feeding strategy characterises the extant leopard and crabeater seals, which employ their front teeth to grasp prey and their back teeth to strain out smaller prey, predominantly krill.

Another fossil, this time from Australia, had widely spaced teeth and was likely to have been a benthic suction feeder, using its muscular snout to ingest organisms from the sea floor.[12] Yet other species had evidence of both teeth and baleen. CT imaging of one 25-million-year-old skull fossil showed structures compatible with proto-baleen, while several fossils had grooves and holes on the side of the palate for blood vessels that nourish the baleen-producing tissue.[13]

The remarkable diversity of recently discovered fossils reveals that the early mysticetes experimented with various feeding methods as the Earth's climate changed. However, by the end of the Oligocene, the winner was declared: baleen filter-feeding, a successful method that has persisted for the last 25 million years.

Then, around 4.5 million years ago, the mysticetes exploded in size, prompting the question of why this occurred. In general, body size increases in evolutionary lineages over geological time, an observation known as Cope's rule, after the famed palaeontologist Edward Cope. Animals also enlarge in response to predation and competition, with better survival odds for bigger animals. However, Graham Slater, a leading palaeobiologist at the University of Chicago, found these explanations unsatisfactory and proposed an alternative theory for cetaceans.[14]

Slater's team realised that the sudden increase in the whales' bulk coincided with climate cooling and the formation of large ice caps in the northern hemisphere. Before the emergence of polar glaciers, krill was never abundant and remained evenly spread around the oceans. But polar ice brought seasonality to the Earth. During the spring and summer, ice-trapped nutrients were released into open water and accumulated near the coasts, causing plankton and krill to blossom. Further out to sea, the new wind patterns and altered ocean currents led to areas where nutrient-rich water rose from the depths, resulting in swarms of krill, which can reach densities of 30,000 individuals per cubic metre.[15] However, such upwellings were widely dispersed across the oceans, sometimes separated by thousands of kilometres, as they still do today. The size of gulp-feeders became a critical factor subject to intense selection pressure. Larger and bulkier individuals had the advantage of endurance and could travel the large distances required to exploit the newly formed feeding grounds. Conversely, smaller individuals struggled to compete and ultimately perished.

Loss of gene function

Whales have undergone profound anatomical and physiological transformations to survive life in the oceans. Many such modifications affect soft tissues and biochemical pathways involving skin, blood, saliva, taste and smell, which are not evident from fossil analysis. We have gained insights into how such fundamental evolutionary changes occurred through the recent availability of high-quality genome sequences. Surprisingly, most of these adaptations stem from mutations that create inactive genes or pseudogenes. Some examples of how the so-called pseudogenisation of genes has acted as engines for evolutionary change are discussed below.

My encounter with grey whales off Baja California fascinated me because of their unusual skin texture. Unlike any other mammal, it is smooth, thick, spongy and tough, with a thick fat layer and no sweat or sebaceous glands. The absence of hair reduces friction and drag, enhancing their swimming speed. What's intriguing is that most of these skin-related innovations result from gene loss. For instance, all the genes controlling sebaceous gland secretions and several fibroblast growth factor genes that regulate hair growth are pseudogenes.[16] Even the *Hr* gene, which regulates hair morphology and growth, is non-functional in toothed whales.[17] This genetic loss extends to hair keratin genes and genes essential for skin barrier function. As expected, such losses occurred early in the lineage's evolution, around 46.5 million years ago, pre-dating the emergence of the last common ancestor of extant cetaceans.[18] Furthermore, the same study revealed that the aquatic skin traits of whales and hippos evolved independently, and that gene losses in the hippo lineage occurred significantly later than in the cetacean lineage. These genetic findings align with the results of skin histology. Unlike whales, which lack sweat glands, hippos possess a highly specialised type of sweat gland that produces 'blood sweat', an orange-coloured substance with unique natural antimicrobial and sunscreen properties. In other words, the last common ancestor of hippos and cetaceans was likely a land-dwelling mammal.

Other examples of non-functioning genes involve the loss of teeth in mysticetes. Initial studies revealed inactivating mutations in three enamel-related genes, although no mutation was common to every species.[19] Then, in 2011, a team led by Mark Springer reported that another gene critical for enamel production, enamelysin (*MMP20*), was inactivated in all baleen species.[20] Notably, the mutation was the same in every taxon, implying that the loss of the genetic blueprint for enamel occurred in their last common ancestor, around 25 million years ago, a finding that accords with the fossil record of mysticete evolution.

Sperm whales dive to greater depths and for longer than any other mammal, submerging for up to an hour and reaching depths of nearly 3,000 metres in search of squid. The loss of the gene *AMPD3*, soon after the species diverged from other toothed whales, helped sperm whales achieve such amazing feats. The lack of the gene's coding protein causes a reduction in haemoglobin's affinity for oxygen, ensuring that more oxygen is delivered to the vital organs.[21] Furthermore, the phenomenon of decompression sickness, or the 'the bends', can arise during such dives due to nitrogen bubbles forming in the bloodstream, potentially causing blood clots. Interestingly, two genes associated with initiating clotting have become non-functional in cetaceans, presumably reducing the likelihood of clot formation. Nonetheless, the remainder of the clotting process remains operational, so whales and dolphins can still heal wounds when injured.[22]

Cetaceans, including modern whales, have experienced a loss of four genes crucial for melatonin synthesis and signalling. Melatonin is a hormone pivotal in regulating sleep patterns in vertebrates. Because of the need to surface to breathe, it is theorised that early whales could not shut down their brains for extended periods. This limitation led to a unique sleeping pattern in modern whales, where one brain hemisphere remains active while the other sleeps. This adaptation suggests a diminished reliance on traditional sleep mechanisms, potentially rendering melatonin production redundant.[23]

The examples presented support the view that pseudogenes are not always 'genomic fossils' but can be powerful adaptive forces. In essence, natural selection favours species with non-functional or missing genes if they are better suited to their environment than individuals with functional versions. The findings for skin, teeth and blood genes validate the intriguing concept, proposed by the geneticist Maynard Olson of the University of Washington in 1999, that 'less is more'.[24]

However, loss of gene function can also arise from reduced or relaxed selection pressures, summed up by the pithy phrase 'use it or lose it'. In reality, most gene knock-outs will be neutral, with no fitness consequences for the individual. However, a change of environment, from terrestrial to aquatic in the case of cetaceans, can remove or weaken a source of selection that previously maintained an individual trait. We have already encountered this concept when discussing the loss of eyesight in mammals living perpetually in the dark (Chapter 5). As we will discuss, several such 'use it or lose it' mutations are scattered throughout cetacean genomes.

Toothed whales have little need for smell, as their nasal passages remain closed during dives, and they echolocate to find their prey. Consequently, Odontoceti has a complete loss of genes involved in olfaction, associated with a lack of parts of the nervous system required to detect and process odours – a classic example of 'use it or lose it'.[25] However, baleen whales have retained limited olfaction to detect dimethyl sulphide (DMS) produced in regions of feeding zooplankton.[26] Taste is also redundant as whales gulp their prey whole, don't encounter toxic food, and exist in a high-salt-concentration environment. As a result, their taste receptor genes for sour, sweet and umami are non-functional, making them the only group of mammals to have lost most of their gustatory sensory system.[27] Genes linked to feeding, such as those involved in saliva secretion and sodium reabsorption by the kidneys, are no longer helpful in a marine environment and have disappeared.[28] As with selected gene loss, the sensory genes became non-functional early, sometime before the Odontoceti –Mysticeti divergence.

Gigantism

Baleen whales are among the largest animals on Earth, providing an ideal model for studying the evolution of gigantism. Cetaceans dramatically increased in size around 4.5 million years ago, helped by the boom in food supplies in the high latitudes and the freedom from gravity in the buoyant marine environment.[29] However, the genes underpinning their size remained a mystery.

Then, in 2023, Mariana Nery and colleagues from the State University in Campinas, Brazil, found four relevant genes all involved in the insulin and growth hormone pathways.[30] One, called *IGFBP7*, controls cell growth and division. A second, *GHSR*, has various effects, including stimulating growth hormone release, increasing food intake, and modulating glucose and lipid metabolism. The other two control body weight and were already known from studies in cows and pigs.

The very same genes act as tumour suppressors and help reduce cancer risk. As with elephants, the gigantism and longevity of cetaceans theoretically increase their risk of malignancy. Not only are baleen whales large, but they live a long time. Indeed, the bowhead whale boasts a remarkable lifespan of up to 200 years, making the species an excellent model for investigating Peto's paradox (Chapter 7). Indeed, work spearheaded by Marc Tollis from Arizona University shows accelerated evolution in specific regions of whale genomes compared to their counterparts in other mammals.[31] These segments encompass crucial genes governing cell cycle regulation, division, and cellular DNA repair, all of which are vital for maintaining cellular integrity. Moreover, whales have amassed far fewer DNA mutations over time than other mammals. Given that the mechanisms guarding against cancer in whales differ from those observed in elephants, nature has devised different strategies to reduce cancer risk. These findings are exciting, as they offer promising possibilities for innovative anti-cancer therapies in the future.

CHAPTER 12

The Buffalo's Story

ADAPTABILITY AND DOMESTICATION

Biologically the species is the accumulation of the experiments of all its successful individuals since the beginning.
H. G. Wells (1866–1946)

Bovids (family Bovidae) are the most recently evolved even-toed ungulates or ruminants, comprising 143 species in 50 genera. They emerged after the divergence of deer and musk deer (family Cervidae), which evolved from ancestors similar to extant duikers approximately 30 million years ago, during the Oligocene. They were small-bodied ungulates with simple antlers and prominent upper canine tusks that inhabited the tropical forests of the Old World.[1]

Bovids consists of two subfamilies: Bovinae, with somewhat primitive teeth, represented by bison, buffalo, cattle, the nilgai, the four-horned antelope, the nyala, and kudus; and Antilopinae, with more advanced teeth, including antelopes, gazelles, goats, sheep, the muskox and their relatives (Figure 12.1). Bovidae exhibit significant variation in size and colouration, with all males possessing two or more horns (in many species, females can also have horns). While the size and shape of horns vary greatly, the basic structure is always the same: one or more pairs of simple bony protrusions without branches covered in a permanent keratin sheath, often having a spiral, twisted or fluted form.

The earliest known bovid, *Eotragus*, was identified from fossilised horn cores and lived in Eurasia approximately 18 million years ago during the Miocene. *Eotragus* was a small, solitary forest-dweller that relied on vegetative cover for protection, with its closest modern relatives likely being Africa's duikers and dwarf antelopes. Early in their evolutionary history, bovids diverged into the Bovinae of Asian origin and the Antilopinae of African origin, which occurred after the two continents split. Later, both groups ventured into each other's territory when the land masses rejoined. Finally, bovids reached North America during the

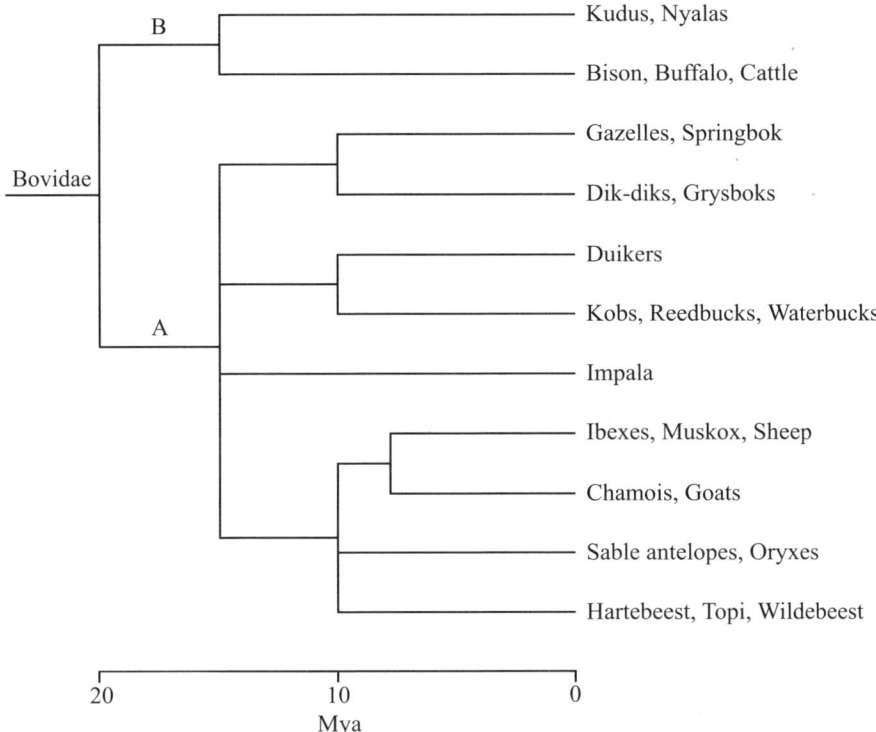

Figure 12.1 Phylogeny of Bovidae. B, Bovinae; A, Antilopinae. Modified from Yang *et al.* (2012)[2] and Calamari (2021).[3]

Pleistocene by crossing the Bering Land Bridge, with the American bison being the only survivor.

The diversification of Bovidae coincided with the spread of grasses, which followed a climate shift in the Middle Miocene from subtropical to cooler, more seasonal conditions. As landscapes gradually opened up and precipitation decreased, extensive grasslands replaced the less productive subtropical woodlands. Browsing artiodactyls, or ruminants, evolved into various species adept at extracting nutrients from fibrous diets, exploiting specific ecological conditions more efficiently than less specialised animals. Bovids, in particular, excelled in adapting their size, shape, digestive systems and social structures to specific habitats. By partitioning ecosystems into numerous zones and occupying narrow niches, they became the most diverse and abundant large herbivores, playing a vital role in shaping grassland ecosystems.

Over millions of years, bovids evolved to include a wide range of sizes, from the massive 1,000-kilogram gaur or wild oxen of Asia to the diminutive 3-kilogram royal antelope and dik-diks of West Africa. This size diversity highlights their adaptability and evolutionary success, and today they occupy nearly every niche accessible to terrestrial herbivores across Africa, Eurasia and North America. Their range spans diverse biomes, from the tropical rainforests of Africa, home to species such as duikers, to the Arctic tundra, where muskoxen thrive in the harsh polar winters. However, the largest concentration of wild bovids today is found in the savannas of East Africa, which host vast populations of migrating blue wildebeest and white-eared kob.

Foregut fermentation

The bovids' extraordinary success is intricately linked to the evolution of foregut fermentation, a solution to surviving on indigestible vegetation that is low in fat and protein and where most nutrients are locked in the plant's rigid cell walls. Such digestive processes are not unique to ruminant herbivores (species with specialised multi-chambered stomachs, such as cattle, goats and sheep). Indeed, foregut fermentation has evolved independently or convergently in several unrelated lineages, including some marsupials and rodents, sloths, colobine monkeys and the hoatzin (a South American bird). All these taxa have a feature in common: they lack the enzymes needed to break down plant cellulose and lignin, and they can only access a fraction of the necessary nutrients by themselves.

To solve the dietary problem, ruminants and other foregut fermenters have harnessed the power of microorganisms – bacteria, archaea, fungi and ciliated protozoa – that provide the missing cellulolytic enzymes. This salmagundi of microbes converts the plant's starches, cellulose and lignin into carbon dioxide, methane and hydrogen, as well as short-chain volatile fatty acids (VFAs). These VFAs – mainly acetate, propionate, butyrate – and microbial proteins provide most of the animal's nutrition. How ruminants acquired their microbial helpers is uncertain, as fossils cannot give us the answer. It's conceivable that early tetrapods ingested contaminating microbes after eating some rotting vegetation or plant-eating insects, and that some species survived in their guts. Over time, a symbiotic relationship developed with the host, as individuals housing enzyme-releasing microbes could extract more nutrients, enhancing their survival rates.

Fermentation takes place before digestion occurs, with the collection of microorganisms contained in specialised stomach compartments. There are four chambers divided by sphincters, each dedicated to a different stage. The first two compartments are the rumen and reticulum, or reticulorumen, representing one functional space that hosts the crucial fermentation process. The reticulorumen is followed by the omasum, where water and minerals are absorbed, and the abomasum, which secretes enzymes and functions like most mammals' 'true stomachs'. Ruminants are named after the process of rumination (from Latin *ruminare*, to chew over again), as the products of the reticulorumen are regurgitated into the mouth for a second regrinding, or 'chewing the cud' – a process essential for efficient digestion. Although this four-stage process, coupled with rumination, enables foregut fermenters to obtain more nutrition from their food, the longer digestion times mean that ruminants cannot bulk-process food, a fact that imposes an upper limit on their body size.[4] It is for this reason that hindgut fermenters such as elephants and rhinoceroses, with faster intestinal transit times, can have larger body masses.

The early foregut fermenters faced a formidable challenge: the loss of crucial nutrients assimilated by the vast number of microbes. Nature, therefore, required a means of breaking down and digesting the microbes once they had served their vital biological function and recycling their valuable constituents. The solution, a significant evolutionary adaptation, was the co-option of an enzyme called lysozyme, a protein that evolved initially to protect against infections, acting as a natural antibiotic.

Unlike most mammals, ruminants secrete large amounts of lysozyme into their stomach lumen (abomasum) to help digest and recycle bacterial components. Bovids have around 10 lysozyme genes, with at least four being expressed in the stomach that are uniquely active at low pH and resistant to pepsin cleavage. Modifying existing lysozyme to act as a digestive enzyme enabled the evolution of foregut fermentation. Colobine monkeys, woodrats and the hoatzin also exhibit high levels of stomach-type lysozyme. The independent emergence of gastric lysozyme in distantly related species to facilitate the emergence of the same digestive strategy is another classic example of the convergent evolution of a protein. However, the convergence is functional rather than genetic, indicating that phenotypic convergence may result from different mechanisms even though taxa are exposed to similar sources of selection pressure.[5]

Diets consisting of grass are contaminated with significant amounts of grit and soil, which can wear down teeth. Evolution has addressed this issue through hypsodonty, a dental pattern characterised by high-crowned teeth and enamel extending beyond the gum line, providing extra material to resist wear. However, in bovids, the fermentation process in the rumen removes most of the grit from ingested vegetation before it is subjected to more intensive chewing after regurgitation. According to Valerio and colleagues, this process gives bovids an inadvertent advantage over hindgut fermenters like horses and zebras, as the most intensive chewing involves grit-free food. This internal cleaning mechanism may explain why ruminants have a lower degree of hypsodonty than other herbivores, and likely contributes to their success in terms of species diversity.[6]

Extreme adaptations

Several bovid species have evolved unique adaptations to aid survival in the most inhospitable environments, notably the cold winters of the Arctic tundra and the low oxygen levels of the Tibetan plateau.

The modern muskox is the last member of a lineage that initially appeared in the temperate regions of Asia around 12 million years ago and only adapted to life in the high latitudes late in its evolutionary history (Plate 18). Initially, the early ancestors were sheep-like animals with high-positioned horns (muskoxen have low-positioned horns) that successfully spread throughout Eurasia during the Pliocene. Modern muskoxen, however, appeared around 1 million years ago in Europe, where they inhabited the tundra south of the Scandinavian ice sheet. After spreading along the ice's edge into Asia, they crossed from Siberia to North America around 200,000 and 90,000 years ago. Along with bison and the pronghorn, muskoxen are among only a handful of species of Pleistocene megafauna in North America that survived the Pleistocene–Holocene extinction event. Today, these iconic mammals are restricted to northern Greenland and the Canadian high Arctic, with successful reintroductions to Alaska and Russia.

To survive in some of the coldest, darkest and longest winters on the planet, where temperatures can plummet to minus 40 degrees Celsius, muskoxen have developed unique adaptations. Their shaggy coat features a matted underlayer known locally as *qiviut* 15 centimetres thick, windproof, snowproof, and eight times warmer than sheep's wool. This extraordinary undercoat, which helps keep the animals alive, is more durable than wool and finer than cashmere. Muskoxen grow qiviut yearly before winter, shedding it in the spring to prevent overheating during the summer. Covering the qiviut layer are long, shaggy outer guard hairs that reach almost to the ground.

Muskoxen have barrel-shaped bodies supported by short legs, reducing their body volume-to-surface-area ratio to minimise heat loss. Their limbs are also kept cooler than the rest of their body to help conserve heat. Muskoxen possess another key trait: their haemoglobin is three times less sensitive to temperature changes than human haemoglobin. This temperature insensitivity enables the animals' protein to maintain a high oxygen affinity in icy conditions, ensuring efficient oxygen diffusion into their cold peripheries.

To investigate the evolutionary journey that led to the muskox's current distribution, a team led by Patrícia Pečnerová, a researcher at the University of Copenhagen, analysed the whole genomes of over 100 individuals.[7] This research involved arduous fieldwork collecting samples from some of the most remote areas in Canada and Greenland, including DNA from one 21,000-year-old specimen preserved in Wrangel Island's permafrost.

By reconstructing changes in population sizes over time, the researchers discovered that climatic changes, refugia, recolonisations and population bottlenecks have shaped the muskox's evolution. Surprisingly, the animals have lost considerable genetic diversity over the past 20,000 years, which coincides with the maximum extent of the ice sheets covering Canada and Greenland. Even the most diverse present-day population has only about 30 per cent of the diversity of animals that lived 20,000 years ago. In East Greenland, the last place the animals colonised, genetic diversity is so low that only two variable positions occur for every 100,000 base pairs.

These findings reveal that muskoxen are the least genetically diverse of any ungulate, a degree of variation typically restricted to highly endangered species like the white rhinoceros and snow leopard with population sizes of around 1,000. In contrast, muskoxen are thriving, with an estimated population of approximately 170,000. The paradox of a stable population with a near absence of genetic heterogeneity can be best explained by a gradual and sustained decline in genetic variation over 20,000 years, accompanied by removing the most deleterious mutations from the population's gene pool. According to Pečnerová, muskoxen managed genetically 'to draw a winning lottery ticket. It is almost surprising that they survived until the present, but the way their evolutionary history unravelled, things just worked out for them.'[8]

However, recent genetic research on a museum specimen of the blue antelope of South Africa, a species driven to extinction by human activity during the colonial era, suggests that the survival of small populations is not purely down to luck. These studies reveal a potential for high levels of purging of deleterious mutations in long-term small populations. Harmful mutations are more frequently present in a homozygous state, and thus more likely to be selected against than in larger populations. From a genomic point of view, if a population survives at low numbers for an extended period, specific adaptations may counteract the effects of genomic erosion. This conclusion suggests that small populations may become less disadvantageous for survival over time, offering hope for conservation efforts.[9]

The long-term stability of the muskox population, however, remains uncertain. As global warming increases, Arctic species are encountering novel challenges. For muskoxen, mild and humid winters pose a genuine threat, as they limit access to food when freezing rain forms an impenetrable barrier over their food source. Furthermore, there is documented evidence of recent disease outbreaks, with parasites such as lungworms extending their range further north. The muskoxen's limited genetic diversity is compounding these challenges, diminishing their capacity to adapt to rapidly emerging threats.

The Tibetan antelope, or chiru, inhabits the Tibetan plateau at elevations ranging from 3,600 to 5,500 metres above sea level. These altitudes are characterised by low temperatures and significantly reduced oxygen levels – approximately half of those at sea level. In such hypoxic conditions, aerobic exercise is typically severely restricted in most mammals, including humans. However, the Tibetan antelope exhibits the extraordinary ability to sustain running speeds exceeding 70 kilometres per hour over distances of up to 100 kilometres. Their exceptional high-altitude endurance is attributed to a unique modification in the oxygen-carrying proteins within their red blood cells, a characteristic not observed in other mammals.

During the developmental stages of many bovids and primates, the type of haemoglobin expressed transitions from a high-affinity variant that facilitates oxygen transfer across the placenta in utero to a lower-affinity form in adulthood. However, such a switch does not occur in the Tibetan antelope. In contrast, fetal haemoglobin expression is maintained into adulthood, enhancing oxygen uptake and increasing its delivery to the tissues. Furthermore, due to a significant chromosomal deletion, the antelope can no longer produce adult haemoglobins. Research has confirmed the biological significance of these findings by means of in vitro investigations. The experiments revealed that the Tibetan antelope's haemoglobin has a higher oxygen affinity than any other bovid haemoglobin and helps explain the species' adaptation to altitude. Interestingly, the retention of a juvenile protein – a form of biochemical paedomorphosis – represents a unique genetic mechanism for the emergence of a novel mammalian phenotype.[10]

Migration

Today, half of ungulate species migrate to some extent, with the behaviour evolving independently in 17 lineages. Seasonal movements first emerged during the Miocene when bovids became obligate herbivores.[11] The long-distance migrations likely arose after the animals extended their smaller annual movements while searching for better food and breeding opportunities. Individuals whose movements enhanced survival would have passed that migratory behaviour, including cognitive and navigational abilities, to their offspring. Such species were more likely to evolve large body sizes and bigger populations, as the blue wildebeest and white-eared kob demonstrate.

The most well-known migration involves over a million blue wildebeest travelling from Serengeti in Tanzania to the adjacent Maasai Mara reserve in Kenya. This annual spectacle showcases some of the most breathtaking wildlife behaviour in the world. Driven by primal instincts and a fierce determination to survive, the wildebeest follow the climatic rainfall patterns, moving towards greener pastures within the Serengeti–Maasai Mara ecosystem during the wet season. Although the number of participants can vary each year, up to 1.5 million wildebeest have been known to embark on this lengthy and challenging journey, which spans several weeks. Along the way, they face significant dangers, particularly when crossing the crocodile-infested Mara and Talek rivers, where many fall prey to the waiting giant Nile crocodiles. The benefits of such a dramatic journey obviously must outweigh the many hazards involved, a testament to the resilience and adaptability of these magnificent creatures.

However, the largest and most spectacular land-mammal migration on Earth occurs across South Sudan and Ethiopia, involving approximately 5 million white-eared kob and other antelopes, including 1 million tiang, a subspecies of topi. The astonishing scale of the 'Great Nile Migration' was only recently documented after an extensive aerial survey of the remote, war-torn area in 2024. Two planes, equipped with cameras programmed to take photographs every 2 seconds, took 330,000 images, which scientists at the University of Juba studied using software to count the wildlife.[12]

Domestication

According to fossil records, wild cattle (tribe Bovini) originated in Asia south of the Himalayas during the Late Miocene, a conclusion supported by the region's high diversity of wild cattle. The critically endangered saola, a shy and secretive bovid from Vietnam and Laos first described in 1992, is the most basal of the living species and supports an Asian origin. Later, the cattle lineage migrated from Asia to Africa, diversifying into many species. Large grazers, including buffalo (Plate 19), aurochs, yak and bison, emerged during the cold Pliocene climate and the associated further extensions of open grasslands.

Despite their similar appearance, the African and Asian water buffalo are only distantly related, having separated around 8 million years ago. Failed cross-fertilisation experiments and unsuccessful attempts at implanting Asian water buffalo embryos into the African species underscore their distant evolutionary relationship. The African species is the largest bovid in the continent's savanna and is of great ecological importance, given their role as bulk feeders in the grazing hierarchy. Their large size enables them to process taller and coarser grasses than other species, playing an important facilitative role for the many smaller grazers. However, the species is unpredictable temperamentally, which may explain why it has never been domesticated. Indeed, they are considered one of the most dangerous mammals in Africa, as old and grumpy bulls may persistently charge for no apparent reason.

In contrast, the domestic Asian water buffalo occurs across all five continents, with a global population of approximately 200 million. More people's livelihoods depend on water buffalo than on other domesticated animals. They supply draught power, milk and meat, and millions of smallholder farmers in Asia rely on them for sustenance and income. There are two distinct types of water buffalo, river and swamp, which originated from different wild populations that separated around 900,000 years ago and evolved in different geographic regions. The river buffalo was domesticated in the western part of the Indian subcontinent around 6,300 years ago and subsequently spread westward to areas including Egypt, the Balkans and Italy. In contrast, the swamp buffalo was domesticated in the China–Indochina border region between 7,000 and 3,000 years ago and spread throughout Southeast Asia and China, reaching as far as the Yangtze River valley. Recently, scientists have discovered that human domestication has shaped the water buffalo's genome similarly to global cattle breeds, suggesting that critical genetic loci are linked to domestication across different species.[13]

The aurochs, a large extinct cattle species, evolved in northern India and migrated west and north during the interglacial periods, reaching Denmark and southern Sweden during the Holocene. Greek and Roman writers depicted them as highly aggressive creatures. Julius Caesar, who observed them in the Black Forest region of southern Germany, documented in Book 6 of his *Commentarii de Bello Gallico* (Commentaries of the Gallic War) that:

> *these animals are a little below the elephant in size, and of the appearance, colour, and shape of a bull. They spare neither man nor wild beast ... not even when taken very young, the animals cannot become accustomed to human beings and be tamed.*

The aurochs declined during the late Holocene due to increasing farming activities, including grazing and intensified hunting and poaching. By the thirteenth century, they had become exceedingly rare, confined primarily to eastern Europe, with the last known animal dying in Poland in 1627.

Current genetic evidence shows that all modern cattle derive mainly from two episodes of aurochs domestication during the Neolithic Revolution: one in the Middle East 10,500 years ago that gave rise to taurine cattle, and one 8,000 years ago in what is now Pakistan that gave rise to the indicine or zebu cattle. Taurine cattle are primarily confined to temperate and cold climates and are mainly distributed across the northern hemisphere. By 2022, there were over 250 breeds and approximately 1.5 billion domestic cattle in the world, with the dairy industry alone valued at over 500 billion pounds. Zebus differ from taurine cattle in having a fatty hump on their shoulders, a large dewlap, and often drooping ears. They are well adapted to the hot tropical savanna climates and have a greater tolerance to heat, drought and sunlight exposure. Adaptations include gene selection for keratins, heat-shock proteins and heat-resistance genes.[14]

Wild yaks inhabit the remote high-elevation alpine meadows and steppes of the Tibetan plateau and exhibit many unique adaptations to survive the region's low oxygen levels and cold. They possess large hearts and lungs, compact bodies, thick outer hair, and non-functional sweat glands. Physiological and biochemical adaptations include lower respiration rates, sweating and metabolism, with efficient nitrogen utilisation. Additionally, long-term natural selection has resulted in a distinctive genetic makeup, including the expression of several hypoxia-response genes, similar to those acquired by *Homo sapiens* from the Denisovans (Chapter 25).

Bison emerged in Asia approximately 2.6 million years ago at the end of the Pliocene. Subsequently, their progeny migrated to North America via the Bering Land Bridge in two distinct waves: the first occurring between 195,000 and 135,000 years ago and the second between 45,000 and 21,000 years ago. These migrations resulted in several species, of which only one, the American bison, survived to become the most populous hoofed mammal in the New World, reaching populations of over 60 million. They roamed the continent's vast prairies, subsisting on grasses and migrating seasonally from north to south in response to winter conditions.

The westward expansion of European settlers necessitated the subjugation of Native American populations, for whom the bison was a crucial resource, providing meat for sustenance and hides for shelter. As a result, the United States Army initiated a campaign to cull the bison, effects compounded by the construction of railroads and the establishment of settlements. By 1870, the slaughter of bison exceeded 1 million a year, and by 1884 only 541 individuals remained.

Conservation efforts were instigated by a few concerned ranchers in 1910, and contemporary populations have rebounded to over 500,000. They include a wild, free-ranging population of approximately 2,000 individuals in Yellowstone National Park, which can trace its lineage back to prehistoric times. Initial attempts often involved crossbreeding bison with cattle, although, since the resultant hybrids exhibited diminished vigour, the practice was soon abandoned. Today, the proportion of cattle DNA in bison is low, around 1 per cent, with many ranchers employing DNA testing to reduce the levels further by eliminating animals with introgressed cattle genes from their herds. In 2016, the American bison was designated the United States' national mammal, symbolising the Wild West's untamed spirit and exemplifying the success of conservation initiatives.

The goat–antelope group (Figure 12.1) emerged during the Miocene, although maximum diversity awaited the arrival of the recent ice ages. Many species adapted to marginal, often extreme environments during their early evolution: mountains, deserts and subarctic regions. In the mountainous areas of Eurasia, two lineages gave rise to sheep and goats. The sheep ancestors moved into the foothills and nearby plains, relying on flight and flocking for protection against predators. In contrast, the goat ancestors adapted to high, steep terrain where predators are disadvantaged. Eventually, the Asiatic mouflon and the Bezoar ibex arose, respectively, with each species being domesticated around 10,500 years ago in the same region of the Middle East: southeastern Anatolia and the Iranian Zagros Mountains. Favoured traits included tameness, rapid growth, stamina, thick fleeces and increased milk production. Since then, humans have spread modern sheep and goats beyond their native range and, ultimately, worldwide.

Unlike other common livestock, domesticated sheep and goats have not hybridised with their native wild relatives, providing scientists with opportunities to determine the genetic basis for the mammalian 'domestication syndrome'. Recently, a comparative study reported 20 common genomic regions related to domestication, but the patterns of gene selection were not the same in sheep and goats. According to Florian Alberto, one of the team's scientists, different genetic solutions have given rise to similar characteristics in both lineages. In other words, sheep and goats have used different genes to reach similar phenotypic endpoints.[15]

A novel speciation mechanism

The genus *Raphicerus* includes three small antelope species: Cape grysbok, Sharpe's grysbok and steenbok, all endemic to sub-Saharan Africa. The three species present a unique case for evolutionary study due to a recently discovered mechanism of speciation revealed through the analysis of their chromosomes. The rapid radiation of *Raphicerus* is explained by periods of isolation in refugia followed by expansion during Pliocene–Pleistocene climatic oscillations. During isolation, structural alterations occurred in the X chromosomes of each ancestral population, leading to genetic incompatibilities and subsequent speciation.

Specifically, regions of the X chromosome known as heterochromatin, characterised by being highly condensed, gene-poor and transcriptionally silent, underwent varying degrees of expansion. This lengthening resulted in significant heterochromatin length and sequence differences among the separate populations. Consequently, female hybrids (XX) exhibit mismatched X chromosomes during cell division or meiosis. Such misalignment initiates a gene silencing program, causing apoptotic cell death and the non-viability of female embryos.

Before this came to light, scientists generally assumed that Haldane's rule applied, which posits that divergent X chromosomes typically drive speciation through males rather than females. However, the *Raphicerus* case challenges this assumption by demonstrating that X-chromosomal divergence can also drive speciation through females. This novel mechanism underscores the complexity of evolutionary processes and, importantly, highlights the crucial role of chromosomal structure in speciation events.[16]

CHAPTER 13

The Giraffe's Story

COMPARATIVE GENOMICS

It is altogether exceptional, novel, and specialised.
Sir Roy Lankester (1847–1929)

The giraffe holds a special place in the annals of evolutionary history. With the longest neck of any mammal, the giraffe was proffered by both Jean-Baptiste Lamarck and Charles Darwin in support of their opposing ideas on the mutability of species. Lamarck proposed that acquired characteristics are inherited and play a role in evolution. In his 1809 work *Philosophie Zoologique*, he speculated that as giraffes stretched to reach higher branches, their neck lengthened from the effort, resulting in the inheritance of taller necks by their offspring.[1] After many generations, the result was the emergence of the long-necked giraffe we recognise today. Indeed, he believed his idea was so irrefutable that it required no amassing of facts or trial by experiment to confirm it.

In contrast, Darwin proposed a different scenario. He suggested that ancestral giraffes would have exhibited minor neck and leg length variations. Those with longer necks and legs would have been more successful in obtaining the highest foliage, resulting in the selection of these traits over time through the improved survival and reproduction of the 'fittest' (those individuals who are best adjusted to their environment). Eventually, longer-necked individuals evolved. This view is now universally held, and is frequently rehashed in children's books, textbooks and internet descriptions of the species.

While Lamarck's explanation is now rejected, the question of how such long necks evolved remains a subject of intense debate. Definitive proof that they arose solely by natural selection to enable the highest leaves to be reached is still lacking. According to Stephen Jay Gould, 'We only prefer this explanation because it matches current orthodoxy.'[2] While giraffes clearly benefit from their ability to feed where others can't, it's uncertain whether this is the sole reason for their iconic phenotype. Other factors could have contributed, and their ability to reach the canopy might be fortuitous.

Over 150 years later, the questions of how giraffes evolved and the drivers behind their iconic necks continue to intrigue researchers.

Why and how?

Challenging Darwin's long-standing theory is Robert Simmonds and Lue Scheepers' novel hypothesis of sexual selection.[3] They argue that field observations fail to support the notion that long necks evolved to reach the canopy. For example, in the dry season, giraffes often feed from low shrubs, not tall trees; males and females feed faster and more frequently with bent necks; and there is little foraging height partitioning with other browsing species. Furthermore, compared to the okapi's neck-to-leg ratio, the giraffe's neck has increased proportionally more than its leg length, an unexpected and physiologically expensive means of gaining height.

In support of the 'necks for sex' hypothesis, male giraffes engage in vigorous neck-to-neck combat to establish mating or territorial dominance, a behaviour unique to the species. This strategy involves the males swinging their necks, their most powerful and manoeuvrable weapon, and striking their opponent's neck, ribs or chest with their well-armoured skull and horns (or ossicones, to use the technical term). These clashes are intense and violent, often resulting in severe injury and, in some cases, even death. The victor, usually the bigger male with the longer neck, stands a better chance of siring offspring and passing his genes to the next generation. Simmonds and Scheepers also stress that larger males demonstrate positive allometry, a prediction of sexually selected traits, by investing more in their necks than smaller males. However, not all scientists agree. Graham Mitchell of the University of Pretoria challenges the evidence, stressing that the giraffe's neck remains an evolutionary enigma.[4]

Then, in 2022, a strange-looking early relative of the giraffe, dubbed *Discokeryx xiezhi*, was reported to support Simmonds and Scheepers' hypothesis.[5] The 17-million-year-old species, which roamed the grasslands of northern China, was adorned with a bony disc-like shield on the top of its skull, covered in a protective layer of keratin. The distinctive headgear exhibited similarities to the horn-like, bony ossicones seen in living giraffes and used for bashing their opponents. But the massive joints connecting the animal's skull and neck revealed its head-butting prowess – thick cervical vertebrae that locked together into a column perpendicular to the head dome, forming a remarkably effective battering ram. Writing in the academic journal *Science*, the researchers conclude that the ancient giraffe's penchant for fighting made their necks longer, eventually enabling them to reach the highest foliage.

The habitat of *Discokeryx* was relatively barren, with less food than the continent's earlier forested landscape, and the resulting increased competition could have resulted in the animal's aggressive behaviour and corresponding phenotypic changes. Intriguingly, around 7 million years ago, the environment in eastern Africa was similar, with forests giving way to grasslands. During this transitional period, the ancestors of modern giraffes faced similar adaptive pressures, potentially culminating in the evolution of the males' combat strategy involving necks and heads. In other words, sexual selection, rather than Darwin's view of reaching for the sky, led to the long-neck giraffes we see today.

However, Edgar Williams argues against both sexual selection and the advantage of reaching higher leaves as aetiological factors. Instead, he proposes that the benefits of 'horizon-scanning' underpin the emergence of the giraffe's long neck.[6] Vigilance, especially while feeding, is essential for the survival of all herbivorous ungulates. Unlike the okapi, the giraffe's neck gives the savanna-dwelling herbivore a heightened sense of its surroundings. This ability offers several advantages, such as better food availability assessment, monitoring family members' proximity, and detecting potential predators more effectively.

While the debate continues about the evolutionary drivers – Darwinian gradualism, sexual selection, or horizon-scanning – what do we know about how the giraffe's neck evolved?

A team led by Nikos Solounias from the New York Institute analysed fossil and extant specimens, employing computational techniques to track evolutionary shifts.[7] They concluded that the giraffe's neck elongated in two distinct stages. Initially, around 7 million years ago, the cranial end of the cervical vertebrae stretched to produce transitional forms like *Samotherium*, found in Greece.[8] Then, approximately 1 million years ago, a second stage of elongation occurred on the back or caudal portion, but only in the modern giraffe lineage. The result is that the giraffe's cervical vertebrae are nine times longer than their width. The third cervical bone is the same length as an adult human's humerus, a bone that stretches from elbow to shoulder. However, as the modern giraffe's neck was lengthening, the neck of the okapi was shortening. Before Solounias' findings, researchers believed the okapi, native to the Democratic Republic of the Congo, was 'more primitive' than the giraffe. However, it now appears that its neck shortened later, placing it on a different evolutionary pathway.

As in all mammals except sloths and manatees (Chapter 8), the giraffe's neck consists of just seven cervical vertebrae, albeit large ones.[9] Until recently, it was assumed that the thoracic vertebrae are held fixed relative to each other and that the head and neck movements result solely from cervical bone articulations. However, two Japanese researchers noted something unusual after carefully dissecting zoo carcasses. The animal's eighth vertebra, the first thoracic vertebra, is highly flexible and functions as a fulcrum in neck movement, extending the giraffe's reach by an additional 50 centimetres.[10] This finding suggested that the first thoracic vertebra could be an extra or eighth cervical bone. However, this is unlikely as the bone is attached to a rib, as do all thoracic vertebrae. Furthermore, there is a strong evolutionary constraint for the number of cervical vertebrae in all but a tiny handful of mammals.[11]

Neck elongation occurs after birth, probably because females would have difficulty giving birth to young with adult neck proportions. By adulthood, the giraffe's head and neck weigh approximately 300 kilograms. To counterbalance the weight, they have a sizable nuchal ligament running the length of the neck and secured by elongated thoracic vertebral spines. Composed primarily of collagen and elastin, this fundamental structure acts like a strong rubber band, exerting an upward pull on the neck. As a result, very little muscular energy is spent when the neck is held upright. Instead, the neck muscles are mainly used to lower the head by stretching the ligament.

In recent years, there has been a trend towards producing robots modelled after biological organisms. Given the impressive attributes of the giraffe's neck – such as its power, flexibility and resilience to damage – it is not surprising that robotic engineers and designers have taken notice. Sure enough, a Tokyo Institute of Technology team has recently developed a giraffe-inspired robotic arm. This innovative design strikes a balance between conventional rigid robots and entirely soft tentacle-like robots. Employing a sophisticated arrangement of beams, joints, arrays and actuators, the team successfully engineered a robot with 18 degrees of freedom, mimicking numerous behavioural aspects observed in living giraffe necks.[12]

Adaptations for a tall life

The neck of a giraffe, which can reach up to 2.4 metres in length, presents significant cardiovascular challenges: how to generate sufficient blood pressure to supply the brain with blood and oxygen while averting damage to tissues and organs situated below the heart due to fluid infiltration and oedema (giraffes require a mean pressure at the heart of 220/180 mm Hg, almost twice that of humans). Even routine actions like bending the neck to drink or standing up from a prone position require solutions to prevent dizziness and loss of consciousness.

Among the distinctive adaptations, giraffes have valves approximately every 3 centimetres in the large veins of the limbs to protect the capillaries when walking or running. In contrast, limb arteries are thick-walled with tiny lumens that provide haemodynamic resistance, whilst those supplying the brain appear normal. Additionally, a sphincter-like structure in the arteries below the knee, coupled with the thick walls, further shields the lower limbs from damage. Despite possessing a heart-to-body weight ratio akin to other mammals, giraffes exhibit low cardiac output, resulting in a comparable cardiac energy expenditure. The kidney is more robust, strengthened with fascia and thickening to reduce interstitial pressure, while the renal veins are anatomically adapted to tolerate higher pressures.

The mystery of how the circulation of the giraffe's head is protected during drinking remains unsolved. One observable behavioural measure is the animal's drinking posture: it awkwardly spreads or bends the forelegs, thereby bringing the heart closer to the ground and reducing the vertical distance between heart and head, and lessening the hydrostatic pressure in the neck and head vasculature. Other potential protective measures include blood storage in the neck and head veins, control of vascular tone, non-collapsible veins and valves in the jugular veins.[13]

The giraffe's long neck imposes other challenges. Although they may drink only once a day, giraffes can go without water for weeks, obtaining most of their needs from dietary moisture. However, the mechanism by which the world's tallest mammal draws water and propels it to the stomach had long been a mystery, until two physicists, Phillipe Binder and Dale Taylor, proposed a solution – plunger pumps.[14] According to their model, the giraffe's lips act as one valve of the 'pump' while the animal's epiglottis, situated at the back of the mouth, serves as the other. The giraffe dips its pursed lips into the water, then retracts its jaw, allowing water to rush into the mouth while keeping the epiglottis 'valve' closed. Subsequently, the giraffe tightens its lips, relaxes the epiglottis, and moves its jaw in a pumping action, pushing the collected water into the oesophagus. This sequence repeats, progressively moving more water into the oesophagus. Eventually, the giraffe lifts its head, allowing gravity and the rhythmic muscular contractions known as peristalsis to get the water to the stomach.

A towering genome

Recently, comparative genomics has emerged as a valuable tool to probe deeper into our understanding of species-specific adaptations, including how giraffes cope with their range of cardiovascular challenges. In 2021, an international team led by Qiang Qiu from China generated a high-quality genome sequence of the giraffe and compared it to those from other ruminants, including the okapi.[15] Mutations specific to giraffes were frequently associated with the cardiovascular system, bone development, vision, hearing and circadian rhythms. Notably, the giraffe's *FGFRL1* gene exhibited seven amino acid substitutions that were not found in other species. Since mutations in *FGFRL1* lead to serious cardiovascular and skeletal issues in humans, the researchers wondered if this gene might protect giraffes from high blood pressure. The answer was an emphatic yes. When the mutated gene was introduced into laboratory mice using sophisticated gene editing techniques, they suffered less cardiovascular and organ damage when their blood pressure was artificially raised. Genes involved in several pathways involving tissues affected by elevated blood pressure, such as blood, blood vessels, heart and kidneys, were also different to those of other ruminants. These unequivocal findings led Qiu and colleagues to speculate that *FGFRL1* might be a promising target for the prevention and treatment of human hypertension. Significantly, the mutated gene also affected the skeletal system and likely functions to maintain the giraffe's normal bone density despite its accelerated growth rate.

The unique effects of the giraffe *FGFRL1* gene offer intriguing insights into evolutionary mechanisms, notably the concept of evolutionary pleiotropy. Pleiotropy refers to genes that fulfil numerous functions, exemplified by *FGFRL1*'s involvement in both the cardiovascular system and bone structure. Alterations in such genes could underpin the sudden appearance of multiple phenotypic adaptations and help explain how rapid evolutionary change, including what Eldredge and Gould called 'punctuated equilibrium', occurs.[16] However, this phenomenon is probably not widespread, as mutations in multifunctional genes are more prone to be harmful and eliminated from a species' gene pool.

The giraffe's genome is also characterised by positive selection for genes contributing to eye development and vision. In contrast, over 50 genes related to olfaction have been lost, indicating a marked degeneration of smell. These findings suggest a 'sensory trade-off' whereby the giraffe has gained improved vision at the expense of its sense of smell. This observation makes sense, since the animals rely on accurate horizon-scanning vision from their tall vantage point, whereas olfaction may be less relevant 5 metres above ground. In other words, natural selection acts against resource-demanding structures that carry only a slight survival advantage.

The extreme anatomy of giraffes increases their vulnerability while asleep by increasing the time required to stand upright. Given the need for vigilance and the long periods spent feeding, giraffe sleep patterns are unusual, with sleep duration among the shortest in the mammalian world at around four hours a day. They sleep in cycles, as short as half an hour, standing or lying, day or night. It was no surprise, therefore, when Qiu's team found evidence of positive selection for genes known to control circadian rhythm and sleep arousal, such as *PER1* and *HCRT*. Although not proven, these genetic changes likely underpin the giraffe's short and fragmented sleep patterns.

An earlier study by Douglas Cavener and colleagues also used comparative genomics and found approximately 70 genes with unique adaptations in the giraffe.[17] Intriguingly, four of these involved homeobox genes that control development, and each possessed a different mutation predicted to alter protein function. Homeobox genes generally encode proteins that specify the characteristics of 'position', ensuring that the right structures develop in the correct places, including the spine and legs. All the genes described in these studies naturally occur in all mammals, including humans, but as Cavener puts it, what made giraffes unique 'was just to tinker with them a bit and alter them in subtle ways.'[18]

The advent of high-grade genome sequencing has revolutionised the field of evolutionary biology. It equips researchers with a powerful tool to explore the evolutionary changes in organisms and identify conserved genes and those that give each animal its unique traits. The two highlighted studies in giraffes are a testament to the value of comparative genomics, showcasing its potential to unveil the genetic basis for many more species' adaptations. This technique opens up a world of exciting possibilities for future research, promising to deepen our understanding of the evolution of life on Earth.

The species problem

Since Linnaeus's first description in 1758, most biologists believed that all giraffes belonged to a single species, *Giraffa camelopardalis*. This species, and its smaller relative the okapi, are the only two living genera in the family Giraffidae, one of the subgroups of even-toed ungulates (Artiodactyla). However, recent studies have concluded that there are, in fact, four distinct species – the northern giraffe (Plate 20), the southern giraffe, the Masai giraffe and the reticulated giraffe.[19] But not all scientists agree – some suggest the existence of as many as nine species – a controversy that reflects the complexity of defining a species.

The 'species problem', as it is prosaically called, has taxed scientists for over 150 years. Indeed, Darwin (1859) even recognised the difficulty when he wrote in *The Origin of Species*, 'No one definition has satisfied all naturalists; yet every naturalist knows vaguely what he means when he speaks of a species.' Today, at least two dozen definitions have been proposed, although most are variations on one of two themes: the 'biological species concept' and the 'phylogenetic species concept'.

The biological species concept, introduced by evolutionary biologist Ernst Mayr in 1942, characterises a species as a population that is reproductively isolated and unable to breed with other groups.[20] In other words, no genes are exchanged between species. If they do mate, they are unable to produce viable or fertile offspring. Distinctive geographical forms of the 'same' species are usually lumped as one, assuming they would interbreed if given the opportunity. One problem, of course, is knowing whether such isolated populations can interbreed.[21]

In contrast, the phylogenetic species concept defines a species as a group with a shared and unique evolutionary history, distinguishable from other such groups through various characteristics, phenotypic or genotypic. It is less restrictive than the biological species concept, as breeding between members of different species is not a problem. Put another way, a species defined according to the biological species concept only exists when the lineage separation is complete. In contrast, species fulfilling the phylogenetic species concept are 'evident' as soon as the evolutionary paths begin to diverge. Of course, a significant limitation of the phylogenetic species concept is the lack of emphasis on reproductive isolation, rendering the dividing line between species highly subjective. For example, species may appear identical morphologically but differ in their DNA sequences.[22] Consequently, DNA studies recognise many more taxa than the biological species concept. However, while some genetic analyses indicate that thousands of additional mammals may be waiting to be discovered, no consensus exists on how much variation is required to separate individual species. Indeed, in the most extreme interpretation, a single DNA base change represents a uniquely derived trait, meaning that every individual is potentially a different species.

The findings have huge implications for conservation. Giraffes have suffered a severe decline in numbers across Africa over the last four decades because of habitat loss, with only 117,000 remaining. As a single species, they are currently listed as vulnerable on the IUCN Red List of endangered species. However, if formally recognised as four separate species, then two, the Masai giraffe and the reticulated giraffe, would be listed as endangered. In essence, the reclassification of giraffes as multiple species offers a more nuanced and effective framework for conservation, ultimately enhancing the prospects for the long-term survival of these iconic animals in the wild.

Then, in 2024, the story of the giraffe's speciation became even more complicated.[23] A team led by Rasmus Heller studied whole-genome sequencing data from 90 wild giraffes and concluded that there had been high levels of gene flow, with major hybridisation events in the past. For example, the reticulated giraffe, which occurs in northern Kenya and Ethiopia, is a hybrid species with almost equal contributions from the northern lineage and an ancestral lineage related to the southern and Masai giraffes. Despite strong differentiation, the team also found evidence of further ancient gene flow among other branches of the giraffe tree. Overall, the results support the emerging consensus that many evolutionary lineages lack a clear-cut branching tree structure that satisfactorily describes their taxonomic relationships. Instead, patterns of gene-wide diversity can better be represented as convoluted, complex networks that are often difficult to interpret.

Heller postulates that unsuitable habitats periodically isolated historical populations. In effect, climate fluctuations and the associated changes in rainfall and vegetation likely drove a series of mixing–isolation–mixing events. Since giraffes are highly dependent on open savanna and acacia trees for browsing, repeated interglacial forest expansions during the Late Pleistocene would have fragmented populations, leading to increased differentiation due to genetic drift and limited gene flow.

The widespread mixing between giraffe groups throws a spanner into how species are defined. Interbreeding between giraffe species is well documented in zoos, and there have been anecdotal reports of natural hybrids in the wild. Traditionally, scientists thought extensive gene sharing only happened within a single species. However, Heller's findings suggest that the amount of genetic difference might not be as clear-cut a way to identify new species as was once thought.

The question of what constitutes a species extends beyond giraffes to the okapi, the only other extant member of the Giraffidae. Like the giraffe, this central African species has a rich evolutionary history and a genetic structure shaped by past ecological conditions.[24] Genetic analyses hint at repeated climatic cycles that led to multiple Plio-Pleistocene forest refugia with periods of limited gene flow. In addition, restricted historical gene flow occurred across the Congo River. Today, the okapi is not only a genetically distinct entity but also a highly genetically diverse one, a rare combination for a threatened species. Indeed, the clear genetic differentiation of the extant population southwest of the river raises the question: should it be treated as a separate conservation population? Could there even be two species of okapi?

CHAPTER 14

The Horse's Story

A BUSHY PHYLOGENY

Sit down before a fact as a little child, be prepared to give up every preconceived idea.
Thomas Henry Huxley (1825–1895)

The odd-toed ungulates or perissodactyls are a clade that encompasses three families: horses (Equidae), tapirs (Tapiridae) and rhinoceroses (Rhinocerotidae), along with a few strange extinct relatives (see Figure 9.1).[1] Their roots trace back to a time of sudden global warming called the Palaeocene–Eocene Thermal Maximum (PETM), around 55.8 million years ago. During this period, temperatures soared by 6 degrees Celsius, and carbon dioxide levels mirrored those seen today. Paradoxically, the PETM was a cauldron of frenzied invention and innovation that spawned two other modern groups – artiodactyls and primates. The nearly simultaneous appearance of all three lineages across North America, Europe and Asia has made it difficult to pinpoint their origins and dispersal routes. While perissodactyls were traditionally believed to have arisen in North America or Asia, some studies propose that they may have evolved on the Indian plate as it gradually moved northwards, subducting beneath the Eurasian plate along its edge.[2] However, the latest fossil discoveries favour an Asian origin, while the earliest horses, or equids, likely originated in Europe before dispersing to North America via the Greenland land bridge.[3]

Perissodactyls, in contrast to artiodactyls, have a unique feature which sets them apart: the primary axis of their feet passes through the third toe, while other toes are diminished to varying extents. For example, tapirs, adapted to walking on soft ground, have four toes on their front feet and three on their back feet. Modern rhinoceroses have three toes on both forelimbs and hindlimbs, while modern equines have a solitary toe on each foot, surrounded by two vestigial splints as remnants of side toes, largely encased by a hoof. In contrast, the hooves of rhinoceroses and tapirs cover only the front edge of the toes, leaving the underside soft.

Another intriguing characteristic of perissodactyls is their digestive strategy. Unlike artiodactyls, which are foregut fermenters and ruminants, perissodactyls solved the problem of plant digestion differently. They utilise an enlarged caecum – a specialised segment of the large intestine housing many plant-eating bacteria. This digestive strategy allows hindgut fermenters to efficiently process food, even of low quality, and attain impressive sizes. However, the arrangement comes at a cost. Dietary cellulose is not broken down until it reaches the fermentation chamber, and the vital volatile fatty acids must be absorbed in the colon rather than in the small intestine. Consequently, hindgut fermenters have a shorter passage time than ruminants and are less efficient in cellulose digestion. They consume large quantities of food to compensate, compelling some species to cover vast distances in search of nourishment, sometimes even resorting to eating their own dung to maximise digestive efficiency.

Equids, of which *Equus* is the only surviving genus, were the first to evolve, and their eventual domestication reshaped the trajectory of human history. Given their durability and traction, donkeys became indispensable for labour and trade and revolutionised overland routes between Africa and Asia. Yet it was the domestication of the horse around 5,500 years ago that left the most indelible mark. Their remarkable speed and strength bridged vast distances, facilitating the exchange of ideas and propelling advancements in science, art and religion. They also revolutionised how our genes, diseases and languages crisscrossed the planet and effectively globalised the world for the first time. Quicker communications enabled vast empires to be stabilised, such as the Chinese Tang dynasty and the Great Mongolian Empire established by Genghis Khan. Moreover, horses transformed warfare, fuelling Europe's colonial expansions and the downfall of states and empires.[4] Thus, the story of the horse's 55-million-year odyssey holds profound significance and, like a gripping thriller, reveals layers of complexity beyond initial assumptions.

Copiously branching bushes

One of Stephen Jay Gould's gripes was the notion that species evolve from basal simplicity to higher complexity along a linear path or ladder. In characteristic style, he expressed his views in an essay entitled 'Life's little joke', focusing on the evolution of the horse.[5] Gould argued that 'evolutionary genealogies are copiously branching bushes – and the history of horses is more lush and labyrinthine than most.' I vividly recall memorising the supposed unbroken continuity from the four-toed *Eohippus*, via three-toed species, to the one-toed *Equus* for a high-school biology examination. Today, what

Gould called the equine 'iconography of an expectation' has been replaced by an intricately branching nexus.[6]

The exceptionally rich fossil record of the Equidae has provided remarkable insights into their macroevolutionary changes.[7] By 55 million years ago, the first members of the clade had reached North America, inhabiting the forests that covered the continent. The earliest known species was the dog-sized *Hyracotherium*, more commonly called *Eohippus* ('dawn horse'). Their legs ended in padded feet, with four functional hooves on each forefoot and three on each hindfoot. Most family members remained small forest browsers with low-crowned teeth for eating leaves. However, changing climate conditions allowed grasslands to expand, and about 20 million years ago a minimum of 19 new species appeared quickly, only for most to go extinct. As stressed, there was no steady, gradual progression but numerous evolutionary experiments and dead ends. Many species remained in the dwindling forests, content with their lineage's long-standing diet of leaves. During this time of flux, high-crowned teeth for grazing appear suddenly in the fossil record, dentition that evolved to cope with cellulose-laden grasses and contaminating grit. Some species, but not all, became larger and evolved the familiar hoofed foot (monodactyly) and diet we associate with the genus *Equus* today. In addition to size, muzzles became longer, legs lengthened, and their brains increased in volume and complexity. Although small and large species lived side by side for millennia, only the large species survived to give rise to the common horses, zebras and asses we know today.

But why did certain species evolve a single-toed-foot 10 million years ago? Monodactyly is often perceived as an 'evolutionary' improvement, but such a view doesn't account for the persistence of horses with three-toed feet for millions of years. In fact, tridactyl horses underwent the most successful radiation during the Late Miocene. According to evolutionary biologists Christine Janis and Raynor Bernor, the single-toed foot enabled equines to do something different in a world where tridactyl odd-toed ungulates dominated: the adoption of a novel foraging strategy.[8] Monodactyl horses are faster and more efficient and can cover longer daily distances to reach suitable food sources. Such species often trot, a form of locomotion found by the US cavalry and endurance equestrians to be the most efficient. Indeed, the trot gait enables greater storage of elastic energy in the flexor tendons than when walking, and this enhanced proficiency drove the evolution of monodactyly. Essentially, a one-toed foot produces a 'spring foot' that maximises energy efficiency compared to a three-toed foot.

Then, around 4 million years ago in North America, the equine population of larger-sized species split into two lineages: the stenonines, which would give rise to the asses and zebras, and the caballines, or true horses (including the domestic horses of today). Both lineages spread to Asia via the Bering Land Bridge, a vast, flat, grassy landmass formed at times of low sea levels. The first to cross were the stenonines, around 2.6 million years ago, followed by a caballine migration around 0.9–1 million years ago.[9]

The caballines also crossed into South America via the Panama isthmus, but by 8,000–12,000 years ago all equids in the Americas were extinct. They disappeared during the Quaternary Extinction Event, along with most other megafauna, due to climate change and overexploitation by newly arrived humans.[10] However, at the end of the fifteenth century, the Spanish reintroduced horses to the New World during their brutal conquests. Inevitably, some escaped, and they soon returned to their wild state, proliferating on the continent's plains. Some of the escapees were integrated into indigenous cultures across western North America long before the arrival of Europeans in that region.[11] By the 1800s, vast herds of wild horses roamed freely across North America, the land of their ancestry.

A recent study led by Beth Shapiro of the University of California has brought the equid story full circle.[12] By sequencing many mitochondrial and nuclear genomes from ancient horses from both North America and Asia, the team constructed a phylogram to reveal how all the samples were related. With a location and an approximate date, they could track the movements of the different lineages. The results were a surprise. After the split from stenonines, the caballine horses moved back and forth across the Bering Land Bridge each time it formed, interbreeding multiple times until around 25,000 years ago. In other words, rather than separating into species once they reached Asia, they continued to exchange genes with their North American relatives for hundreds of thousands of years. In effect, the North American and Asian horses were the same species. These novel findings will likely intensify the heated debate surrounding the management of the sizeable population of wild horses or mustangs remaining in the United States. While some view them as an invasive species, others now argue that they are an integral part of the native fauna of North America and should be reclassified as a reintroduced species.

The caballine horses were successful and rapidly spread across Asia to replace most stenonines except in Africa. While the original migrants were large, about the size of a draught or shire horse, between 25,000 and 15,000 years ago they became smaller, to resemble Icelandic horses. During the Last Glacial Maximum, the horse population declined, only to rebound, possibly due to a convergence of climatic shifts and human impact. The late Neolithic witnessed a cooling trend across much of northern Eurasia that could have boosted the population size and expanded the viable habitat of the cold-adapted equids. Such favourable climatic changes may also have been enhanced by the behaviour of pastoralists migrating into new areas of Eurasia, who likely engaged in deforestation and the creation of grasslands across a vast region.

In 1879, the Russian geographer and explorer Nikolai Przewalski embarked on a journey across central Asia, where he discovered a remarkable species of horse that now bears his name. Renowned for its gregarious nature, this rare equine has a sturdy physique, short legs and a sandy-red coat. Despite these distinctive traits, the taxonomic classification of the Przewalski's horse remains a subject of debate, with no consensus reached on whether it is the last surviving wild horse, a subspecies, or a descendant of horses domesticated by the ancient Botai culture of central Asia some 5,500 years ago that later reverted to a feral state.[13] While molecular evidence supports early domestication, recent archaeological discoveries suggest the Asian horse might be the last remaining wild equine species.[14] This ongoing uncertainty underscores the importance of its conservation. Przewalski's horse, once considered extinct in the wild, has been reintroduced through meticulous conservation efforts, particularly in Mongolia and other regions, from a founding population of fewer than a dozen individuals. However, modern domesticated horses do not trace their lineage back to Przewalski's horse, leaving the quest to pinpoint the time and place of their initial domestication ongoing.[15]

What is certain, however, is that the genetic diversity of the modern horse has significantly decreased during the last 250 years. The introduction of closed stud breeding and the excessive influence of specific bloodlines have eroded their ancestral diversity, resulting in reduced fertility, a decline in overall fitness, and the emergence of disadvantageous traits.[16]

Zebra stripes

In Asia, the stenonines underwent divergence leading to the ass and zebra lineages around 1.85 million years ago. The asses split into Asian and African populations approximately 250,000 years later. Presently, there are two Asian species: the Tibetan kiang, found across the expansive Tibetan plateau, and the critically endangered onager, which inhabits the border region between Mongolia and China. In Africa, there is only one species, the critically endangered African wild ass, with a remaining wild population estimated at 600 individuals.

The donkey traces its roots back to the Somali and Nubian subspecies of African wild ass and was probably first domesticated by the pastoralists of Nubia.[17] Evidence from archaeological sites near Cairo, Egypt, dating back to around 4000 BCE, supports this view. The Nubians preferred using the donkey as their chief pack animals, as ruminants like oxen have the disadvantage of having to chew their cud.

Zebras underwent parallel radiation in Africa approximately 1.3 million years ago, resulting in the emergence of the three extant species: the plains zebra (Plate 21), Grévy's zebra and mountain zebra. The three species feed primarily on low-quality vegetation and are well adapted to gazing. Their social dynamics differ: plains and mountain zebras organise into stable harems led by a stallion, accompanied by several mares and their foals. In contrast, Grévy's zebras tend to live alone or in loosely connected herds. Among harem-structured species, mating occurs solely between mares and their harem stallion, whereas male Grévy's zebras establish territories to attract females, displaying promiscuous behaviour.

Strikingly, multiple instances of hybridisation occurred throughout the equid lineage despite chromosomal numbers varying from 16 to 31 pairs. These observations challenge the concept of chromosome rearrangements driving reproductive isolation and the origin of some species. Equids, therefore, offer a valuable model for investigating the complex interplay between chromosome structure, gene flow and the process of speciation.[18]

The adaptive significance of the zebra's striking stripes is one of the oldest problems in evolutionary biology, taxing scientists since Darwin and Wallace debated the issue. While many theories have been proposed – social cohesion, thermoregulation, individual identification – the idea that stripes act to bewilder or dazzle remains the most plausible. Supporting the concept, neuroscientists Martin How and Johannes Zanker have shown that stripes function as a form of motion camouflage.[19] When zebras move, their coat patterns create a highly perplexing array of signals so that the observer's visual system becomes flooded with erroneous motion signals corresponding to two well-known visual illusions: the wagon-wheel effect, resulting in perceived motion in the opposite direction to actual movement, and the barber-pole illusion, where stripes seem to move vertically instead of horizontally. These illusions confound mammalian predators during hunts, especially when many zebras are observed moving together. However, Tim Caro believes these optical illusions evolved to thwart the attacks of bloodsucking flies – African horseflies and tsetse flies – which carry diseases fatal for equids. By filming zebras and horses dressed in zebra-striped coats, Caro showed that flies would near the animals and then, for some reason, fail to land.[20]

But how did the zebra's stripes evolve?

Animal skin pigmentation patterns represent one of nature's most amazing observable phenomena and have intrigued scientists from many disciplines, including the renowned British mathematician and polymath Alan Turing. Turing gained fame for his groundbreaking contributions to computing and codebreaking during World War II, and for his eponymous test, which is used to determine whether a machine demonstrates human intelligence. Less recognised, however, is his work on chemical diffusion gradients.[21] Turing realised that when two diffusible substances with different diffusion coefficients interact, they can generate a spatially periodic pattern even when the starting condition is homogeneous and uniform. Turing's publication was highly mathematical and abstract, and it was only later that his seemingly arcane ideas gained traction to explain how zebra stripes, jaguar spots and human fingerprints can arise naturally.[22]

In terms of the biological world, Turing modelling envisages that when short-range activators and long-range inhibitors, known as morphogens, are combined, a patterned distribution of cells should result. Although direct experiments on zebras are lacking, observations of how their stripes vary at different anatomical positions provide indirect evidence of a Turing effect. For instance, the observed chevron shapes at the zebra's shoulder, where two stripes intersect orthogonally, agree with the model's predictions. The only problem is that the Turing effect produces shapes with indistinct edges rather than the sharp boundaries observed in zebra and leopard skins. However, two physicists, Benjamin Alesso and Ankur Gupta, added a new dimension to the mix – diffusiophoresis, or the movement of particles in response to concentration gradients.[23] Using their upgraded Turing model, the two scientists showed that specialised pigment cells would respond diffusiophoretically to physiological conditions and create a robust patterning of cells with fine detail that matches nature's clarity.

Turing effects also account for the arches, loops and whorls of human and koala fingerprints.[24] In the womb, fingerprint-defining ridges expand outward in waves starting from three points on each fingertip. The raised epithelium produces a striped pattern under the influence of three interacting proteins: Wnt, EDAR and BMP. As required for a Turing system, the protein messengers provide opposing signals for skin cells. Wnt informs cells to multiply, forming cutaneous ridges and releasing EDAR, further enhancing Wnt activity. In contrast, BMP blocks these actions. The combination of all three protein messengers results in individually patterned fingerprints.

The researchers turned to mice to prove these effects, as their toes have striped skin ridges like human fingerprints. Increasing EDAR produced thicker, more spaced-out ridges, while reducing the protein resulted in spots rather than ridges. The scientists proposed that the switch from spots to stripes was to be expected if governed by a Turing reaction-diffusion system. Mouse digits, however, are too tiny for the elaborate shapes of human fingerprints. Instead, the team used computer models to simulate the effects of spreading from the three known ridge initiation sites on the fingertip.

By altering the relative timing, location and angle of the starting points, the fingerprint's characteristic arches, loops and whorls were produced, which, together with the inherent randomness of a Turing system, delivers an individual uniqueness to each fingerprint.

Evolutionary stasis

Tapirs and rhinos belong to the suborder Ceratomorpha and are more closely related to each other than they are to horses. Tapirs fascinate evolutionary biologists because of their size and unusual geographical distribution. They are bulky creatures, with the largest of the four extant species, the Malayan tapir, reaching 2.5 metres in length and weighing up to 320 kilograms. All four taxa possess a muscular proboscis, an extension of the upper lip and nose that forms a mobile, tactile soft-tissue structure that allows the animal to grab leaves, seeds and fruit that would otherwise be out of reach. The evolution of the proboscis, which lacks any internal skeleton, gives the tapir's skull a unique shape amongst the perissodactyls, features that palaeontologists can easily identify. Tapirs play a vital ecological role as 'gardeners' or 'architects' of the forest since they disperse seeds across vast distances while foraging for food. Such dispersal is crucial for the regeneration and diversity of their forest ecosystem.

The geographical distribution of tapirs, especially the Malayan tapir, native to Southeast Asia (Myanmar, Malaysia, Thailand and Sumatra) has long been a puzzle. Why is this species such an outlier, with the remaining three taxa found in the lush forests of Central and South America? This anomaly has sparked speculation since the nineteenth century, with Alfred Wallace, the 'father of biogeography', offering his insights in his book *Darwinism* (1889).[25] Wallace's conclusion was that the tapir must have once had a contiguous distribution, with their kin perishing in the intervening regions. This view is supported by the fossil record, which has yielded numerous discoveries from Europe, Asia and the Americas.

One of the candidates for the earliest tapir is *Heptodon* from northern British Columbia, a species about half the size of modern tapirs that lacked a proboscis. Instead, it probably possessed an elongated fleshy upper lip. The northernmost fossils are from Eocene deposits in Ellesmere Island in the Canadian Arctic, an area considered 'en route' for land animals dispersing across Holarctic continents during the Palaeogene when intercontinental links were restricted to high latitudes. Indeed, Asia has been hypothesised as a likely source of many faunas that utilised this route to reach the mid-latitudes of North America during the Eocene. However, the palaeontologist Jaelyn Eberle holds that the opposite applies to tapirs, and that their lineage represents a rare instance of counterflow, wherein a North American taxon successfully invaded Asia.[26]

After the formation 3 million years ago of a land bridge connecting North and South America, the Great American Biotic Interchange commenced, marking the entry of tapirs and numerous other species into Central America and the southern hemisphere. Throughout the late Quaternary period, speciation occurred as populations ventured into the cloud forests of the Andean foothills or became isolated due to fragmentation of lowland forests caused by a cooling climate.[27] Additionally, the Amazon acted as a north–south barrier to gene flow. Among the various species that evolved, only three survived the Pleistocene megafauna extinction: Baird's tapir in Central America and the lowland and mountain tapirs in South America.

While many families of perissodactyls achieved very high levels of diversity, there have never been more than a few species of tapirs. They are also morphologically conservative, with teeth and skeletons resembling those of their ancestors. Their lack of diversification is known as evolutionary stasis, a feature frequently associated with the theory of punctuated equilibrium, in which most evolutionary change is concentrated during the phylogenetic branching of lineages during rapid bursts of speciation. Much longer episodes of relative morphological invariance, or stasis, follow such speciation events. Over the past 35 million years, tapirs have remained mainly unaltered, likely due to their consistent dietary requirements and reliance on ancient, densely forested habitats.[28] Indeed, their unwavering habitat preference has led scientists to use tapir fossil distributions to infer historical rainforest conditions and changes over geological time.

The rhinoceros family was once a diverse clade, but today only five species remain, all highly endangered and global priorities for conservation. The rhinos diverged from tapirs approximately 55 million years ago in North America or Eurasia.[29] Afterwards, they radiated into over 100 species across Africa, Eurasia, and North and Central America, including some of the largest terrestrial mammals that ever lived.[30] The earliest species, the hornless *Hyrachyus*, was widespread and resembled a tapir or small horse rather than a modern rhino. Nearly all the species went extinct before the Pleistocene, although some lingered on, like the woolly rhinoceros, until the Late Pleistocene.

Given that there are only five extant species – African black, African white, Indian, Javan and Sumatran – it has been challenging to elucidate their evolutionary history. However, a recent international team of scientists compared the genomes of all extant and three extinct species. They showed that the main divergence did not result in one- and two-horned phenotypes, but rather occurred geographically, producing African and Asian lineages.

The Afro-Asian split occurred around 16 million years ago after the formation of a land bridge linking the two land masses.[31]

Another important finding was that the living species show lower genetic diversity than the extinct species. Such a finding is not unexpected, given the huge population declines that modern species have undergone due to human activity. Surprisingly, however, when all the data from extant and extinct species was lumped and compared to other mammalian families, rhinos showed the lowest genetic diversity of any group. Only cats (family Felidae) approach similarly low levels, probably due to the natural small size of predator populations, coupled with human pressures. These findings are important, as they indicate that rhinos have historically had low genetic diversity, and that this may be their natural state. Therefore, the rhino's small gene pool is partially a consequence of their biology, which may have contributed to their evolutionary stasis. Furthermore, given the species' size, ferocity and lethal defensive horns, modern and extinct rhinoceroses lacked predators, and their hardships and selection pressures remained relatively constant.

With the exception of equines, the perissodactyls have exhibited some remarkable evolutionary stability. As a result, these 'beasts of the slow lane' are invaluable for the unique insights they can provide for understanding evolutionary stasis. Surprisingly, we know very little about long-lived, phenotypically conservative, and species-poor lineages, colloquially dubbed 'living fossils'.[32]

CHAPTER 15

The Bear's Story

INTER-SPECIES GENE FLOW

The grossest blunder in sexual preference which we can conceive of an animal making, would be to mate with a species different from its own.
Ronald Aylmer Fisher (1890–1965)

Backed by a glacial breeze, I scrambled up the low ridge that marked the edge of one of Svalbard's remote northern inlets. As I crouched down and peered cautiously over the crest, clutching my trusted Lee-Enfield, I was met by the sight of an ice-choked bay with a vast Arctic wilderness beyond. On the ice, over 150 metres out, was an adult polar bear, already upright on its hind feet, with its front legs and paws hanging down and its nose lifted skywards as it sniffed my presence. Emboldened by the knowledge that it had a fresh kill, I motioned for my six clients to join me, and we settled down to observe our number one target species. Our majestic quarry, the undisputed 'King of the Arctic', had already returned to its bloodied meal, its face and chest stained a vivid crimson.

We marvelled at its features, which were so distinct from those of brown bears. Its thick, cream-coloured fur blanketed its entire body, exposing only its high-set eyes and prominent black nose. Less bulky than its southern relatives, it had a longer neck, small rounded ears, and a short, stubby tail. Its snout, with a distinct curve reminiscent of a 'Roman' nose, added to its striking appearance. Its large paws, heavily furred for traction and warmth, completed the picture. Once the bear had eaten its fill, it paused to survey the surroundings, casting a final glance in our direction. After several more sniffs of the crisp air, it strolled across the ice, disappearing into the encroaching mist.

Once back in the safety of our old fishing trawler, we huddled around the heater with mugs of coffee, still buzzing with euphoria and exhilaration from our experience. Fulfilling my role as guide, I regaled them with details of the bear's many adaptations that allow survival in one of the coldest environments on Earth. I explained how they are incredibly well insulated, with a thick layer of fat below two layers of fur, and how their pelt provides camouflage amid the snow and ice. Surprisingly, the bear's fur lacks white pigment, and the hollow hairs reflect the entire spectrum of light. Moreover, the underlying black skin absorbs what little warmth the sun offers, although the mystery of the black tongue remains unexplained. Bumps or papillae cover the black footpads to grip the ice and prevent slipping. Polar bears have an acute sense of smell, as our encounter vividly confirmed, and their noses have such a large surface area that they can detect prey over a kilometre away. After mating, male bears lead a solitary life, while females prepare dens within deep snow drifts and give birth, typically to twin cubs, during the harshest winter months.

However, I could not discuss the polar bear's complex evolutionary history, for that storyline had yet to be written. It would be another two decades before rapid genome sequencing and powerful computational analyses added this fascinating chapter to their story.

Carnivora

Polar bears are laurasiatherians (see Figure 9.1) and belong to the order Carnivora, the fifth-largest order among mammals. The 296 existing species can be traced back to a common ancestor that lived in North America around 60 million years ago, some 6 million years after the K–Pg extinction event. This early species, which would have resembled a small weasel or marten that likely foraged in trees or on the forest floor at night, marked the start of one of the mammals' most fascinating and prolific evolutionary journeys.

The term Carnivora, meaning meat-eater, can be misleading, as not all the clades are carnivores. Consider the giant panda, classified as a bear but living a herbivorous lifestyle, with bamboo making up about 99 per cent of its diet. The success of the carnivorans is reflected in their wealth of habitats across all continents and seas, as well as their vast range of diets, hunting methods and social organisations.

Carnivora is made up of two suborders: Caniformia and Feliformia. Caniformia encompasses nine families: bears and pandas, dogs, raccoons, the red panda, skunks, weasels, the walrus, eared seals and true seals. The Feliformia are represented by seven families: cats, mongooses, hyenas, Malagasy carnivorans, African linsangs, civets and genets, and the African civet. While the crown carnivorans first appeared during the Early Eocene Climatic Optimum, the Caniformia diverged in the Eocene and the Feliformia more recently during the Oligocene.[1]

A bamboo specialist

The lineage that gave rise to the giant panda split from other bear species around 19 million years ago, making it the earliest ursine divergence.[2] Fossil discoveries show that early pandas evolved in Europe, and that by 11.6 million years ago they had already developed adaptations for a herbivorous diet. Their teeth were designed for processing tough, fibrous plants that require shearing rather than crushing, implying that the panda's specialist plant-chewing behaviour persisted throughout their evolution. Early pandas also had proportionally larger teeth than their modern descendants, which provided good protection against predators. As a result, scientists believe their dietary switch was driven by competition from carnivorous species.[3]

Over time, climate change forced pandas out of Europe. Approximately 6 million years ago, a salinity crisis occurred in the Mediterranean Sea, causing it to dry up entirely and become a massive basin resembling the Grand Canyon in parts. Europe's swampy forests transformed into warmer, arid lands, rendering the areas inhospitable for pandas. However, during the Early Pliocene, rising sea levels from glacial melting breached the Straits of Gibraltar, refilling the Mediterranean basin. But by then it was too late: the pandas had already migrated eastwards on their journey towards China.

The panda's ability to thrive on a diet of mainly bamboo necessitated a raft of adaptations, one of which was an 'extra thumb.' Unlike most mammals, with five digits on each hand, the panda has a sixth formed by an enlarged wrist bone called the radial sesamoid. The extra digit helps the animals grip and manipulate bamboo, although it is not as dextrous as a primate's opposable thumb. Recently, palaeontologists found this unique feature in a panda fossil from the Late Miocene in China, a significant find as it shows that the panda's specialised bamboo diet started as early as 7 million years ago.[4] Despite its usefulness in bamboo handling, the pseudo-thumb has yet to evolve further, probably because a panda must balance the demands of gripping bamboo and supporting its weight while walking.

Interestingly, while the giant panda is closely related to bears, the red panda is more closely related to ferrets. However, the two species share a common trait, a reliance on bamboo for sustenance that led to the independent development of an extra digit. This example of convergence in lifestyle and anatomy of two species that diverged over 40 million years ago from a common meat-eating ancestor continues to intrigue evolutionary biologists.

Research by the Chinese Academy of Sciences has recently shed light on this mystery.[5] Through decoding the red panda's genome and comparing it with the giant panda's, the researchers identified two genes (*PCNT* and *DYNCZ2H1*) involved in limb development that were similarly mutated in both species. In humans and mice, mutations in these genes cause abnormalities in bone and muscle, including the development of extra fingers and toes. Crucially, the mutations altered protein structure, indicating their potential contribution to developing the pandas' pseudo-thumbs. The Chinese team also found the umami taste receptor *TAS1R1* to be a non-functional pseudogene in both species, although this occurred from different 'knock-out' mutations. The team hypothesised that after the species changed diets, there was no longer a need to detect savoury, meat-like flavours, allowing subsequent gene mutations to persist unchecked. Overall, the giant and red pandas represent a remarkable case of genetic convergence, a scenario that sheds light on the complex relationship between evolution and environmental adaptation.

Surprisingly, giant pandas haven't evolved the ability to digest bamboo completely. Genes related to cellulose degradation are absent from their genomes, and bamboo digestion relies on intestinal microbes. However, the result is inefficient, compounded by the inheritance of a short, straight carnivore-like gastrointestinal tract that prevents the colonisation of the more efficient flora found in other herbivores.[6] To compensate, pandas cut back on energy expenditure and live life in the slow lane. They move less than other bears, and their energy-draining organs, such as the brain, liver and kidney, are relatively small compared to those of other placental mammals.

Phylogenetic networks

The spectacled bear, a reference to its light facial fur, which can look like eyeglasses in some individuals, is the only native bear in South America. It is the most arboreal and herbivorous of bears, sometimes building leafy platforms in trees, which it uses to sleep and feed. The species is characterised by its stocky build, short snout and distinctive markings, which vary from individual to individual. However, spectacled bears occupy a remarkable range of habitats on the slopes of the Andes, from the coastal desert of Peru to the cool páramo grasslands above the timberline, at around 4,200 metres.

An iconic medium-sized South American species, the spectacled bear belongs to the subfamily Tremarctinae or short-faced bears. They diverged after the pandas and before the Ursinae, a group of six extant species (polar bear, brown bear, American black bear, sun bear, sloth bear and Asiatic black bear). The short-faced bears, of which only the spectacled bear survives, evolved in North America around 7 million years ago and are named for their disproportionately short snout compared to modern bears. Around 5.5 million years ago, the lineage underwent a rapid diversification, as did the Ursinae, which coincided with the spread of grasslands and open habitats, alongside marked cooling and increased seasonality.

The early species were relatively small, but some became truly colossal. Among them was the giant short-faced bear, the largest land-living mammalian carnivore, almost twice the size of today's brown bear. Unlike modern bears, it was lightly built, with long legs and feet that enabled it to run fast. This huge species roamed across the expanse of North America, especially in the west, from Alaska to the Yukon, up until 12,000 years ago. The earliest people to reach North America, perhaps as early as 15,000 years ago, must have encountered these intimidating animals.

The spectacled bear's ancestors also arose in North America as part of the marked radiation of bears, but their subsequent evolutionary history is so convoluted that scientists are only beginning to unravel its complexity.[7] Hybridisation, a process where two different species interbreed, played a significant role in its evolutionary history. It appears that during the Pleistocene, the ancestral spectacled bears hybridised with a species of Ursinae, probably the American black bear. The two populations overlapped, and hybridisation was possible because reproductive barriers had presumably not yet evolved. In support of this hypothesis, offspring between spectacled bears and Asiatic black bears have occurred in zoos.

Another significant hybridisation occurred between the early spectacled bears and a species of now-extinct short-faced bear. This second encounter likely occurred in Central America as both populations migrated into South America. Such discoveries have profound implications for our understanding of mammalian speciation. Until now, most reports on mammals have focused on the post-speciation exchange of genes, or introgression, rather than the emergence of new species through hybridisation. But spectacled bears are not the only member of their family to have evolved through this genetic process.

The omnivorous Asiatic black bear, a medium-sized, mainly arboreal species, has a wide range that extends from Iran to Japan. In a seminal study led by Tiantian Zou from Yunnan University, China, this species was shown to have evolved from a historical hybridisation between an ancestor of the northern clade of Ursinae (polar bear/brown bear/American black bear) and an ancestor of the southern clade (sun bear/sloth bear) (Figure 15.1).[8] Speciation occurred soon after the two parental lineages diverged in Eurasia due to climate-driven population expansion and dispersal. The northern clade are large bears, while the southern clade are much smaller species. Complex genetic analyses showed that gene mixing from the two parental lineages led to the Asiatic black bear's intermediate size. Furthermore, genomics indicated that matings were likely between males from the ancestral southern group and females from the ancestral northern group.

However, the initial hybrid population would have required reproductive isolation to permit independent evolution, as back-crossing would have

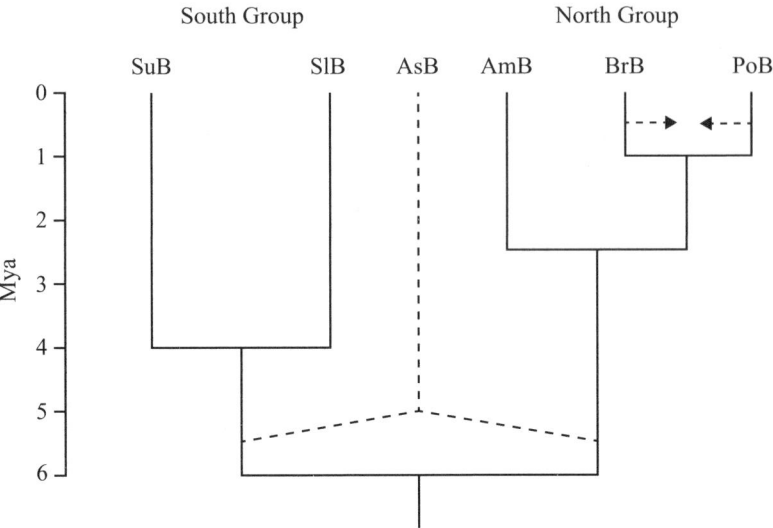

Figure 15.1 Hybrid origin of the Asiatic black bear resulting from a cross between the ancestors of the south and north groups 5.6 million years ago. SuB, sun bear; SlB, sloth bear; AsB, Asiatic bear; AmB, American bear; BrB, brown bear; PoB, polar bear. Mya, million years ago; dotted lines, genetic exchange. Modified from Zou *et al.* (2022).[8]

caused early speciation to collapse. Indeed, Darwin was so concerned that this scenario could negate speciation by natural selection that he devoted a whole chapter to the subject in *The Origin of Species*. Therefore, many biologists saw hybridisation as a natural barrier that maintained each species' integrity and, as such, was of little evolutionary significance. The renowned English polymath Ronald Fisher encapsulated this view when he wrote, 'the greatest blunder in sexual preference which we can conceive of an animal making, would be to mate with a species different from its own.'[9]

However, botanists subsequently demonstrated that hybridisation underpins over a third of plant speciation events. In most cases, both progenitor sets of chromosomes are amalgamated, resulting in a hybrid with double the chromosome complement: a condition known as allopolyploidy. This novel chromosome number makes it impossible for the hybrid's genetic material to pair up with either of its parental species, leading to reproductive isolation.

Hybrid speciation with the same chromosome number, termed homoploid hybrid speciation (HHS), is less frequent but has been documented in flowering plants and animals, such as fish, insects and birds.[10] In contrast, HHS is rare in mammals, with the only convincing examples being the speciation of bats and a fur seal.[11] However, all such examples present an intriguing paradox: gene flow between parental species is necessary for the hybrid to form, yet there must be some genetic isolation for the hybrid to maintain its distinctiveness. In other words, there must be a reduction of gene flow with their parental populations to prevent the formation of hybrid swarms, where the hybrid phenotype is continually eroded through parental back-crossing. If intrinsic reproductive barriers are weak, as may occur initially, other isolating mechanisms, such as morphological, behavioural or ecological differences, are assumed to be necessary.

Geneticists are now beginning to explore whether hybridisation itself can attenuate gene flow between hybrid and parental populations through the effects of sorting and recombining parental alleles. Indeed, recent simulation and empirical modelling predict that recombinational speciation can occur when parental incompatibilities are sorted, with a subset acting as barriers to gene flow against either parental species.[12] The genetic and evolutionary implications of such unions are complex and challenging for biologists to understand fully, and no responsible genes have yet been identified.

Studying Asiatic black bears provides us with valuable insights into the mechanisms of HHS. Their intermediate body size appears to have played a key role in promoting immediate isolation through sexual selection. Moreover, as noted earlier, the fixation of several genes that had diverged between the parental lineages likely reinforced the hybrid's reproductive isolation. Notably, this research identified potentially relevant genes for the first time associated with reproductive processes such as sperm development, egg production and fertilisation.

The extinct cave bear, named for its habitat preference, resembled a brown bear but was larger and herbivorous. Around 25,000 years ago, cave bears vanished, likely due to ice-age-driven food shortages and competition with early humans. However, cave bears and brown bears coexisted in Europe, and research led by Axel Barlow, a palaeontologist at the University of Potsdam in Germany, suggests they interbred.[13] Barlow showed that varying segments of cave bear DNA persist in the genomes of present-day brown bears. This finding is significant for several reasons. Firstly, it challenges our understanding of a 'species', highlighting that species aren't necessarily isolated entities but can interbreed. The study also raises intriguing questions about what it means to go extinct, since species may survive at the genetic level and influence evolution for tens of thousands of years after disappearing.

Introgressive hybridisation

Polar bears owe their origins to the brown bear, a species that also includes grizzly bears (Plate 22). Brown bears are a genetically diverse species ranging across Europe, Asia and North America. They live primarily in forested habitats and are generalised foragers, with plants making up nearly

90 per cent of their diet. However, in coastal Alaska, some populations also feed on spawning salmon that migrate upriver to lay their eggs.

Around 400,000 years ago, the polar bear lineage diverged from brown bears. In a mere 300,000 years, they had fully adapted to a marine diet and life in the Arctic, a remarkably short time for a large mammal.[14] This rapid evolution, occurring in just over 20,000 generations, led to the development of a suite of adaptations that allowed polar bears to thrive in their unique environment. These adaptations include maintaining their metabolism under low temperatures, occupying a hyper-carnivore niche, subsisting on mainly seal meat and blubber, using fat as a primary energy source, and developing white fur for camouflage. But how did such rapid and multiple adaptations come about?

Brown bears in the northernmost regions of their range would have faced intense selection pressures to adapt to colder and harsher environments. Any minor genetic advantages would have been fixed, with subsequent selection pressures shaping the genome further, facilitating increasingly northward expansions. Two distinct genetic mechanisms would have underpinned the bear's Arctic adaptations: selection acting on pre-existing variation (standing genetic variation), and newly arising mutations. The former, with its readily available alleles, is anticipated to drive evolutionary change more quickly, as the alleles are readily available. At the same time, with its random mutations, the latter takes longer to evolve and to become evenly distributed. Given the large population size of the ancestral brown bears and their interconnectivity and interbreeding, their standing genetic variation would have been significant, offering a broader array of genetic options for natural selection to operate upon. A recent investigation by Eline Lorenzen and colleagues from the Globe Institute at the University of Copenhagen found that both mechanisms were involved.[15] Furthermore, a more extensive follow-up study deduced that selection on existing variation was more widespread and prolonged than selection for *de novo* mutations, conclusions that help explain why the species evolved so rapidly.[16]

Polar bears have been under a stronger selection pressure than brown bears, especially concerning genes associated with adipose tissue, fatty acid metabolism, heart function and fur pigmentation. These findings are unsurprising, given their lipid-rich diet and need for camouflage against the ice. One gene, *APOB*, encodes a part of low-density lipoprotein (LDL), and functional mutations may explain how polar bears can tolerate lifelong elevated LDL levels that cause severe cardiovascular disease in humans. Moreover, polar bears have evolved alterations to genes controlling cutaneous melanin production, which likely explains the lack of pigmented fur.[17] This unique adaptation, along with body shape and subcutaneous fat layers, helps reduce heat loss. Furthermore, a recent study discovered that polar bears have modified their genes to produce more nitric oxide than other bear species.[18] Nitric oxide is a signalling molecule with diverse functions, including controlling whether cells utilise available nutrients for metabolic energy production or conversion into body heat. This adaptation is significant, as it allows polar bears to allocate more of their body's resources to generate heat and maintain their body temperature during extreme Arctic conditions.

Despite their striking differences in appearance and ecology, brown and polar bears are not reproductively isolated. Instances of hybrid offspring, with a blend of their two distinct phenotypes, referred to as grolars or pizzlies, have been observed both in captivity and in the wild. While the ranges of brown bears and polar bears overlap along the northern coasts of Arctic lands, polar bears are usually offshore hunting seals during their mating season, and are separated from their land-based cousins. However, as temperatures rise, it has become increasingly common to observe brown bears north of their traditional range, sometimes more than 500 kilometres from the mainland coast and well within the range of polar bears in the Canadian Arctic.

In 2006, a hunter shot an odd-looking polar bear in Canada's Northwest Territories, believing it to be a typical polar bear. However, the authorities were interested in the animal after noticing that although it possessed the thick, creamy-white fur typical of polar bears, it also had numerous features of grizzly bears – long claws, a humped back, a shallow face, and brown patches around its eyes, nose, back and feet. The need for accurate identification extends beyond academic curiosity. Had the bear been declared a grizzly, the hunter would potentially have faced a fine or even a jail sentence. However, DNA testing confirmed the bear's hybrid status, with a polar bear mother and grizzly bear father – the first documented case of interbreeding in the wild.[19]

Hybrids can act as 'go-betweens' that enable genes to transfer from one species to another and increase the genetic and phenotypic variation of one of the parental populations – for example, brown bears possessing cave and polar bear genes. Such genetic events, known as introgressive hybridisation or simply introgression, occur mainly through male hybrids, as the female sex chromosome contains a range of infertility factors. Intriguingly, the evolutionary outcome does not depend on the long-term survival of the hybrids. Most fertile first-generation hybrids will be too rare to breed with each other and far more likely to back-cross with one of the parents. Additionally, some species harbour 'fossilised' genes transferred from another species that no longer exists, such as the cave bear in the case of brown bears.

Introgressive hybridisation is not a recent phenomenon. An investigation of the brown bear population of the Admiralty, Baranof and Chichagof

Islands in Alaska showed that these bears possess approximately 6–8 per cent of polar bear DNA from interactions with female polar bears during the last glacial period, between 17,000 and 14,000 years ago. According to Beth Shapiro, the study's lead researcher, it is likely that a population of polar bears became trapped, and that male rather than female brown bears swam to the islands from Alaska and bred with the polar bears.[20]

Subsequent genomic studies hinted that a similar admixture may have occurred around 110,000 years ago. However, genetic shuffling or recombination reduces the amount of admixed DNA in each successive generation, making it challenging to prove ancient DNA mixing. Then, a team from the University of California, Santa Cruz, successfully extracted ancient DNA from the 125,000-year-old skull of a juvenile polar bear found on the coast of the Beaufort Sea in Alaska.[21] The skull was nicknamed 'Bruno', although subsequent analysis showed it was female. Bruno's palaeogenome confirmed an ancient admixture that impacts the genetics of all brown bears today. During the last warm interglacial period in the Pleistocene, the polar bear lineage leading to Bruno and the ancestral brown bear lineage overlapped and hybridised. As a result, polar bear DNA now accounts for as much as 10 per cent of extant brown bear genomes. Such a discovery would not have been possible without Bruno's legacy, since all brown bears today have this admixture as part of their genomes.

In 2018, James Cahill obtained similar results using DNA extracted from an extinct Irish brown bear population, noting that the proportion of admixed polar bear DNA was maximal during milder interglacial periods when populations interbred, and subsequently declined, as predicted by biologists.[22] In total, polar bear admixtures have occurred in at least four brown bear populations between 25,000 and 15,000 years ago. However, the smaller admixture in living polar bears suggests that brown bear genes could reduce a bear's fitness to survive in the Arctic.

The frequency of such admixtures is projected to rise with the Earth's warming climate. The melting ice is pushing polar bears onto land, while the northward expansion of brown bears' range creates more opportunities for interaction. This scenario, driven by climate change, has far-reaching implications. Climate-facilitated admixtures can lead to widespread and long-term evolutionary consequences, potentially serving as a significant mechanism for generating and maintaining diversity.

Lastly, a comprehensive study by Viktar Kumar and colleagues from Goethe University Frankfurt, involving the entire bear family, revealed that gene flow was more widespread than previously thought, extending beyond closely related species.[23] Additionally, some cases cannot have occurred directly, since the species exist in different habitats, and may have involved intermediate or vector species. Introgression has also occurred during the evolutionary histories of other mammalian taxa, including horses, gibbons and humans, significantly impacting our understanding of speciation. If gene flow across taxa is frequent and can persist for several hundred thousand years after divergence, then speciation should not be viewed solely as achieving genome-wide reproductive isolation. Instead, it should be regarded as a selective process that maintains species divergence despite gene flow – a conclusion that adds a layer of complexity to our understanding of evolution.[24]

CHAPTER 16

The Cat's Story

DISPERSALS AND BOTTLENECKS

One cannot teach a cat not to catch a bird.
Albert Einstein (1879–1955)

Cats (family Felidae) evolved in northern central Asia, although the scarcity of fossils, and the morphological similarity of many of them, means that an understanding of their evolutionary history had to await the genomic era. In 2006, Stephen O'Brien and colleagues addressed the challenge after procuring blood samples from all 37 extant species known at the time.[1] This was no small feat considering the rarity of certain species and the remote, inaccessible habitats of others. The team then extracted DNA from the samples and compared identical stretches of DNA from 30 genes across each species. The resultant phylogeny mapped the branches and the order of their emergence. By incorporating fossil calibrations from accurately radiocarbon-dated specimens, they could determine the approximate timings of the phylogeny's bifurcations. The results indicated that the Felidae consist of eight distinct clusters or lineages, encompassing two subfamilies: a single 'big' cat lineage (Pantherinae) and seven 'small' cat lineages (Felinae) that include bay cats, caracals, ocelots, lynxes, pumas, leopard cats and the wildcat (including the domestic cat).

The 41 extant species recognised today are similar anatomically, with rounded skulls, large orbits, noses that project beyond the lower jaw, actively retractable claws, lithe muscular bodies, and strong, flexible forelimbs.[2] As obligate carnivores, their teeth and facial muscles are designed for powerful bites, while their tongues are covered with horn-like papillae to rasp meat from prey. Cats have modified molars, known as carnassial teeth, for tearing and cutting flesh. Their whiskers are well developed and highly sensitive and aid in capturing and holding prey and navigating at night. Except for lions, they are solitary hunters and showcase remarkable predatory skills, including acute senses of hearing, sight and smell, traits that underscore their apex-predator status.

Due to their carnivorous diet, all cats have lost the ability to detect sweet tastes. This trait results from the pseudogenisation of a gene that codes for part of the sweet taste receptor.[3] A small deletion in the *T1R2* gene renders this receptor non-functional across the entire cat family. This 'knock-out' mutation likely arose because the selection pressure to maintain the receptor was relaxed when cats became obligate carnivores. Interestingly, this gene is also non-functional in other meat-eaters, including otters, sea lions and dolphins.[4] However, because these disabling mutations occur in different areas of the gene, the loss of sweet taste function must have occurred independently and repeatedly in carnivores due to convergent evolution.

Feline chromosomes are remarkably stable compared to other mammalian groups, with minimal differences between the chromosomes of lions and domestic cats. In contrast, the genomes of great apes and humans have undergone frequent duplications, rearrangements and other variations that led to the evolution of different species, including ourselves. Primate genomes tend to break and rearrange because of the frequency of chromosomal segmental duplications – blocks of DNA that share more than 90 per cent sequence identity and occur at many sites within the genome. In other words, the more segmental duplications a mammal possesses, the more likely it is that chromosomes will rearrange, leading to genetic isolation and speciation. Primates have seven times as many segmental duplications as cats.

Since feline genomes are so stable, how do new species arise? Researchers at the Texas A&M School of Veterinary Medicine, have recently shed light on the problem.[5] The research team, led by William Murphy, discovered that cats still exhibit significant genetic variation, primarily concentrated in the centromere of the X chromosome, an area that has long been challenging to sequence. Within this region, a segment known as DXZ4 stood out as the most rapidly evolving part of the feline genome. Although DXZ4 doesn't code for a physical trait, it consists of an extensive array of tandemly repeated, non-coding DNA that shapes the three-dimensional structure – and therefore the function – of the X chromosome. Given its rapid rate of evolution, Murphy believes that DXZ4 is likely to play a novel role in driving cat speciation.

'Big' cats

The subfamily Pantherinae was the first lineage to evolve, around 10.8 million years ago, and by 4.6 million years ago it had diverged to produce the ancestors of the seven species alive today. They include five large cats (genus *Panthera*): the lion, leopard, tiger, snow leopard and jaguar (Plate 23), together with two smaller species (genus *Neofelis*), the mainland and Sunda clouded leopards.

All *Panthera* cats, with the exception of the snow leopard, have an incompletely ossified hyoid bone and a specially adapted larynx with large vocal folds to enable them to roar. This unique feature sets them apart from other felines. In contrast, the snow leopard possesses shorter laryngeal folds that provide a lower resistance to airflow, resulting in a variety of other sounds like mews and growls. Also, clouded leopards cannot roar or purr, and their vocalisations consist of growling and hissing. However, all seven Pantherinae can 'prusten', a short, low-intensity sound, also known as a chuffle, used for non-threatening communications, such as between mother and cubs.

In 2017, a multinational research consortium led by PUCRS, a Brazilian university, unveiled the entire sequence of the jaguar genome and conducted a comparative analysis with the genomes of other big cats. Their findings revealed that all five *Panthera* species diverged from a common ancestor resembling a leopard around 4.6 million years ago.[6] The study also examined the historical demography of each species, showing that all populations had declined over the past 300,000 years. These declines occurred in two distinct phases: the first, affecting all five species, took place between 300,000 and 100,000 years ago, while the second phase, following the Last Glacial Maximum 10,000–20,000 years ago, primarily impacted lions and leopards. Notably, the jaguar experienced the most significant population reduction, stabilising at the lowest level among the big cats. Overall, these changes led to a considerable decrease in genetic diversity across all species.

The investigators also looked for evidence of admixture and found a complex network of ancestral hybridisation amongst many *Panthera* lineages. The most affected was the lion, probably due to its broad historical range throughout much of the Holarctic, which overlapped with other big cat species. While introgression has been documented between the lion and snow leopard, the above study reveals a much more complex history of post-speciation admixture than was previously appreciated.[7]

Until recently, the lion was one of the most widely distributed terrestrial mammals. During the Pleistocene, their ancestors roamed widely across the northern hemisphere, using transient land bridges. Around 500,000 years ago, this global population diverged to give rise to the modern lion in Africa, the Middle East and India, the cave lion in Eurasia and North America, and the American lion, which extended as far as Mexico. However, like all big cats, the lion populations declined, and by 14,000 years ago the cave lion was extinct, followed by the American lion. By the early twentieth century, modern lions had disappeared from southwestern Eurasia and North Africa due to anthropogenic factors. The decline continues, and over the last 150 years populations in North Africa, South Africa and the Middle East have become extinct. Today, only fragmented populations remain in sub-Saharan Africa, with a small population of around 600 individuals in the Gir forest sanctuary in the Gujarat province of western India.

Recent genetic studies have revealed a reduction in lion diversity, especially among the Indian population, due to centuries of persecution and historic genetic bottlenecks.[8] As is usually the case in small, isolated populations, the Indian lions exhibit the detrimental effects of inbreeding, including depleted sperm counts, skeletal defects, testosterone reduction and reduced mane size. Genetic analysis has also confirmed that the Gujarat lions are native to the region, dispelling the belief that they were brought over from Africa in the pre-colonial era. This finding has significant implications and underscores the imperative for urgent conservation measures.

During the Pleistocene, the jaguar lineage evolved in Asia and migrated into North America via the Bering Land Bridge around a million years ago. The modern species arose approximately 400,000 years ago and spread throughout the Americas, crisscrossing the Panama isthmus several times before reaching as far south as Patagonia.[9] However, the fluctuating climatic conditions, characterised by glacial cycles, caused significant range contractions and expansions, with small populations surviving in southern tropical refugia. Today, the species is distributed from Mexico to central Argentina, with a 50 per cent contraction over the last century, including probable extirpation in the United States.

Unlike other large, solitary cats, jaguars typically hunt smaller prey rather than large herbivores. However, this behaviour was not always the case. In the past, jaguars were much larger and competed with American lions and cheetahs for large prey. Over time, selection pressures drove the three lineages down different evolutionary paths, each developing its own unique characteristics. Lions grew bulkier and developed as ambush predators, hunting cooperatively to bring down the grassland's large herbivores. Meanwhile, American cheetahs became leaner and faster, filling the role of chase predators and preying on fast-running herbivores.

The large jaguars, however, took a different trajectory and became smaller, with shorter legs, and retreated to the forests. They gave up hunting the megafauna and instead preyed on species they could closely approach and pounce upon. To do so, jaguars evolved the ability to crush skulls with strong jaws rather than throttling their prey as lions do. Their jaw muscles were reconfigured, and jaw length was slightly reduced, to enhance biting leverage. Consequently, jaguar jaws can easily penetrate the skull of a capybara and pierce the thick skins of caimans using their robust, cone-shaped canines. According to Matt Haywood and coworkers, jaguars survived the megafaunal extinction event by preferentially preying on small species such as these.[10] A comparative genomic analysis of big cats hinted at how this phenotypic advantage may have evolved. Unlike other *Panthera* species, jaguars have protein-altering mutations in two genes (*ESPR1* and *SSTR4*) that control craniofacial development.[11]

The charismatic tiger originated and diversified in Asia. Like other *Panthera* species, their population contractions and expansions in the Pleistocene contributed to the low genetic diversity of modern populations. Once spanning 70 degrees of latitude of Asia, they now occupy only 6 per cent of their former range. However, despite these recent contractions, tigers continue to inhabit diverse habitats, from mangrove forests and rainforests to cold temperate forests, in 11 Asian nations. Although their evolution remains unclear, recent genetic analyses suggest that the six subspecies diverged within the last 20,000 years and experienced substantial population bottlenecks.

Comparative genomics has revealed that tigers possess the most olfactory receptor genes among all felids. This adaptation is crucial for tigers, who live solitary lives and rely on detecting scent marks and locating potential mates over large distances. In contrast, many of the lion's olfactory receptor genes are non-functional due to their social lifestyle, which involves living in closely knit cooperative groups. The increased pseudogenisation in lions stems from the relaxed selection of chemical cues for determining sexual status and finding partners.[12] Additionally, comparative genomics has shown Siberian tigers, native to the Russian Far East, to have undergone selection pressure for genes related to metabolic adaptation to the cold. Moreover, Sumatran tigers show a weak selection for a range of genes related to morphological development, a find that might explain their relatively small size.[13]

The snow leopard is adapted anatomically and physiologically to the high altitudes of the Himalayas and Altai Mountains. It is a scansorial species that hunts alone, ambushing prey from above at speed, rather than employing the typical stalking technique of other large cats. They have evolved enlarged scapular and pectoral muscles that provide stability for the shoulder joint when grappling with large prey and support while jumping and climbing. The clavicle is small and unarticulated to provide both support and flexibility.[14] The paw muscles are broad and fleshy and act as snowshoes to prevent them from sinking into the snow.

At altitudes above 4,000 metres, where oxygen levels fall to around 60 per cent of those at sea level, the species faces formidable environmental selection pressures. The resulting adaptations include long body hair, dense woolly underfur and a long thick tail used to wrap around the body to aid in thermoregulation and insulation against harsh conditions. Snow leopards also have well-developed chests to help draw in the rarified air and enlarged nasal cavities that warm the cold air before it reaches the lungs. Furthermore, two genes (*EGLN1* and *EPAS1*) that regulate cellular responses to hypoxia are uniquely mutated in snow leopards. Intriguingly, similar genetic changes occur in humans inhabiting the Tibetan plateau, protecting them against altitude sickness and preventing dangerous increases in blood viscosity.[15]

'Small' cats

After the divergence of the Pantherinae, the next split occurred around 9.4 million years ago, producing the bay cat lineage. This branch gave rise to three poorly known, small-sized species with long tails and short heads that inhabit the forests of Southeast Asia, including the marbled cat and the Asian golden cat. A million years later, the caracal lineage arose, and it is represented today by three medium-sized species, including the serval. The early caracals undertook the first of many 'small' cat migrations, crossing into Africa when sea levels fell by 60 metres and land bridges formed across the Red Sea. Indeed, all cats, big and small, are hard-wired to disperse. On reaching adolescence, young males, and occasionally females, leave their place of birth and seek new territories, often travelling considerable distances. Such innate behaviour, combined with their instinct to track prey, probably accounts for the many feline dispersals in the past.

While caracals migrated to Africa, other cats took the opportunity of the low sea levels to disperse across the Bering Land Bridge into North America. Once the sea rose again, continents became isolated, leading to the evolution of scores of new species. The precursors of the ocelots and lynxes diverged in North America from the original migrant population 8 and 7.2 million years ago, respectively.

The ocelot lineage spread throughout Central and South America, crossing the Panama isthmus in several waves from around 3 million years ago. The lineage comprises eight similar species with 36 instead of 38 chromosomes, as in all other cats, confirming their monophyly.[16] Today, these small cats occupy a broad range of habitats, from the coastal thicket areas of Texas to the shrublands of Patagonia, from high-altitude Andean salt flats to lowland rainforests. Like many South American mammals, their evolution has been shaped by past climatic upheavals. For instance, the subspecies of pampas cat from Chile experienced a marked bottleneck, from which it never fully recovered at the time of the Last Glacial Maximum, when sizeable glacial ice sheets covered the southern Andes.[17]

The puma lineage diverged 6.7 million years ago and gave rise to the puma (cougar), the jaguarundi and the now-extinct American cheetahs, differing in size, range and coat pattern. The puma's ancestors migrated from

Asia across the Bering Land Bridge to North America. Subsequently, the cheetah returned to Asia and eventually reached Africa, while the puma and jaguarundi spread to South America around 2–3 million years ago. Once across the Panama isthmus, the ocelot and puma lineages encountered a continent teeming with marsupials, including some successful carnivores like the sabre-toothed metatherians. However, the marsupials could not compete and were soon replaced by the fierce and territorial placental cats that spread throughout the continent.

The African cheetah, a marvel of adaptation, is the fastest animal on Earth, sprinting after prey at over 100 kilometres per hour. They have evolved unique physiological traits over several million years to achieve this incredible feat. Their claws, a vital tool in their hunting arsenal, are non-retractable and, together with tough foot pads, provide maximum traction. Instead of a tail for balance, the cheetah's tail is long and acts as a rudder, enabling sharp turns at high speed. Their legs are elongated, with very light bones, their skulls are slim and aerodynamic, and their hearts are enlarged. Cheetahs have binocular vision, allowing them to spot prey up to 5 kilometres away, and their eyes are specially adapted to enable sharp, wide-angle vision. Amazingly, they can keep their eyes fixed on their prey despite their entire body moving while running at speed. This remarkable ability is achieved through a modified cochlea or inner ear. In vertebrates, the inner ear's balance system comprises three semicircular canals filled with fluid and sensory hair cells that detect head movements. Each canal is oriented at a distinct angle, making it particularly sensitive to various types of motion: vertical, lateral and angular. However, cheetahs possess uniquely structured inner ears, with a more extensive vestibular system and longer semicircular canals than other cats. These adaptations increase the animal's response times to head movements, allowing cheetahs to keep their sight fixed on prey, even during high-speed chases.

Cheetahs have the lowest genetic diversity of all the cats, with their genomes exhibiting 95 per cent homozygosity. In other words, cheetahs have inherited identical versions (alleles) of most genes from each parent. Such an extreme lack of genetic variation explains their ability to accept reciprocal skin grafts from unrelated cheetahs.[18] Furthermore, excessive inbreeding has resulted in increased juvenile mortality, abnormalities in sperm development, and a marked vulnerability to infection. Consequently, cheetahs are heading towards extinction today. However, this isn't the first time they have faced annihilation.

The first bottleneck event occurred around 100,000 years ago when cheetahs expanded into Asia, Europe and Africa. This dispersal from North America likely occurred rapidly, thus restricting their ability to exchange genes.[19] The second bottleneck occurred about 10,000–12,000 years ago when the last ice age ended. Many other large mammals disappeared forever, including mammoths, mastodons, short-faced bears, American lions and sabre-toothed tigers. During this tranche of worldwide megafaunal extinctions, the cheetahs of North America and Europe disappeared, leaving only those that had escaped to Asia and Africa. However, their numbers dwindled, leading to extreme inbreeding. Since then, climate change, habitat loss and human activities have all put further pressure on the remaining animals. Approximately 8,000 wild cheetahs remain, confined to several regions in Africa and a small area in Iran. Only time will tell if this iconic species, a paradigm of physical prowess, can survive the current genetic bottleneck.

During their migration across the Bering Land Bridge back to Asia, cheetahs were accompanied by the progenitors of the leopard cats and the domestic cat. The leopard cat lineage swiftly underwent diversification, giving rise to the Asian leopard cat and four smaller species: the rusty-spotted cat in India, Pallas's cat in Mongolia, the flat-headed cat in Indonesia, and the more widely distributed fishing cat.

Domestication

Linnaeus first classified the ubiquitous domestic cat as *Felis catus* in 1758. While his binomial is still widely used, recent genetic studies indicate that it is no longer accurate and should be replaced by *Felis silvestris catus*. Let's explore the reasoning behind this statement.

Scientists have long suspected that all breeds of domestic cats have descended from a single species – the wildcat – but this theory remained unconfirmed. The wildcat is widespread throughout the Old World, encompassing five subspecies that inhabit different areas across Eurasia and Africa. Furthermore, all subspecies look alike and are hard to distinguish from the feral, tabby-coated, domesticated cat, as they all exhibit similar fur patterns. Like most felines, they also interbreed freely, blurring population boundaries.

Then, in 2007, comparative genomics resolved the scientific impasse. Carlos Driscoll and coworkers compared DNA from nearly 1,000 wildcats and domestic cats. They found that the domestic-cat sequences were virtually indistinguishable from a subspecies of wildcat, *Felis silvestris labia*, that inhabits the remote deserts of Israel, the United Arab Emirates and Saudi Arabia. In contrast, the DNA of the remaining four subspecies, from China, Europe, central Asia and Africa, were less closely matched, proving that the domestic cat arose from a single locale, the Middle East.[20]

Determining the timeline of domestication is not possible using molecular clocks, as these genetic tools are only suitable for dating divergences in the distant past. The likely timing for cat domestication is around 10,000 years ago, which is too short for any informative genetic changes to have evolved. Instead, biologists resorted to archaeology. In 2004, a team from the National Museum of Natural History in Paris unearthed an eight-month-old cat deliberately buried with its owner in a Neolithic grave in Cyprus, some 9,500 years ago, along with other offerings.[21]

This discovery marks the earliest record of cat domestication. Since no wildcats ever reached Cyprus, the islanders must have brought them over by boat, probably from the eastern Mediterranean coast. Cats were likely first tamed when permanent settlements were established in the Middle East, around the area known as the Fertile Crescent. Here, humans began to gather and cultivate cereals in more and more elaborate ways. It is not hard to imagine that their grain stores encouraged the proliferation of mice, which then attracted wild cats. Individuals more tolerant of humans would have come close to settlements, and this self-selection process likely played a crucial role in their domestication. While selection pressures favoured tameness, it also ensured that competition limited how submissive cats became. Since the early animals were left to fend for themselves, their hunting and scavenging skills remained highly tuned. Even today, most pet cats are free spirits and can survive without humans, as evidenced by the large populations of feral cats in most cities and towns. Nevertheless, domestic cats have altered genomes. A study led by Wesley Warren from the Genome Institute, Washington University School of Medicine, revealed a strong selection for genes affecting memory, fear-conditioning behaviour and stimulus-reward learning, hinting at likely molecular modifications by which cats became domesticated.[22]

Cats remain one of the most beloved and widespread pets globally, with a population of nearly 500 million strays and 350 million living as household pets. By the nineteenth century, cat owners started selectively breeding their pets, and the International Cat Association now recognises 73 different breeds. Today, the domestic cat is almost the only cat species not considered vulnerable or worse by the International Union for Conservation of Nature (IUCN).

Domestic cats, much like their wild ancestors, possess an intrinsic desire to maintain territories and prefer a solitary existence. However, when living under the same roof, cats typically tolerate one another – primarily to access essential resources such as food. Moreover, as illustrated by Einstein's observations of his pet cat 'Tiger', the visual and auditory stimuli associated with prey activate a cat's deeply ingrained hunting instincts. This anecdote from one of the most eminent scientists of all time highlights a fundamental aspect of feline behaviour: they are irresistibly drawn to the fluttering of bird wings or the scurrying movements of a mouse.

CHAPTER 17

The Bat's Story

POWERED FLIGHT AND ECHOLOCATION

Without an understanding of development, we cannot fully understand life on Earth.
Karen Sears

One summer's evening, we gathered among the strolling couples and local barbecues at a picturesque location – a boating and fishing lake near the Hungarian village of Felsőtárkány. Peter, a bat enthusiast employed by the National Parks, had set up a mist net to show us some of the local bat species. As we waited, watching the occasional fluttering shape in the moonlight, my colleague Jim, equipped with his bat detector, diligently detected bats as they flew somewhere above our heads. As I viewed the screen, he pointed out several species' display patterns or spectrograms: a soprano pipistrelle, a *Myotis* species, and a likely common noctule.

Bat calls are produced by the passage of air over vocal cords and occur typically at frequencies beyond the range of human hearing. However, they can be recorded and analysed using modern bat detectors. Each species emits high-frequency sounds with distinct characteristics determined by size and flight behaviour. Moreover, they adapt their calls to their surroundings, producing different sound patterns, for example, in woodlands and over lakes. Specialised circuitry in the detectors captures these inaudible sounds and transforms them into a form discernible by human hearing. By fine-tuning the detector, we were able to immerse ourselves in a bewildering cacophony of ultrasonic chirps, smacks and clicks, a testament to the echolocation skills of these remarkable creatures.

Meanwhile, a few bats had inadvertently become trapped in the net, hanging silently upside down in the still evening air. In Peter's capable hands, the species were quickly identified, and he unveiled their most remarkable feature: their wings. These delicate structures, which gave the order Chiroptera its name (literally 'hand-wing'), are not just wings but marvels of evolution. They enable powered flight, a defining feature of all bat species.

Examining an outstretched wing, I could see that the bones were similar in structure to those in the human hand and arm (Figure 17.1; Plate 24). The four finger bones in bats are greatly elongated, light and slender, supporting the wing's soft and stretchy membrane, known as the patagium. The second digit (index finger) and the outer end of the third digit (middle finger) form the wing's stiff leading edge, while the third digit also creates the wing tip. In contrast, the fourth and fifth digits act like umbrella ribs, keeping the patagium taut without needing large muscular forces, while the wing's trailing edge is unsupported. The first digit (thumb) protrudes from the wing as a small claw, which bats use to climb, handle food, and fight. Closest to the body lies the humerus, a longer and thinner bone than in other mammals, which meets the bones of the lower arm to form the 'elbow'. Between the humerus and wrist region lies the radius, a long, thin, but strong bone supporting the wing, and the diminutive ulna fused to the radius. Like the wings of the ancient pterosaurs, the bat's patagium connects to its ankle, forming a huge surface to maximise lift. Amazingly, every bone mentioned has its counterpart in the human body, except with markedly different proportions. In other words, the skeleton of the bat's flying wing and our upper limb are homologous, a reflection of their different descendant lineages from a common ancestor.

During my brief introduction to bat watching, I encountered two of the most remarkable and successful mammalian adaptations: powered flight and echolocation. Bats are the second most diverse group of mammals,

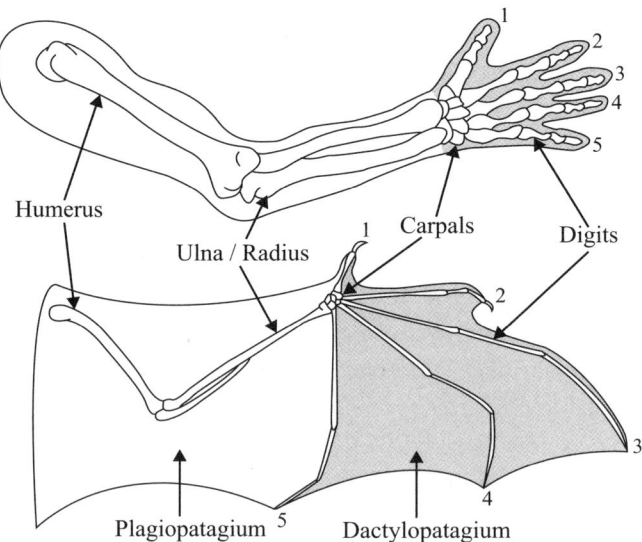

Figure 17.1 Comparison of a human upper limb with a bat wing. Numbers refer to digits: 1, thumb; 2, index finger; 3, middle finger; 4, ring finger; 5, pinkie. The grey areas show the evolutionary relationship of the human interdigital skin and the bat wing's dactylopatagium.

with 1,498 living species, surpassed only by rodents.[1] They are ecologically diverse and can be found in tropical, subtropical and temperate regions. Indeed, bats are found across the globe apart from a few remote islands, high mountains and polar areas. They even reached Hawaii around 500,000 years ago from America, marking the longest transoceanic dispersal event known for terrestrial mammals. No matter how remote an island, a bat will find it if it has plentiful insects and fruit. Yet, despite their distinctiveness, their widespread distribution, and the availability of exquisite fossils, bats remain one of evolution's greatest puzzles. Why is this?

The earliest bats

Stem bats emerged a few million years after the Palaeocene–Eocene Thermal Maximum (PETM) period of warming. The oldest known fossil was recently recovered from 52.5-million-year-old limestone from the Green River Formation in western Wyoming. The specimen consists of a wonderfully complete skeleton that reveals all the bat's bones in a lifelike position.[2] Palaeontologists can even deduce from the arrangement of its foot bones that it probably roosted upside down, like many modern bats. We only know of this 40-gram, walnut-sized species because it lived around a lake that favoured exceptional preservation, known as a fossil-lagerstätte. The lake's fine sediment and oxygen-depleted water, with few bacteria, allowed the bat carcass to sink to the bottom and leave a durable impression in the powder-like matrix before decomposing.

Further specimens, dating to around 50 million years ago, include mainly teeth and bits of jawbone but also stunning, fully articulated skeletons in places as far-flung as Australia, Europe, North America, Africa, and the limestone mines of Gujarat in India. At the time, Australia, Africa and India were islands – and bats, unlike other mammals, could disperse unchecked across oceanic barriers to become the first cosmopolitan placental mammals. This significant radiation occurred during the global 'hothouse' conditions of the Early Eocene, a period marked by rapid diversification of plants, insects, and other mammals. As the global climate transitioned to cooler and less stable conditions, around 49 million years ago, bats may have adapted by roosting in caves, a behaviour observed in almost half of all extant species.

Another exquisitely preserved fossil, recovered from 50-million-year-old cave deposits in France, enabled scientists to obtain three-dimensional images of its near-complete, uncrushed skull.[3] Crucially, its inner ear revealed echolocating adaptations, indicating that advanced laryngeal echolocation evolved very early, predating the crown radiation of bats. Moreover, the ancestral bat probably used echolocation for hunting insects, as its mandible and teeth resemble those of extant insectivorous bats. Such fossils can even reveal that their fur was reddish-brown.[4] Remarkably, these early bats, which had spread over the entire globe by the early Eocene, had already evolved wings and echolocation systems, although less advanced than extant species. So where did these early bats come from?

Genomic studies reveal that bats belong to the extensive and diverse laurasiatherian clade, which evolved in the northern hemisphere (see Figure 5.1). Additionally, time-calibrated molecular phylogenies suggest that they evolved shortly after the K–Pg boundary, with their two suborders, Yinpterochiroptera and Yangochiroptera (see below), diverging approximately 63 million years ago.[5] However, no fossils have been found from this early period of bat evolution, resulting in a critical 10-million-year gap during which bats diverged from an unknown terrestrial mammal and developed powered flight. According to Emily Brown, a palaeontologist from the University of Birmingham, bats from this era may have inhabited forests, an environment not conducive to fossil preservation.

Palaeontologists have faced similar intriguing puzzles before. As previously discussed, scientists were baffled for a long time by the origins of whales and sirenians (Chapter 11). However, genomics and numerous fossil discoveries subsequently revealed how terrestrial ungulates took to the water and evolved to produce the world's largest mammals. Birds posed a similar challenge, as their reptilian ancestry bewildered experts. Eventually, new theories about dinosaurs and exquisite fossil discoveries from China confirmed Thomas Huxley's view that birds are living dinosaurs.

Unlike whales and birds, the earliest stages of bat evolution remain unclear, in the absence of transitional fossils. However, two teeth from northern Xinjiang, China, might offer some insights.[6] These tiny fossils, dating back to the beginning of the Eocene around 55 million years ago, display dental characteristics typical of both bats and early eutherian mammals. While interesting, they confirm what we might have anticipated and offer no clues about the species or even the branch of eutherians from which bats evolved. Nonetheless, they suggest a possible central Asian origin for bats, similar to the proposed origins for even-toed ungulates, lagomorphs, rodents and primates.

Given the absence of fossils, it is possible to envision a plausible scenario. Bats likely evolved from tree-dwelling mammals, like squirrels, that leapt from branch to branch. Whether these ancestors were nocturnal or diurnal is debated. Imagine some members of this unknown species developing, by chance, a small area of loose skin between their arms and body, which provided extra lift when jumping through the air. Individuals with this adaptation would have been more likely to survive by evading predators or reaching food sources more efficiently, thus passing on their modified genes. Repeated cycles of mutation and selection would lead to individuals with more extensive skin membranes and improved gliding abilities. This scenario is not so far-fetched, as gliding has evolved independently many times in mammals, including in colugos, flying squirrels, lemurs and marsupial possums. Each group possesses membranes between the body and legs, acting as an aerofoil to generate lift and allow directional movement during flight.

Then, as first proposed by Darwin in 1859, the evolution of flight rested on the transformation of the hand into a webbed and elongated 'hand-wing' that became integrated into the pre-existing glider body plan. In support of Darwin's theory, all bats have the glider's membranes, but they also evolved an extra area of patagium around their elongated digits, termed the dactylopatagium. Over time, natural selection favoured greater gliding efficiency, causing the digits to lengthen and the membranes' surface area to increase. Eventually, flapping became possible. However, powered flight is the most demanding animal locomotion mode and requires the wing to be as light as possible, especially at the tip. To this end, the limb muscles migrated towards the body, leaving the tendons extending the entire length of the digits, crossing many hand joints. Given the force exerted on the wings during repeated wingbeats, the arrangement posed a risk of tendon slippage.

The bat's solution was sesamoids, small bones embedded in tendons that function as pulleys to reduce stress between tendons and joints. We have a sesamoid in each of our legs, the patella or kneecap, but bats have a forelimb equivalent – the ulnar sesamoid – which is not present in any other mammal. Moreover, bats have numerous sesamoids, with some species having over 20, primarily concentrated in the joints of the hand-wing. These small, easily overlooked structures are crucial, allowing bats to perform repetitive wing movements without damaging their tendons. Like the development of gliders, natural selection refined the early wings of bats to enhance their aerodynamics, increasing lift and manoeuvrability. Such adaptations, coupled with the emergence of echolocation, allowed bats to hunt insects on the wing. Although the hypothesis that bats evolved from arboreal ancestors via gliders and flyers is widely accepted, fossils from the 10-million-year gap are still needed to confirm this sequence of events.

Wing evolution

Scientists now have a basic understanding of how the delicate hand-wing evolved, with its unique dactylopatagium. The complex anatomy of vertebrate limbs begins from small limb buds of undifferentiated cells on either side of the body. Guided by growth factors, these buds elongate, forming terminal paddle-like structures. Nascent fingers emerge as a series of stripes that fan out from the centre. This arrangement relies on the interaction of two key morphogens, BMP and Wnt, with a third, Sox9, in a three-way mutual activation and inhibition network. Readers will recall from the discussions of zebra stripes and fingerprints that the interaction of an activator and inhibitor, with feedback loops, can function as a Turing system, leading to precise tissue patterning (Chapter 14). Such effects are now believed to occur in the proto-hand. Sox9 molecules, which trigger the formation of bones and cartilage, are guided by BMP and Wnt into a Turing pattern of five radial spokes destined to become the fingers.

Without further intervention, hands would remain paddle-shaped, with fingers bound together by webs of tissue. Instead, the webbing cells are removed *in utero* by apoptosis or 'programmed cell death'. Apoptosis is an energy-consuming process encoded by genes that dismantle cells in a distinct and highly organised sequence of events without releasing toxic molecules that occur when cells die from injury. Under the influence of bone morphogenetic proteins (BMPs), the interdigital cells shrink and form bubble-like protrusions or 'blebs' on their surface. Then, the DNA in the nucleus fragments into small chunks, and specific organelles disintegrate into small pieces. Finally, the entire cell divides into small, membrane-bound packages that emit signals that attract phagocytic cells, like macrophages, to engulf and digest them. Failure of interdigital apoptosis is one of the most common limb abnormalities in humans, occurring in approximately one in every 2,500 births, resulting in fusion of the fingers or syndactyly.

In bats, only the hindlimbs undergo interdigital apoptosis, resulting in feet that are neither webbed nor elongated. In contrast, the webs in the animal's forelimbs are retained and stretched to form the thin dactylopatagium. Apoptosis in the bat's hand is prevented by a unique pattern of growth factors, including fibroblast growth factor 8 (Fgf8), working in conjunction with BMP inhibitors, such as Gremlin. Interestingly, ducks employ the same protein inhibition of apoptosis to produce webbed feet. In addition, the bat's elongated digits result from increased BMP protein within the epiphyseal cartilage that keeps the fingers growing. In other words, increased BMP in the digits and decreased BMP in the interdigital webs appear to underpin the evolution of the bat's forelimb and dactylopatagium.[7] However, as previously discussed, the development of the remaining area of the patagium is controlled by different cellular processes involving the *Wnt5a* gene (Chapter 3). For example, the plagiopatagium arises from novel outgrowths from the body flank that merge with the limbs to form the wing aerofoil.[8]

The protein Prx1 is also involved in finger elongation. Prx1 plays a crucial role in shaping large-scale anatomical features in the early stages of embryonic development. A mutation in its DNA enhancer sequence causes an increase in the protein's production and contributes to the digit lengthening.[9] This finding is significant, indicating that alterations in pre-existing, non-coding regulatory sequences within a population can drive morphological differences between species. We will explore this topic further in later chapters, focusing on the role of gene regulation and phenotypic plasticity in the evolution of primates.

Two biologists, Sophia Anderson and Graeme Ruxton, have suggested that interdigital webbing could have evolved before the forelimb-to-hindlimb patagium. Not only would this have improved jumping between branches in the ancestral proto-bats, but it could also have enabled a better grip in the dark and even allowed the detection of vibrations from predators and prey. If Anderson and Ruxton's hypothesis is correct, then interdigital webbing was the ancestral phenotype from which all bats' morphologies evolved.[10]

The evolution of flight required more than just the development of wings. Like birds, bats expend high energy levels when flying – three to five times the maximum energy used by land mammals of comparable size. They have the fastest recorded flight speed of any vertebrate and undertake energy-demanding curved flight trajectories that quickly shift while chasing insect prey. To meet such demands, they have optimised the efficiency of their mitochondria, the ubiquitous cellular powerhouses. These organelles break down glucose and convert ADP into ATP, the energy molecule essential for various cellular processes, particularly muscle and brain activity. Early bats were under intense selection pressure, and the genes that regulate these chemical reactions were altered to accommodate the significant increase in energy needs for flight. However, producing more energy also creates harmful byproducts like reactive oxygen species (ROS) or free radicals, which can damage cells and DNA. To mitigate these effects, bats have altered various genes involved in detecting and repairing genetic damage.[11] They include the genes *KU80* and *RAD50*, which control the repair of DNA double-strand breaks by recruiting other proteins to process and ligate the broken ends.

Despite these remarkable adaptations, the intrinsic inefficiency of mammalian respiration may have posed limitations on the body size of bats and their ecological niche range compared to the birds and pterosaurs. Like all mammals, bats use a tidal breathing pattern, where air enters and exits via the same airway. In contrast, birds have a unidirectional (one-way) breathing system, with the air moving mainly in a loop. As size increases, flying becomes more energetically demanding because mass grows faster than the wing surface area, which generates lift to counteract gravity. Although bats are diverse, none have ever been significantly heavier than the largest existing flying foxes. For example, the heaviest extant species, the great flying fox endemic to New Guinea, may reach only 1.6 kilograms. In contrast, several birds and pterosaurs have reached masses orders of magnitude greater than this.

Echolocation

Bats use echolocation by creating intense high-frequency pulses with their larynx and emitting these ultrasonic sounds through their mouth or nose. Each species utilises specific frequencies depending on their environment and whether they are searching, feeding or socialising. By analysing how the echo changes, bats can interpret their surroundings, including an object's size, shape and texture. Remarkably, some species can resolve objects less than half a millimetre in size and even determine whether their prey is flying and in what direction.

To achieve this, bats emit sounds up to 140 decibels, equivalent to a jet engine's noise during take-off. Even the so-called whispering bats, like the brown long-eared bat, will emit 110-decibel shrieks, comparable to power saws and pneumatic drills. Peng Shi, a Chinese biomedical engineer from the Kunming Institute of Zoology, was puzzled why bats didn't go deaf from their squeaks, since the fine hairs in their inner ears should be damaged at such levels. To find out, Shi's team attached electrodes to five species of echolocating bats, non-echolocating fruit bats and laboratory mice, and subjected them to continuous high-intensity noise. The echolocating bats were unaffected, while the mice and fruit bats lost their inner ear hairs, resulting in hearing impairment. Genome analysis revealed an overexpression of several genes, including *ILS1*, which produce noise-cancelling proteins that protect echolocators from their sounds.[12]

Another intriguing puzzle concerns when echolocation evolved. To answer this query, we must first highlight the latest classification of Chiroptera. Initially, the phylogeny of bats appeared straightforward, with living species categorised into two suborders based on size and behaviour. Bats that used echolocation were classified as 'small' bats (Microchiroptera), while the non-echolocating Old World fruit bats were placed in the suborder of 'big' bats (Megachiroptera).

However, recent research presents a more complex picture. Extensive genetic studies have uncovered varying relationships among bat families. For instance, the horseshoe bat family, comprising small insect-eating bats, is genetically more akin to the 'big' fruit bats than other families traditionally classified as 'small' bats. This discovery means that microbats and megabats are not monophyletic groups, making the subdivision into Microchiroptera and Megachiroptera obsolete. The most recent and widely accepted phylogeny now classifies bats into two new suborders: Yinpterochiroptera, which encompasses the fruit bats and five microbat families, and Yangochiroptera, consisting of all the remaining small bats.[13]

While debates on bat phylogenies may seem esoteric and arcane, the updated family tree has profound implications for our understanding of the evolution of laryngeal echolocation. It poses an intriguing question: did echolocation evolve independently in the Yinpterochiroptera and Yangochiroptera suborders, or did it originate once in a common ancestor and then disappear in some yinpterochiropterans? According to Peng Shi, the answer lies hidden within the amino sequence of an inner ear protein.[14]

Prestin is an essential protein for sensitive hearing in mammals. Located in the hair cells of the inner ear, it functions like a tiny motor, quickly contracting and expanding to help the cochlea amplify sound waves. Peng Shi and colleagues performed molecular convergence analyses using hearing-related genes from 10 bat species across different phylogenetic branches. Non-bat mammals were used as controls, including the echolocating common bottlenose dolphin. Their findings revealed enriched molecular convergences in all echolocating bats but not fruit bats.

Furthermore, a distinct mutation in the *Prestin* gene (leucine to methionine at position 497) mirrored the same phylogenetic pattern, appearing exclusively in echolocating bats. The modified protein is functionally significant, as demonstrated in both *in vitro* and *in vivo* experiments. Cultured cells expressing the mutated protein exhibited enhanced electrical properties upon stimulation, while transgenic mice with the mutation displayed heightened high-frequency hearing compared to control animals. These findings could even have practical implications, potentially leading to advancements in the treatment of hearing loss.

The same *Prestin* mutation in echolocating bats was found in dolphins, which also use echolocation. However, the mutation is absent in other mammals, including the non-echolocating common minke whale. Despite toothed whales and echolocating bats using different echolocation methods, both groups need high-frequency hearing to detect echoes and form a 'picture' of their surroundings. This parallel adaptation highlights evolution's ability to find similar solutions to physiological challenges, providing another striking example of convergent evolution at the molecular level.

Peng Shi believes that the above findings support a one-time emergence of laryngeal echolocation in an ancestral bat that was later lost in the fruit bat lineage. However, not all studies agree with this conclusion, as distinguishing between genomic changes that directly cause a phenotypic trait and those that occurred secondarily after the trait evolved remains a significant challenge.

Fruit bats may have subsequently lost laryngeal echolocation because they evolved acute vision and feed on fruit during the day rather than insects at night. Nevertheless, they still use echolocation but with a different method of sound generation – tongue-clicking. Hyper-short clicks are produced by rapid tongue movements that are then directed towards objects of interest. While mainly required to navigate around caves during roosting, some species use echolocation during the day to supplement their vision, including the Egyptian fruit bat. A similar mechanism is used by some blind people, which can result in the activation of their primary visual cortex. This phenomenon, known as neuroplasticity, hints at how evolution can repurpose structures for new functions.

Genetic adaptations for sanguivory

Vampire bats, the only mammals to subsist solely on blood (sanguivores), are a captivating anomaly. The three extant species, which emerged from their insectivorous counterparts approximately 26 million years ago, are found in Central and South America. The evolution of such a peculiar diet remains a mystery, but one can envision insectivorous bats being lured to insects around animal wounds, eventually progressing to feeding on the wound's blood. In a mere 4 million years, the vampire bat lineage had developed all the essential adaptations for sanguivory, marking it as one of the most rapid examples of natural selection among mammals.

Vampire bats, for example, evolved a specialised ability to detect infrared radiation, which allows them to locate warm-blooded animals and find blood vessels near the skin surface. Infrared awareness is a rarity among vertebrates, otherwise found only in three types of snakes – pythons, boas and pit vipers. The vampire bat's adaptation involves heat-sensitive nerve channels that innervate specialised pit organs around the animal's nose. In most bat species, the receptor channel becomes activated at temperatures above 43 degrees Celsius. However, in vampire bats, the channel activates at a much lower temperature of 30 degrees, making them sensitive to the infrared radiation emitted by their warm-blooded prey. This specialised sense evolved through 'alternative splicing' of the messenger RNA that codes for the excitatory ion channel TRPV1, resulting in a truncated protein which lacks 62 amino acids from one end.[15]

The hypothesis of 'one gene, one protein' was shown to be an oversimplification in the 1970s with the discovery of 'split genes'. Most genes contain non-coding regions called introns, which are removed or spliced out after transcription to mRNA. However, this process can vary, and coding regions, or exons, may be shuffled to produce proteins with different sequences or lengths. This phenomenon, known as alternative splicing, enables the creation of different proteins from a single gene, significantly expanding a species' proteomic repertoire. Recent research has underscored the importance of this process in evolution, as the resulting phenotypic plasticity can lead to population divergence, adaptation and speciation.

Remarkably, in bats, the alternative splicing of TRPVI only occurs in the nerve ganglion that supplies the pit organs on the bat's face. The full-length protein is expressed in the rest of the nervous system, maintaining normal heat and pain sensation. Humans express TRPV1 nerve channels in the tongue, skin and eyes that sense pain when we encounter capsaicin, the chemical responsible for the heat sensation in chilli peppers. However, vampire bats have repurposed the nerve channels around the nose to function as heat sensors, allowing them to detect thermal radiation instead of feeling pain.

While most bats have lost the ability to move effectively on land, vampire bats can walk, jump, and even run. They employ a unique bounding gait that relies on their forelimbs rather than their hindlimbs, as their wings are significantly stronger than their legs. The ability to run evolved independently within the vampire bat lineage, allowing them to manoeuvre around prey animals while feeding.

The vampire bat's adaptations for a diet of blood are unique and highly efficient. When a suitable blood source is found, their razor-sharp upper incisors effortlessly slice through the prey's skin, and anticoagulants in their saliva prevent the blood from clotting during feeding. However, blood is a challenging diet due to its high fluid content and low caloric value, requiring vampire bats to consume up to 1.4 times their body weight in a single feeding session. Consequently, their stomachs have adapted more than any other mammal's, expanding quickly to store blood and absorb large amounts of fluid. Additionally, their kidneys begin excreting urine within minutes of feeding. Microbes in the intestinal tract aid digestion by preventing blood clotting while simultaneously releasing essential vitamins, such as carotenoids, that the bats cannot obtain from blood alone.

The microbiomes of vampire bats' guts include hundreds of bacterial species, many transmitted by fleas and blood-sucking ticks, which can cause illness in other mammals. However, vampire bats remain unaffected due to their unique immune system, which features a constantly active first-line defence called the interferon response. Zepeda Mendoza and her colleagues also found that vampire bats carry a greater number of transposons – mobile genetic elements – than other bat species. One particular 'jumping gene', known for its strong ability to cause mutations, was found in regions of the bat's genome linked to essential traits for a blood-based diet, such as antiviral defence, fat metabolism, and vitamin processing. This finding suggests that the inserted DNA accelerated genetic changes in the surrounding areas, enabling the vampire bat to adapt rapidly to a sanguivorous diet.[16]

Blood contains toxic levels of iron, but vampire bats have adapted by deactivating their *REP15* genes. This adaptation allows them to shed excess iron from their gastrointestinal tract instead of absorbing it. Additionally, several genes necessary for insulin production have also been lost. Because blood has minimal sugar content, these insulin-related genes are no longer needed. Consequently, vampire bats have low carbohydrate (glycogen) stores, making them more susceptible to starvation. To mitigate this risk, these bats are highly sociable, and share food with starving colony members by regurgitating their meals.

CHAPTER 18

The Rat's Story

EXTREME EVOLUTION

If you want to know something about evolution, rodents are perfect to look at.
David Thybert

The success of the brown rat is reflected in numerous newspaper headlines highlighting its fecundity and adaptability. Articles with such eye-catching titles as 'The truth about Paris's rat problem' and 'New York City appoints its first ever rat czar' are not historical accounts but recent stories in reputable broadsheets. Such articles underscore a growing concern about rising rat populations in major cities, especially in the post-COVID era. But this is not a new phenomenon, and their long-standing association with human societies reflects their gamut of distinctive behaviours and habits: agility, nocturnal activity, high fertility, ability to swim, and capacity to exploit diverse food sources. Brown rats easily navigate through sewers, cavity walls and dark crevices, securing shelter and sustenance in the most improbable locations. They are gnawing machines with teeth as strong as steel that can chew through wood, plaster, and even concrete. Their extraordinary spatial awareness and intelligence allow them to map out pathways through our homes. At the same time, their highly flexible bodies enable them to squeeze through seemingly impossible gaps, even negotiating the U-bend of our toilet pipes. Their tendency to appear 'from nowhere' and their ability to spread diseases like plague, typhus, leptospirosis and hantavirus have fostered our widespread dislike of their kin – views reinforced by the frequent negative cultural portrayals in media and literature.

Despite our familiarity with the brown rat, little is known about its evolutionary history (Plate 25). Even its name is misleading, as the species is not always brown and didn't originate from Norway, as its binomial name, *Rattus norvegicus*, suggests. Genetic studies indicate that its ancestors, along with those of its sister species, the black rat, emerged in East Asia, between northern China and Mongolia, approximately 2 million years ago. They likely inhabited grassy, savanna-like environments near streams, feeding on seeds, fruit and insects. With the advent of agriculture in China around 11,000 years ago, some rat populations began living alongside humans, taking advantage of cereal harvests and stored grain. This commensal relationship persisted for millennia. Only recently have they expanded their range significantly, aided by the southward migration of humans across China between 800 and 1550 CE. After colonising Southeast Asia, rats rapidly expanded via maritime trade routes to reach Europe around 500 years ago. They then spread to Africa, the Americas and Australia during the 'golden age' of exploration and conquest.

Black rats were the first to arrive in America, stowing away on the ships of early European visitors and colonists, including Columbus's arrival in the Caribbean in the late 15th century, before spreading further afield. Recent archaeological research, including shipwrecks and land sites, shows that brown rats followed closely and were established in North America by 1731. Once ashore, the more aggressive latecomers spread rapidly and replaced their smaller cousins within decades. The exact reasons for their success remain unclear, with possible explanations including competition for food, territory, nest space, and even their predation on black rats.[1]

Brown rats are arguably the poster child of evolutionary success in urban environments, yet their genetic adaptation remains largely unexplored. To investigate further, an American multi-centre study compared the genomes of brown rats from New York City to those inhabiting their ancestral range in rural northeast China.[2] Using large-scale genomic data and powerful bioinformatic tools, the team sought evidence of 'selection sweeps' as signatures of genetic adaptation. When natural selection increases the frequency of a novel and beneficial mutation in a population, the surrounding genomic sequences can also be affected. This situation most likely occurs when allele selection is rapid, with little time for gene shuffling or recombination. The result is that large stretches of adjacent DNA containing other genes can be swept along with the advantageous allele to reach high frequencies in the population.

The study found that New York City rats have signs of selection sweep for genes associated with the nervous system, locomotory behaviour, diet and metabolism. One such gene, *CACNA1C*, is already known to influence social interactions and communication in rats. In humans this gene is linked to psychiatric disorders, and it could potentially affect anxiety behaviour in rats, including anti-predator responses or reactions to novel stimuli. Other genes (*FGF12* and *CHAT*) have been linked to rodent locomotion, suggesting that gait or movement could also have undergone adaptive change. This conclusion makes sense, given that urban rats must negotiate highly artificial, constructed environments rather than their ancestor's naturally occurring vegetative habitats. Urbanisation has also exposed rats to human diets that contain an increasing proportion of highly processed sugars and fat, foods that have led to human public health issues. This dietary shift may have modified the rat's genome, as several highlighted genes (*CHST11* and *BGTG1*) appear linked to carbohydrate metabolism. Thus, the success of urban rats seems to be partly due to the rapid genetic adaptations brought about by natural selection in response to their changing environments, adaptations that have occurred within only 500 rat generations.

News stories such as 'Health experts issue super rodents warning' highlight another aspect of the urban rats' adaptability and success: their tolerance to rodenticides. Since the discovery of anticoagulants as rat poisons in 1944, their use has sky-rocketed globally. These poisons interfere with vitamin K recycling, leading to blood clotting failure and death from internal bleeding. However, over the last 20 years, most rats and mice in the UK have developed a tolerance to these agents, making infestation control more challenging. The reduced efficacy is primarily due to genetic mutations that alter the structure of proteins vital for vitamin K activity. In the Netherlands, one-third of rodents now carry the same mutated *Vkorc1* gene, which prevents its encoded protein from binding to the poisons.[3]

Rats are members of the Rodentia, one of five orders collectively known as Euarchontoglires, arguably the most significant superorder of mammals to have evolved. In the remaining sections of this book, we will explore their evolutionary paths and extraordinary adaptations, starting with the clade that contains rats and rabbits.

Dispersal and adaptation

Approximately 90 million years ago, the Euarchontoglires diverged from the Boreoeutheria in the northern hemisphere, forming two distinct clades: Glires (lagomorphs and rodents) and Euarchonta (treeshrews, colugos and primates) (Figure 18.1). Despite the lack of unifying anatomical traits among Euarchontoglires, their phylogeny is robustly supported by nuclear and mitochondrial DNA analysis.[4]

Glires consists of two orders, Lagomorpha and Rodentia, and is the largest and most diverse clade of living mammals, with over 2,600 species. Lagomorpha, including pikas, hares and rabbits, is much the smaller of the two orders, with only 109 species, and is the more conservative regarding body size, dentition, feet and brain structure. They arose in northern India and, despite a close evolutionary relationship with rodents, exhibit significant differences. They possess four upper incisors (two large front incisors and two smaller peg teeth behind), while rodents have only two. Additionally, lagomorphs are herbivores, whereas rodents are omnivores, and lagomorphs'

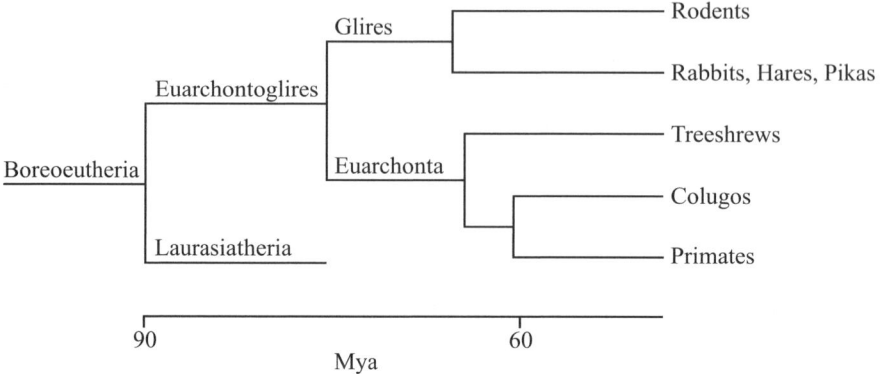

Figure 18.1 Phylogeny of Euarchontoglires. Divergence dates remain uncertain due to the rapid radiation of the superorder. Modified from Doronina *et al.* (2022).[5]

feet are covered with fur and lack foot pads. Their body plan is also more uniform, with only two variations: long-limbed (rabbits and hares) and short-limbed (pikas).

Lagomorphs

Rabbits and hares belong to the family Leporidae, but neither 'rabbits' nor 'hares' constitutes a monophyletic clade, as they are taxonomically mixed throughout the family. Biologists term this situation paraphyly, since neither group includes all the descendants of a single common ancestor. In contrast, pikas (family Ochotonidae) are monophyletic, as they derive from a common evolutionary ancestor not shared with any other group.

The small mountain-dwelling pikas are native to Asia and North America. They resemble their closest relatives, the rabbits, but are smaller and cold-adapted, inhabiting tundra, plateau and steppe environments. The earliest pikas arose in Asia and, from the Late Eocene to Middle Miocene, spread throughout Africa, Asia and North America when the climate was warmer. However, as the Earth began to cool, many species went extinct, and their distribution became restricted to the Tibetan plateau. At altitude, the environmental stresses likely led to the pikas' adaptations to the cold: high resting metabolic rates and non-shivering heat generation due to altered fat and steroid metabolism. These genetic changes gave the pikas' ancestors a great advantage when the Miocene global cooling event occurred around 14 million years ago, enabling them to survive and disperse into North America during the Earth's final transition to an 'icehouse' climate.[6]

Hares primarily inhabit open grasslands but can also be found in various biomes, including deserts, arctic tundra and forests. They thrive at different altitudes, ranging from sea level to the heights of the Himalayas. Their common ancestor originated in North America and spread across most of the northern hemisphere and Africa over the past 4–6 million years after major climatic shifts in the Miocene, Pliocene and Pleistocene. The establishment of the Bering Land Bridge allowed the colonisation of Asia and Europe around 2.5 million years ago. Throughout their evolutionary history, hares have undergone repeated hybridisations, leading to introgressions that have enhanced their adaptation to highly seasonal environments, including genes affecting circadian rhythms, fur colour and thermoregulation. In other words, contrary to the view that hybrids will suffer a reduction in fitness, recurrent introgressions have substantially contributed to hares' adaptation and speciation during their widespread radiation.[7]

Globally, there are many species named rabbit, jackrabbit and cottontail, but for most Europeans, at least, 'rabbit' means the European rabbit or coney. This species, found in heathland, grasslands, meadows, woodlands and dunes, originated in the Iberian Peninsula during the Mid-Pleistocene. Unlike the related hares, rabbits are altricial, the young being born blind and furless in their underground burrows, termed warrens. The rabbit's widespread distribution and colonisation of various regions globally is a relatively recent development driven primarily by human activity. However, genetic studies suggest that domestication involved French populations, though the exact timeline is unclear. However, the process probably began in the Palaeolithic era with hunting and gradually evolved into keeping rabbits in Roman *leporaria* and Medieval pillow mounds or warrens. This practice facilitated their spread across Europe and eventually led to their breeding as pets.[8] Today, over 300 recognised rabbit breeds exist worldwide.

Rodents

However, rodents are the mammals' most remarkable success story, comprising over 40 per cent of all extant species. They occupy virtually all terrestrial ecosystems, from tropical rainforests and deserts to the Arctic tundra, serving crucial functions such as seed dispersal and soil engineering. New rodent taxa are discovered each year, including entirely new genera. For instance, in 2005, the Laotian rock rat was identified as the only living member of a family that is unique both morphologically and phylogenetically. And in 2020, a new genus with three species was documented from the montane forests of Ecuador. Such findings support Bryan Carstens' theory that the diversity of Rodentia is greater than previously thought, and that many cryptic species remain to be discovered (see Prologue).

Rodents evolved in Asia and underwent numerous complex radiations throughout the Cenozoic, eventually spreading to all continents except Antarctica. The superfamily Muroidea, which includes rats, hamsters, gerbils and true mice, was involved in repeated colonisations of Africa, North America, Southeast Asia, South America and Eurasia. Rodents are hardy seafarers, and they reached most islands, with the notable exception of New Zealand. Indeed, almost one in five of the world's rodent species is an island endemic, with such species reaching their highest levels on large or remote oceanic islands such as Madagascar, the Ryukyu Islands, Sulawesi and the Solomon Islands.[9] Rats even reached Galápagos from Ecuador and Peru on rafts of vegetation, although today they are represented by only four species of rice rat. Rodents are particularly species-rich on islands once linked to the mainland by land bridges, such as New Guinea, which has over 70 extant species. However, the most dramatic sea crossings resulted in the colonisation of South America and Madagascar, and their stories warrant further discussion.

In the Mid-Eocene, between 48 and 38 million years ago, a sweepstake dispersal event (see Chapter 5) led to the colonisation of South America by caviomorph rodents originating from Africa, possibly involving just a few individuals from a single species. The current view is that the Eocene immigrants made a rare crossing of the Atlantic Ocean on a natural raft consisting of portions of a riverbank adorned with grasses, trees and their associated faunal communities. These floating islands likely formed during excessive flooding, when debris was swept down rivers and carried out to sea. Although South America and Africa were closer during the Eocene, it is estimated that such a journey would still take 1–2 weeks.

The sudden appearance of rodent fossils in the southern continent approximately 41 million years ago supports the rafting hypothesis. Additionally, no caviomorph fossils have been found along other possible routes, such as through Antarctica or North America. Furthermore, geologists have yet to find strong evidence for past land connections or island stepping stones, and even if such routes existed, more families of terrestrial mammals would be expected to have made the journey, not just a few monkeys (Chapter 21). However, the most convincing evidence for an ocean crossing is provided by molecular studies. Caviomorph rodents are a monophyletic group, meaning they all evolved from a single species, while their closest relatives – the mole-rats, cane rats and the dassie rat – are found in Africa. Therefore, despite the improbable sweepstake scenario, the alternative theories are even less plausible.

From a tenuous beginning, the caviomorphs underwent an explosive diversification, outcompeting other mammals, including some marsupials, to occupy virtually every major habitat. Their dramatic radiation led to an astonishing variety of forms, ranging from rat-sized to bison-sized species, although the largest are now extinct. They evolved a variety of ecologies, including the fossorial tuco-tucos, the arboreal porcupines and some spiny rats, and the semi-aquatic capybara and coypu. The various habitats include high mountains (viscachas, chinchillas and the guinea pig), grasslands (maras), forest edges (prehensile-tailed porcupines), and tropical forests (agoutis, pacas and the pacarana). Caviomorphs also expanded into the Caribbean, reaching the Bahamas and the Greater Antilles by the Early Oligocene, and are today represented by 10 species of hutias. The absence of other rodent groups, like beavers, squirrels, gophers, voles, muskrats, rats and mice, from South America before the Great American Biotic Interchange likely contributed to their rapid diversification compared to their African cousins. Indeed, caviomorphs were the only rodent group on the continent until the arrival of rats and mice from the north in the Late Miocene before the Panamanian isthmus became established. The remaining Neotropical rodents, including squirrels, gophers and pocket mice, were late arrivals from Central America and underwent more limited radiations.

Like the South American caviomorphs, all of Madagascar's endemic rodents are descendants from a single sweepstake invasion, although, in this case, from eastern Africa. Molecular clock studies date the species' arrival during the Miocene, about 20–25 million years ago, somewhat later than the arrival of the tenrecs (Chapter 5) and lemurs (Chapter 19). This monophyletic group includes 27 species and nine genera, including tufted-tailed mice and big-footed mice, all inhabiting diverse native forest ecosystems. Endemic Malagasy rodents display a wide range of forms and sizes and have adapted to arboreal, terrestrial and semi-fossorial lifestyles. However, a recent study on skull morphology suggests that these rodents evolved under unusual insular constraints, as they don't appear to have experienced strong ecological release.[10] In other words, rodent diversity on the island is not as extensive as one might expect, given Madagascar's size and habitat complexity.

Evolutionary success

A key innovation underpinning the general success of rodents was the evolution of a unique dental arrangement. The name Rodentia derives from the Latin *rodere*, to gnaw, and *dens*, tooth. All rodents are characterised by having four long, rootless incisors, an upper and a lower pair, made up of a hard enamel layer in front and a softer dentine layer at the back. Because their incisors grow constantly, they must be worn down by continuously gnawing, which maintains a chisel-shaped cutting edge, given that dentine wears away faster than enamel. Furthermore, they lack canine teeth; instead, a gap called a diastema between the incisors and their molars or cheek teeth separates the two functionally different sets of teeth. The ability of the rodent's masseter muscle to pull the jaw forward ensures that the animal's incisors do not meet when food is chewed and that the premolars and molars do not make contact when gnawing. To achieve such biting efficiency, a portion of the masseter muscle has shifted forward at least seven times, involving different clades, during the 70-million-year history of rodents. The repeated forward movement of the masseter muscle and the increasing prevalence of this trait throughout time shows that convergence has played a critical role in rodents' extraordinary evolutionary and ecological success by opening up resources not available to other mammals.[11]

Most rodents are opportunistic omnivores, feeding on vegetation, seeds and invertebrates, although a few species have evolved specialised diets with major anatomical adaptations to their guts and molar teeth. However, even these taxa retain the crucial gnawing and grinding components, the *sine qua non* of rodents. It is remarkable, therefore, that a worm-eating shrew-rat from Sulawesi should have been described recently without cheek teeth and with bicuspid upper incisors, a dental arrangement not seen in any other of the 2,600-plus species of rodent. By specialising in soft-bodied prey, the shrew-rat does not need to process food by chewing, allowing its dentition to evolve only to procure worms. Jake Esselstyn and colleagues, who first described the species, point out that the new rodent highlights the opportunistic nature of evolution, whereby the loss of key innovations allows organisms to exploit previously unavailable resources.[12] Indeed, we have already encountered such an event: the loss of tetrapod limbs that enabled cetaceans to return to the water (Chapter 11).

The success of rodents also relates to their ability to adapt to different habitats, which is due in part to fast evolution rates. Amazingly, rodent genomes evolve 4–6 times faster than those of primates. Evolutionary biologist David Thybert has calculated that 10 million years of rat evolution is equivalent to 40–60 million years for other mammals. Although the exact mechanisms remain unclear, some rodents rapidly reshape their genomes and chromosomes, even after a single genetic mutation involving transposons and repeat elements.[13] With their vast number of species and impressive phenotypic diversity, rodents serve as valuable models to help explain how mammals evolve and adapt to new and challenging environments. The rest of this chapter will explore the intriguing insights such studies have provided, including novel mechanisms to explain speciation and sex determination.

Predator evasion

African spiny mice (genus *Acomys*) range from small to medium-sized terrestrial rodents native to arid regions of Africa and parts of the Middle East (Plate 26). True to their name, these mainly nocturnal mammals are distinguished by the stiff, inflexible hairs on their upper bodies. All have large eyes and a nearly hairless, scaly tail that can be almost as long as the body.

Spiny mice intrigue cell biologists, as they possess the remarkable ability to regenerate damaged tissue without scar formation, a trait not seen in any other mammal. They possess brittle, easily damaged skin that helps them to evade predators, by shedding patches of skin when caught or bitten. Remarkably, spiny mice can fully regenerate the damaged tissue without scarring, including hair follicles, sweat glands, fur, cartilage, and even complex structures like the tips of their tails. While the exact mechanism remains unknown, a research team from the University of Kentucky suggests that a specific subtype of macrophage is important, a type of immune cell not present in other mammals. In response to inflammatory cues, a subset of the spiny mouse's macrophages releases a selection of proteins, including vascular endothelial growth factor c (VEGFC), that dampen inflammation while promoting the growth of new blood vessels and lymphatic vessels and initiating tissue regrowth. These findings are exciting, as they offer hope that potential therapeutic targets for enhancing healing in humans can be found, especially as our genome contains the relevant genes.[14]

Spiny mice have another highly unusual feature among mammals: tails armoured with bony plates called osteoderms within their dermis. While their function remains uncertain, osteoderms likely support the animal's skin-shedding defensive behaviour by preventing predators from damaging the underlying, non-regenerative tissue. Intriguingly, a similar defence mechanism is present in the fish-scale geckos of Madagascar and the Comoros. All other mammals, having evolved hairs, took a different evolutionary route and modified the epidermis rather than the dermis for protection. Thus, we see the modified guard hairs of hedgehogs, porcupines and echidnas, the ectodermal scales of pangolins, and the keratinised horns of many herbivores.

Osteoderms have evolved independently multiple times in the tetrapods, including in reptiles such as lizards, crocodiles and turtles, and in mammals including armadillos and spiny mice. Although different anatomically, the bony plates of the various lineages share developmental similarities, including groups of cells capable of mineralising within a well-organised dermis and a structure where collagen fibres connect with more rigid elements, providing flexibility without compromising strength. These findings suggest the presence of an underlying regulatory network of genes that nature has repeatedly recruited to produce protective dermal armour on divergent branches of the vertebrate tree.[15]

The African crested rat, marked with skunk-like colouring, possesses one of nature's most remarkable defences by harnessing the potent toxins of the 'poison arrow tree.' This evergreen shrub, *Acokanthera schimperi*, contains lethal levels of cardenolides throughout its parts, making it highly dangerous if consumed. Traditionally, its toxins have been used to poison the tips of hunters' arrows. A key component, ouabain, is a cardiotonic glycoside capable of causing breathing difficulties, convulsions and cardiac arrest, even in small doses.

Astonishingly, the crested rat chews the tree's roots and bark without harm and applies the lethal mixture onto specialised sponge-like hairs that absorb the poison. How it withstands toxin levels potent enough to kill an elephant remains a mystery. Nonetheless, the African crested rat's remarkable defence mechanism, unique among placental mammals, underscores the power of predation to drive the evolution of unusual protective strategies in prey species.[16]

Desert and subterranean adaptations

Cactus mice, native to the deserts of southwestern North America, display a suite of adaptations to their extreme environment, including both anatomical and behavioural traits to avoid and dissipate heat. Their large ears, nocturnal lifestyle, and a state of dormancy known as aestivation during particularly hot and dry periods all contribute to their success. Cactus mice also maintain a low metabolic rate to reduce water loss and increase resistance to heat stress. While many desert species have a decreased urine output, the cactus mouse is essentially anuric due to extreme water reabsorption by the kidneys.

Comparative genomic studies have identified widespread selection for genes related to protein degradation and removal, consistent with the need to protect cellular proteins from heat-induced denaturation. Furthermore, a reduction in the family of cytoskeletal genes may enhance the animal's tolerance to dehydration, as cell shrinkage from lack of water disrupts the network of filaments encoded by these genes, leading to cell death. This concept is supported by experiments on captive cactus mice undergoing acute dehydration, which show gene upregulation that effectively prevents kidney damage. Cactus mice also exhibit gene selection for bitter taste receptors. The thinking is that the dominant alleles code for receptors less sensitive to bitter compounds, making food more palatable in an environment dominated by bitter-tasting plants and insects. Another significant gene selection is a growth factor associated with lipid metabolism that enables the release of metabolic water from fat stores.[17]

However, the top prize for extreme adaptation – and arguably for unattractiveness – goes to the naked mole-rat, a species native to the area around the Horn of Africa (Plate 27). This reclusive, embryonic-looking mammal with wrinkled, translucent skin is the longest-lived rodent, boasting a maximum lifespan of over 30 years, in contrast to the typical two years for rats and mice. Naked mole-rats are strictly subterranean, residing in underground tunnels that they constantly enlarge, remodel, and patrol in search of roots and tubers. Other than the Damaraland mole-rat, the naked mole-rat is the only eusocial mammal living in colonies where only a single queen breeds, while the rest of the clan dedicate their entire lives to working for the colony. Both species belong to a clade of more than 30 species of African mole-rats, all exhibiting a subterranean lifestyle, although with differing degrees of sociality, from solitary to eusocial. Cooperative breeding and eusociality have evolved convergently during the clade's 30-million-year history.

As naked mole-rats don't venture above ground, tunnelling is their only means of finding food. The non-breeding workers excavate the hard-packed earth with their powerful, ever-growing incisors, aided by a quarter of their muscle mass being dedicated to closing their jaws. Their lips abut firmly behind their incisors to prevent soil from getting into their mouths, giving them a strange toothy appearance. The somatosensory cortex of the naked mole-rat's brain, which is involved in proprioception, is unusually large, with nearly a third of the area allocated to the four front teeth: an odd arrangement that reflects their use of incisors to interpret the world around them. In other words, as the naked mole-rat evolved its strange array of adaptations for life underground, its brain evolved in parallel, resulting in an unusually specialised sensory cortex.

Naked mole-rats huddle together in nest chambers in their extensive network of tunnels (totalling up to 4 kilometres). Such living conditions limit gaseous exchange with the surface air, leading to low oxygen and high carbon dioxide levels. This challenging subterranean environment created strong selection pressures for a suite of adaptations over the last 30 million years. Incredibly, naked mole-rats can withstand oxygen deprivation for up to 18 minutes without any evidence of a stroke or cardiac damage. A crucial innovation enabled their tissues, including the heart and brain, to switch from glucose to fructose as an energy source. This novel biochemical pathway is anaerobic and doesn't require oxygen.

High levels of carbon dioxide create painful acids that most mammals cannot tolerate. However, naked mole-rats do not find acids painful, nor are they affected by capsaicin, the chemical that gives chilli peppers their burn. Interestingly, while they have the relevant pain-sensing neurons, or nociceptors, like most other mammals, including ourselves, they are fewer in number and are blocked rather than activated by acids. Additionally, their capsaicin receptors do not produce the neurotransmitters needed to send pain signals to the brain, owing to alterations in their neuronal circuitry. One explanation for this remarkable adaptation is that it evolved to enable a diet of irritant-containing tubers.[18] Another African mole-rat, the highveld mole-rat, is similarly unaffected by allyl isothiocyanate (AICL), the chemical that gives wasabi its intense heat. In this species, the AICL receptor is completely switched off, a unique adaptation that likely evolved to protect the animals from the potent venom of the Natal droptail ant, which induces pain via the same receptor. Natural selection favoured the non-functioning receptor, allowing the species to move into areas inhabited by the ants that other mole-rats avoided.

The southern grasshopper mouse that lives in Death Valley is a highly aggressive predator that will attack pretty much anything that isn't bigger than itself. Favourite prey includes the Arizona bark scorpion, whose venom is potent enough to kill humans. Even if stung several times, the grasshopper mouse appears unaffected. As the scorpion's stinger became more toxic, the mice counter-adapted to tolerate the more potent venom, a classic example of an evolutionary arms race. Such selection pressures led to a novel defence: the paradoxical alteration of a nerve channel not normally affected by scorpion venom. The changes, involving amino acid substitutions, converted the protein to a venom-binding channel. Once attached, the complex acts as an inhibitor to another channel that typically transmits the scorpion venom pain signal to the brain – a unique evolutionary strategy, as most examples of resistance to deadly toxins involve structural modifications of receptors that lead to the inhibition of toxin binding. In other words, grasshopper mice have solved the predator-pain problem by enabling the toxin to bind to a non-target channel to block transmission of the pain signals the toxin would usually initiate through another channel.[19]

While the various mechanisms rodents have evolved to protect themselves against pain may seem esoteric and arcane, they provide vital insights for human pain research. Scientists have pinpointed potential genetic and molecular targets for new pain treatments, hopefully leading to more effective and safer painkillers tailored to human physiology.

Sympatry versus allopatry

The concept that new species can evolve from a surviving ancestral population while sharing the same geographical area has been debated since Charles

Darwin first broached the idea in 1859. Darwin emphasised that species are not static but tend to diverge when faced with unoccupied ecological niches, even without geographical separation, a process known as sympatric speciation. Early geneticists concurred, suggesting that new taxa could arise within their existing ranges through spontaneous mutations. However, in the 1940s, the German-American evolutionary biologist Ernst Mayr asserted that speciation could only occur after geographical, and thus reproductive, isolation. The influential polymath's views held sway, and allopatric speciation remained the dominant dogma for over 40 years.

The 1990s saw a resurgence in the debate, fuelled by theoretical and empirical studies that challenged Mayr's viewpoint and suggested that sympatric speciation was possible despite ongoing gene flow. However, demonstrating sympatric speciation is not easy. Proffered examples must meet four strict criteria: the species must overlap geographically, be sister species, exhibit reproductive isolation, and lack a history of allopatric separation. While the first three criteria can be relatively straightforward, proving the absence of a previous allopatric phase is particularly challenging as many species have shifted their ranges significantly in the past. Given these hurdles, researchers proposed that the best way forward would be to study species inhabiting a small area or an isolated island. Enter the diminutive common spiny mouse, an inhabitant of Israel's 'Evolution Canyon'.

With its two contrasting slopes, Evolution Canyon on Mount Carmel offers biologists a natural laboratory for studying evolution. The south-facing 'African' slope is hot, dry and exposed to high solar radiation. In stark contrast, the north-facing 'European' slope, just 200 metres away, is humid, cool and forested. Recently, a team led by renowned evolutionary biologist Eviatar Nevo from Haifa University analysed the genomes of spiny mice from both slopes. The two populations were confirmed to be sister taxa that diverged from a common ancestor around 20,000 years ago without undergoing an allopatric phase.[20]

Crucially, Nevo's team showed that the genomes of the mice have diverged significantly, with differences in structure and methylation patterns suggesting incipient speciation. For instance, the mice on the south-facing slope have developed adaptations related to immunity, temperature regulation and the ability to concentrate urine, which are best suited for their arid environment and marked temperature difference between day and night. Additionally, selection for genes controlling circadian rhythms may explain why they are active earlier in the day, given their exposure to more prolonged daylight. In contrast, mice on the cooler slope have positively selected genes for metabolism and cell cycle, possibly due to more abundant, regular and different food supplies. Although gene flow continues, it will likely decrease until full speciation, aided by the ongoing selection of olfactory receptor genes contributing to mate choice and reproductive isolation.

Interbreeding causes gene homogenisation and prevents gene drift and speciation, while natural selection favours beneficial genes, leading to reduced gene flow and the emergence of new taxa. Nevo's findings demonstrate that the environment is the driving force behind modifying the spiny mouse's gene pool, even in the presence of interbreeding and gene flow. In other words, the adaptation of spiny mice depends on whether they live on the drier, hotter side of the canyon or the more humid, cooler side. The evidence for sympatric speciation is now compelling and extends beyond the spiny mouse to other canyon fauna and flora, including bacteria, wild barley, fruit flies and beetles. No wonder Evolution Canyon has been nicknamed 'Israel's Galápagos Islands', where microclimate selection overrides gene flow to drive incipient sympatric speciation at a microscale.

Before modern genomic sequencing, assessing the environmental impact on a species' genetic makeup and proving sympatric speciation was difficult. However, this has now changed, shifting the argument from plausibility to frequency and how much of life's diversity it underpins.

Sex determination

Marsupial and placental mammals typically use an XX/XY sex-determination system, where the X and Y sex chromosomes are believed to have evolved from a pair of autosomes. An embryo with an X and a Y chromosome will develop into a male, while an embryo with two X chromosomes will develop into a female. Sexual differentiation occurs because the Y chromosome carries a master-switch gene, *SRY*, which activates 'male' genes on other chromosomes, with *SOX9* being the crucial target. Indeed, *SOX9* triggers the development of a male fetus by producing a protein that binds to DNA, initiating the formation of testes. The male sex organs then produce testosterone, a steroid hormone that stimulates the development of the rest of the male reproductive system.

During early mammalian history, the X and Y chromosomes appeared similar, being X-shaped and pairing up neatly during meiosis. Like autosomes, the sex chromosomes exchanged genes and could repair DNA and eliminate harmful mutations. However, around 166 million years ago, a dramatic change occurred in therian genomes. A large section of the Y chromosome became inverted, preventing the X and Y chromosomes from aligning. As a result, the Y chromosome could no longer exchange genes with the X chromosome, leading to its degeneration as mutations accumulated and genes were lost. In humans, the Y chromosome is now only a third of its original size, containing

3 per cent of the genes it originally shared with the X chromosome. This ongoing degeneration has led some scientists to speculate that the Y chromosome may whither and disappear entirely, potentially leading to the absence of males and the extinction of some mammals, including humans. Others believe that evolutionary forces would strongly act against such a loss despite no known compensatory mechanism.

The Amami spiny rat, which lives on one of the small Ryukyu islands in Japan, has long puzzled biologists, as its sex chromosomes defy the standard arrangement (Plate 28). Unlike other mammals, the male rodent has lost its Y chromosome, while females have just one X chromosome, yet the species has not become extinct. Recently, Asato Kuroiwa and her team at Hokkaido University explained this genetic paradox after generating high-quality genome sequences of both sexes.[21]

Kuroiwa and her colleagues found several male genes, including those controlling sperm production, had translocated onto the X chromosome. This observation suggests that the transfer had happened before the Y was lost, analogous to 'rats fleeing a sinking ship'. But the *SRY* gene – the conductor of the male orchestra – was not among them, and no trace was found elsewhere in the genome. How is it possible, therefore, for male mice to express SOX9 without the *SRY* gene? Intriguingly, Kuroiwa discovered that male rats possess a tiny duplication in one of the two copies of chromosome 3, a small stretch of nucleotides located in the regulatory region upstream of *SOX9*. The team confirmed the significance of this finding by showing that the tiny, duplicated region acts as a potent enhancer for *SOX9* and effectively replaces the role of *SRY*.

Such duplications, known as copy number variations, are challenging to detect, which explains why previous attempts to understand how male spiny rats become male have failed. The duplication likely arose around 2 million years ago, when the Amami spiny rats diverged from a related species that still possessed a Y chromosome. According to Kuroiwa, once this duplication occurred, the loss of the Y chromosome would no longer have resulted in the loss of the species. Surprisingly, this otherwise nondescript rat from Amami Oshima Island demonstrates that the generation of new sex-determining genes may not be that difficult from an evolutionary perspective. This remarkable conclusion should provide reassurance to those concerned about the potential degeneration and disappearance of the human Y chromosome, alleviating fears that such an event would lead to humanity's extinction.

A final thought before we allow primates to take centre stage. In his book *The Earth After Us*, Jan Zalasiewicz, a geologist at the University of Leicester, postulates that rodents, particularly rats, would be the most likely mammals to inherit a post-human world.[22] He based his doomsday view on the characteristics of rodents that we have highlighted: high fecundity, exceptional adaptability, unrivalled dispersal ability, having reached most continents and islands, and persistence despite determined efforts to control their populations. Other animals, like feral cats and pigs, thrive in diverse ecosystems but are less widespread than rats. Whether the next mass extinction relates to human activity or a catastrophic natural event, rats are the likely winners. However, readers should not fear an impending tsunami of fur, as, based on past mass extinction recovery rates, rodents would not inherit the Earth for at least 3–10 million years. Nevertheless, Zalasiewicz suggests his findings serve as an alarm call for humans to recognise their adverse impact on the environment and how our actions could mould the future.

CHAPTER 19

The Lemur's Story

SWEEPSTAKE DISPERSAL

Everything about them is faintly absurd.
John Gimlette

The travel writer John Gimlette was right. You always remember your first lemur, with their teddy-bear bodies, human-like hands, furry pyjamas, and the expression of surprise.[1] For me, however, it was their plaintive song that left the most profound impression. We were birding amid the morning's stillness of Madagascar's eastern lowland forest when the unmistakable wails of a distant family of indris filled the air. We headed quickly towards the sound and were soon rewarded with a male's long call high above us. Almost immediately, his mate joined in, their duet reminiscent of the haunting calls of humpback whales: a series of long notes followed by eerie wails, beginning high and descending the scale. Soon, the whole family joined in, with the chorus filling the forest with a ghostly lament that provoked replies from a distant group.

Despite being the largest of the lemurs, the indris' black and white fur, rudimentary tail, large hands and feet, and round furry ears were surprisingly difficult to distinguish amongst the foliage. But these arboreal choristers were my first encounter with wild lemurs, and the forest's unforgettable soundscape, a symphony of nature, still gives me goosebumps whenever I play my iPhone recording. Indris are known as the 'singing lemurs', and their communications are considered the most remarkable in the animal kingdom – sounds that have evolved to help the group stay in touch and proclaim territorial primacy.

But there is more to the indris' calls than evocative sounds. A recent study by Chiara De Gregorio from Warwick University revealed that the primate's vocal communications, whether calls or songs, exhibit isochrony, meaning there is an equal time interval between their sounds and notes, creating a consistent beat or rhythm – much like music.[2] Since Darwin's era, animal songs have been hypothesised as precursors to human vocal displays. De Gregorio's finding supports such an idea, and the fact that indris have the most vocal rhythms, even surpassing those of songbirds and other mammals, implies that elements of human musicality evolved early in the primate lineage. Furthermore, as alarm calls likely pre-dated more complex vocalisations such as songs, isochrony could be an ancient rhythm from which other rhythmic patterns developed.

At this point, we should step back in time and meet the colugos, the closest living relatives of lemurs, as they provide valuable insights into what our earliest ancestors might have been like and shed light on why humans suffer from some rare genetic disorders of the eyes, muscles and bones. The colugos, also known as flying lemurs, may look like bats, with a stretchable patagium, but unlike bats, they cannot fly. Instead, they jump off branches and glide through the air like flying squirrels. Their gliding membrane is the most elaborate of any living vertebrate, restricting their movement on the ground and confining them to forest environments. This lifestyle, added to the fact that they're found in the remotest parts of Asia, makes studying these fascinating mammals challenging, especially as they fail to survive long in captivity.

Current taxonomy describes their order, Dermoptera, as one of the least speciose of all Mammalia, consisting of just two extant species – the Sunda colugo and the Philippine colugo – and their phylogeny has been a subject of debate for over a century. Previously, some experts posited a close relationship with treeshrews because of shared physical characteristics. However, recent advancements in genetic sequencing have demonstrated that colugos are more closely related to primates, a discovery that significantly enhances our understanding of early primate evolution.

For example, comparative genomics shows that the number of odorant and pheromone receptor genes in colugos falls between treeshrews and primates, suggesting that the decline in olfaction began before the first primates emerged.[3] In contrast, there was a positive selection for genes related to vision and hearing. These findings suggest a genetic trade-off – heightened vision and reduced olfaction – that supported the arboreal lifestyles of colugos and primates. Interestingly, some positively selected genes lead to human hearing disorders and eye conditions like macular degeneration when mutated. Additionally, the adaptive changes in a suite of muscle and bone genes that enabled gliding in these mammals are linked to muscular, joint, and bone disease in humans. In other words, several of our ophthalmological and musculoskeletal disorders can be traced back to our ancestors' genetic adaptations for life in the trees.

The earliest primates

The Malagasy name for the indri is *babakoto* or 'ancestor of man', a fitting moniker given recent genetic studies highlighting their divergence around 60 million years ago, making them part of the earliest primate branch (Figure 19.1). Lemurs share a common ancestor with lorises from India and Southeast Asia, as well as galagos (bushbabies) and pottos from Africa. Together, these species form a superorder known as Strepsirrhini (strepsirrhines), which means 'comma-shaped nose' – like those of dogs and cats. In contrast, the remaining primates, from tarsiers to humans, are categorised as haplorrhines, or 'simple-nosed', because they all have nostrils with a round opening.

As highlighted, the earliest primates likely evolved from small, nocturnal, insect-eating mammals that adapted to life in trees. However, the evolutionary forces that led to their key characteristics – grasping hands, forward-facing eyes and depth perception – remain hotly debated, with several competing theories. One, the arboreal theory, suggests that manual dexterity and improved stereoscopic vision developed to help leap from branch to branch through the canopy. Another, the visual predation theory, posits that these traits, especially forward-facing eyes, evolved as adaptations for hunting insects and small vertebrates in a complex three-dimensional environment. Yet a further theory

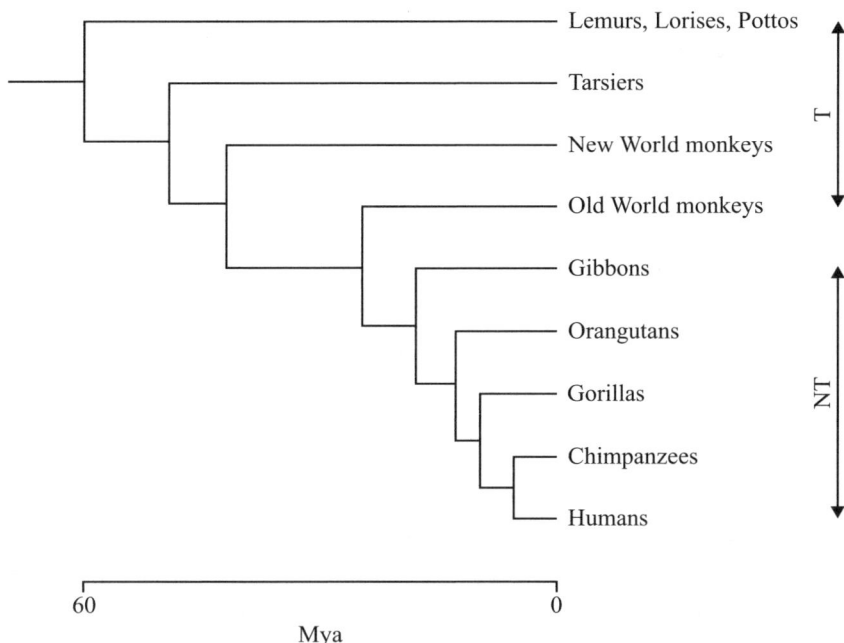

Figure 19.1 Phylogeny of primates. T, tail; NT, no tail; Mya, million years ago.

suggests that the spread of angiosperms may have created new ecological niches, with primates' fine hand movements evolving to aid in exploiting these new food sources. Such differing ideas highlight the complexity of evolutionary biology, suggesting that primate traits likely evolved through a dynamic interaction of diverse ecological and adaptive pressures.

The traditional view of primate origins suggests that the earliest crown primate, or the last common ancestor of all living species, appeared shortly after the Cretaceous–Palaeogene (K–Pg) boundary. This perspective, known as the 'explosive model', proposes that the sudden extinction of dinosaurs and the rise of angiosperms created new ecological niches that primates quickly exploited. The primary evidence for this model is the absence of primate fossils before the K–Pg event. However, molecular genetic studies present a different picture, indicating divergence dates that pre-date any known fossils, placing the origin of crown primates in the Cretaceous alongside most modern placental lineages. Proponents of this alternative view, encompassed by the 'long fuse' or 'short fuse' models (Chapter 9), argue that an incomplete early primate fossil record explains the missing 20–25 million years.

Regardless of which model is preferred, early primates had undergone adaptive radiation by the time of the Paleocene–Eocene boundary, diversifying into various ecological niches across a broad geographic range. Many marmoset-sized species had already spread throughout Asia, Europe and North America, aided by the land bridges connecting the large landmass of Laurasia. However, they did not reach South America, an event which occurred later and from an unexpected direction (Chapter 20). Fossils of the earliest known crown genus, *Teilhardina*, were widely distributed across the northern hemisphere and appeared suddenly during the Palaeocene–Eocene Thermal Maximum (PETM), a brief period 56 million years ago when global temperatures rose by 5–8 degrees Celsius.[4] While the exact geographic origin of *Teilhardina* remains uncertain, they likely dispersed east to west, migrating across the Turgai Strait from South Asia into Europe, eventually reaching North America via Greenland. These primates were better adapted to an arboreal lifestyle than their ancestors, having evolved increased foot mobility and grasping abilities, with claws replaced by flat nails, opposable thumbs, long toes, and more flexible and robust ankle joints. Their brains also became more complex, with an expanded neocortex to process enhanced sensory inputs, while their faces became flatter as their snouts shortened.

The warm climate of the Eocene allowed tropical animals to thrive in the forested northern latitudes. However, as global temperatures later cooled, many species, including early primates in North America and Europe, went extinct. The surviving populations, equipped with new adaptations, migrated southward, spreading through the Arabian Peninsula into tropical Africa.

The centre of primate evolution shifted from Asia to Afro-Arabia. During the Oligocene, these early primates flourished, diverging and evolving into lineages that produced many of the living primate species we know today, including lemurs, tarsiers, New World monkeys, Old World monkeys, gibbons, orangutans, gorillas, chimpanzees and humans.

Strepsirrhines, or lower primates, form a monophyletic group distinguished by two key features: a moist nose tip or rhinarium and a tapetum lucidum – a reflective layer located behind the retina that functions like a tiny mirror. This specialised structure enhances night vision by reflecting light that passes through the retina back toward the photoreceptors. For nocturnal animals that depend heavily on vision, every photon is a valuable resource, and the tapetum lucidum ensures that photons initially lost get a second opportunity to trigger a neural response, thereby improving the clarity of images formed in low-light conditions. They also have smaller brains than all other primates, large olfactory lobes for smell, and a vomeronasal organ that detects pheromones that determine social and reproductive behaviours. However, one of the most distinctive features is their modified teeth on the lower jaw, known as a tooth comb. Formed from thin, long incisors and canines, the comb-like structure protrudes forward and outward, allowing the animals to groom themselves effectively and extract insects and gum from tree bark, which forms part of their diet. To maintain this dental tool, strepsirrhines have a secondary tongue, called the sublingual, used for cleaning the tooth comb after use. In contrast, higher primates lack such dental modifications and instead rely on their hands and fingers for grooming and food acquisition.

Phylogenetic analysis based on 'jumping' gene studies suggests that early strepsirrhines diverged in Africa, giving rise to the lorisiform and lemuriform lineages.[5] The lorisiforms then diversified into various species, including galagos, angwantibos and pottos, which remained in sub-Saharan Africa, and the lorises, which colonised India, Sri Lanka, China and the Philippines. Unlike the isolated lemurs of Madagascar, lorisiforms faced significant competition from the more aggressive and intelligent higher primates, which had evolved by the Oligocene. These ancestral monkeys drove many early diurnal primates to extinction, leaving behind only a few nocturnal and elusive species, such as slow lorises – the only venomous primates. Slow lorises have a toxic bite due to a toxin produced by licking a gland on their arm and activated by their saliva. During the Miocene, both slow lorises and cobras spread across Southeast Asia, and the evolution of primate venom and distinctive fur patterns may have been an adaptive strategy to deter predators, functioning potentially as a form of Müllerian mimicry with spectacled cobras.

Improbable immigrants

Like tenrecs, lemurs reached Madagascar through a random sweepstake dispersal event, floating across on mats of vegetation around 50 million years ago. In support of this hypothesis, large islands of vegetation are occasionally sighted drifting from the mouth of the Zambezi, the largest river to the west of Madagascar. Furthermore, as discussed in Chapter 5, the currents were more favourable in the past, flowing east to west until about 23 million years ago, when they reversed, making such crossings even less likely. The ancestral lemur's small size and ability to alternate between endothermic and ectothermic states likely contributed to their survival during the journey. To paraphrase George Gaylord Simpson, the proposer of the sweepstake route, any event that is not impossible becomes probable, given enough time.

Over millions of years, the founding population of lemurs, in theory a single pregnant female, evolved and diversified into one of the most varied primate groups on Earth, with some species growing as large as gorillas and resembling bears. However, only the smaller species survived, leading to more than 100 lemur species recognised today. This number has significantly increased from just 35 species identified 25 years ago, thanks mainly to advances in molecular genetics that have revealed many cryptic species. Irrespective of the exact number, lemurs are remarkably diverse, accounting for over 15 per cent of all living primates yet residing on an island that constitutes less than 1 per cent of Earth's land area. Their diversity reflects Madagascar's wide range of ecosystems, with each lemur species adapting to a specific habitat – rainforest, dry deciduous forest or spiny forest. As a result, lemurs are often cited as a textbook example of adaptive radiation. Like Darwin's finches of the Galápagos Islands, *Anolis* lizards in the Caribbean, and the cichlids from Lake Victoria, lemurs were thought to have undergone a rapid speciation burst followed by a slowdown as available niches became occupied.

However, recent studies have questioned this scenario, and suggest that lemurs didn't undergo the typical 'early burst' pattern of diversification seen in other adaptive radiations. This observation raises two intriguing questions: when and how did the extraordinary diversity of lemurs emerge? To explore these issues further, a research team led by David Weisrock from the University of Kentucky employed a detailed phylogenetic approach to trace the evolutionary history of strepsirrhines.[6] Their analysis revealed that three separate episodes of speciation occurred around 15 million years ago, with speciation rates continuing to increase up to the present without any signs of slowing down. In essence, lemur evolution resulted from multiple distinct radiations occurring alongside a relatively steady, gradual process of evolution.

High diversification rates, especially following island colonisations, are closely associated with the number of unfilled ecological niches. However, the habitats that lemurs first encountered in Madagascar were considerably less diverse than those of today – a result of Madagascar's spell in the Earth's arid belt as it slowly drifted northwards, after its separation from India. All strepsirrhines, like most primates, prefer warm, tropical environments. Yet, when lemurs first arrived in Madagascar, the climate was drier and cooler than in Africa, and its rainforests, spiny forests and grasslands had yet to develop. Additionally, the arrival of heavy seasonal rainfall patterns and Indian Ocean monsoons had to wait until approximately 23 million years ago. The uplift of the central and northern mountains was in its infancy and would continue for 25 million years, adding further to the variety of available habitats. As the island headed north and encountered the trade winds around 30 million years ago, the temperatures and rainfall gradually increased. The resulting river systems acted as substantial barriers to dispersal, creating isolated refugia and fostering allopatric speciation. Overall, these geological and climatic changes provided new opportunities for speciation. In other words, although limited initially, the ecological niche space for lemurs increased over time, contributing to their continued speciation.

Weisrock's group also observed that lemur clades with the highest diversification rates – specifically mouse lemurs, sportive lemurs and true or brown lemurs – exhibit high levels of introgression. This observation implies that hybridisation is not an evolutionary dead-end but may serve as a speciation catalyst. The underlying genetics are not fully understood but could involve reinforcement, where reproductive barriers are strengthened by selection against hybrids (a topic explored further in Chapter 20). Alternatively, hybridisation could promote speciation by mixing pre-existing genetic variants or introducing new alleles into different populations. Regardless of the underlying mechanisms, the discovery of extensive introgression throughout lemur history challenges the long-standing notion that hybridisation hampers speciation by producing offspring with reduced fitness or by homogenising gene pools, thereby inhibiting speciation.

Another feature of Madagascar's climate – unpredictability – profoundly affected the evolution of lemurs.[7] For most mammals, erratic weather patterns can be devastating, whether through direct effects like droughts or floods or from indirect consequences such as disrupted fruiting and flowering cycles or even the complete collapse of food supplies. Around 10 million years ago, the island's rainfall became unusually variable, presenting lemurs with an evolutionary choice: adopt a 'slow and steady' strategy or embrace 'life in the fast lane and die young'. Natural selection tends to favour slower reproductive cycles when juvenile mortality is higher than that of adults, while rapid reproduction is advantageous when adults face more significant risks than their young. Sifakas chose the 'slow lane', with females reaching sexual maturity at 3–4 years, producing a single offspring after a 160-day gestation period, and often waiting several years before reproducing again. On the other hand, mouse lemurs and the black-and-white ruffed lemur opted for the 'fast lane', characterised by short gestation periods, large litters, rapid sexual maturation and the ability to breed twice a year under favourable conditions. From the perspective of adapting to climate change, the rapid reproductive cycle of mouse lemurs – the fastest among all primates – makes sense. The unpredictability of fruiting trees also helps explain why there are so few medium- to large-sized frugivorous lemurs, and why lemurs eat less fruit than other primates.

The remarkable diversity of mouse lemurs went largely unnoticed until the availability of genetic analyses allowed the identification of many cryptic species. While mouse lemurs share a common ancestor that lived between 18 and 11 million years ago, many species emerged within the last million years. The mechanisms underlying such rapid speciation have intrigued biologists. One factor is the rapid mutation rate of mouse lemurs, the fastest of any mammal.[8] Another relates to Madagascar's past climatic fluctuations. During the Pleistocene, while glaciers in the northern hemisphere advanced and retreated, the southern hemisphere experienced cycles of cool, dry periods alternating with warm, wet ones. These cycles happened quickly in geological terms. As the climate cooled and dried, forests shrank, while they expanded during warmer, wetter periods. Mouse lemur populations would have become isolated during the cooler periods as forests fragmented. When the forests expanded again, allowing the separated populations to meet, they no longer interbred, having evolved into distinct species – a classic case of allopatric speciation. The cryptic, nocturnal mouse lemurs mainly use olfactory and vocal clues to locate potential mates and avoid hybridisation with other species that occupy the same habitat.[9]

Adaptations

The aye-aye belongs to the oldest known lineage of lemurs, which, according to some palaeontologists, may have arrived in Madagascar through a separate dispersal event.[10] Either way, they are one of the most unusual primates on the planet, renowned for their unique foraging technique. The largest of all nocturnal primates, the aye-aye has a penchant for wood-boring beetle larvae, which it locates by tapping its exceptionally long, slender middle finger against tree trunks. It then uses its large ears to detect subtle changes in sound that might betray hidden prey. Once a potential grub is located, the aye-aye uses its forward-facing, ever-growing incisors to gnaw a hole in the wood. It then inserts its middle finger into the hole, twisting it back and forth to hook and extract the grub. The evolution of this 'percussive foraging' strategy has resulted in several unique adaptations not seen in other primates, such as acute hearing, an elongated middle digit, and enlarged, ever-growing incisors. Because the fingers and thumbs are so specialised for finding food, the aye-aye has evolved a tiny extra thumb that can move in three directions to enhance its grip.

Interestingly, the aye-aye's entire dental structure is so similar to that of rodents that early taxonomists mistakenly classified it as a type of squirrel. But the resemblance to squirrels goes beyond its incisors. Recently, Philip Cox and his team from the University of York and Hull York Medical School used CT scanning to create three-dimensional reconstructions of the skulls and lower jaws of the aye-aye, squirrels, and other primates and rodents.[11] The data were analysed using statistical software, allowing the researchers to visualise how the evolutionary paths of the aye-aye and squirrel converged. Despite their entirely different evolutionary histories, the analysis revealed a remarkable similarity in both species' skulls and lower jaws. This finding suggests that the demands of producing a high bite force while eating – in the aye-aye for gnawing holes to obtain beetle larvae and in the squirrel for cracking nuts – have led the aye-aye to evolve not only rodent-like teeth but also a squirrel-shaped skull and jaw. In other words, lifestyle and ecology can strongly influence a species' phenotype and almost override its ancestral history.

While the aye-aye diverged first, the relationships among the remaining four lemur families – indri and sifakas; dwarf and mouse lemurs; true lemurs and bamboo lemurs; and sportive lemurs – are less clear since they diverged rapidly sometime between 42 and 30 million years ago. Over the next tens of millions of years, they evolved and diversified into the most varied group of primates, exhibiting a remarkable array of adaptations to diverse niches.

Madagascar's rainforests, particularly in the island's eastern region, have fostered the evolution of several species uniquely adapted to the canopy-rich environment. The indri, one of the largest living lemurs, is a prime example. It has developed powerful legs and an upright posture, allowing it to leap up to 10 metres between trees in a single bound. Its loud calls and songs, which can be heard over several kilometres, are also well suited to the dense forest where visual signals are less reliable.

Sifakas are a group of tree-dwelling species with remarkable adaptations. Their digestive systems have evolved to handle the fibrous diet of leaves and bark, with intestines 14–15 times their body length and optimised for extracting nutrients. Furthermore, sifakas exhibit convergent evolution with other leaf-eating mammals, particularly of genes related to detoxifying plant toxins and extracting energy from leaves. Their unique mode of locomotion, called vertical clinging and leaping, allows them to cling to tree trunks and leap between them easily, a skill perfectly suited to dense forest habitats. Their elongated fingers and strong grasping toes enable a secure grip on branches. On the ground, sifakas move by hopping on their hind legs, using their arms for balance. Like indris, their social structure is characterised by small, cohesive groups that allow them to defend territories and share resources in competitive forest environments.

Mouse lemurs, the smallest primates in the world, are nocturnal and arboreal and have developed large eyes and excellent night vision, enabling them to navigate the rainforest's dark understory. Their small size allows them to move swiftly and discreetly through the underbrush, evading predators and efficiently exploiting food resources that larger lemurs cannot access. To cope with the fluctuating availability of food, especially in drier regions, they exhibit a remarkable ability to enter torpor, a state of lowered metabolic activity, to conserve energy during periods of scarcity. Scientists believe these miniature primates may soon replace fruit flies, zebrafish, worms and mice as the primary laboratory research animal.[12] The essential advantage is that their genes are a close enough match to humans to enable relevant medical research. They are not endangered and are easy to manage, reproduce quickly, and yield many offspring. Importantly, unlike other biological 'guinea pigs', they have naturally occurring mutations that cause obesity, hypercholesterolemia, pre-diabetes and cardiac arrhythmias. Ageing lemurs also develop dementia associated with the build-up of abnormal protein plaques around their brain cells – just like humans with Alzheimer's disease.

Lemurs inhabiting Madagascar's dry forests and spiny forests have evolved different lifestyles. The ring-tailed lemur, for example, is more terrestrial and social, with adaptations that allow it to thrive in the arid, open habitats of southern Madagascar (Plate 29). Its iconic striped tail, held upright when moving in groups, acts as a visual signal that helps maintain cohesion in the open terrain. They are opportunistic feeders, adapting their diet to the seasonal availability, from fruit and leaves to small vertebrates and insects. Ring-tailed lemurs use river basins during the dry season as dispersal corridors and have moved across Madagascar more easily than the restricted arboreal species. This fact accounts for their high levels of genetic diversity despite severe habitat fragmentation throughout their range. Although possessing smaller brains than monkeys and apes, ring-tailed lemurs can organise sequences, understand basic arithmetic computations, and preferentially select tools based on functional qualities. These experimental results suggest that some of our more complex numerical abilities may be rooted in our ancient evolutionary past, emerging at least 60 million years ago before the divergence of strepsirrhines.[13]

Finally, I need to highlight the remarkable adaptive radiation of the bamboo lemurs. In the forests of Ranomafana, three closely related species – the grey bamboo lemur, the greater bamboo lemur and the golden bamboo lemur – live sympatrically, feeding on bamboo. Coexistence is possible because each lemur targets a distinct bamboo niche. For example, the grey bamboo lemur only eats the leaves of *Cephalostachyum perrieri*. In contrast, the greater bamboo lemur and the golden bamboo lemur feed on *Cathariostachys madagascariensis* (the giant or woody bamboo), with the former eating the pith from mature stalks while the latter only takes new shoots and leaves. Scientists were intrigued by why one species would spend so much time and effort chewing the tough stalks to obtain pith while the other eats the readily available leaves.

The answer was simple: cyanides. These toxins, which contain a cyano group, block the body's ability to generate energy from oxygen. As a result, consuming the tender, cyanide-rich parts of bamboo would be fatal for the greater bamboo lemur. Although bamboo uses this chemical to defend against herbivores, the pith contains no cyanide. However, by eating the fresh shoots, the golden bamboo lemur consumes an astonishing 12 times the lethal cyanide dose for a mammal of its size each day. It has evolved to tolerate such levels by enhancing its detoxification processes and modifying its cellular respiration and thyroid function to negate the toxin's effects.

One can imagine a plausible scenario for how the golden bamboo lemur's extreme adaptations arose. An ancestral population of lemurs developed a less sensitive bitter taste receptor on the tongue and palate, allowing it to tolerate small amounts of cyanide and to eat bamboo shoots.[14] In response, the bamboo increased its cyanide production as a defence mechanism, sparking an evolutionary arms race between the lemur and the plant. Over time, the lemur steadily increased its cyanide tolerance while the bamboo ramped up its cyanide production. The ongoing battle eventually led the golden bamboo lemur to carve out a unique niche, thriving on a toxic food source that no other lemur could stomach. This unique survival strategy is a testament to the power of natural selection to drive speciation. In contrast, those ancestral bamboo lemurs who remained intolerant of the cyanide-rich shoots fed on bamboo pith and evolved into what is now known as the greater bamboo lemur.

The hyper-specialisation of bamboo lemurs provides a classic example of the evolutionary trade-off hypothesis, which posits that as a population adapts to a specific environmental condition, its ability to thrive in other ecological niches diminishes. Most herbivores face a variety of plant toxins, and thus rely on a broad range of detoxifying enzymes, especially from the cytochrome P450 superfamily. However, because bamboo lemurs primarily need to neutralise a limited number of bamboo-specific toxins, their P450 enzyme repertoire has narrowed due to gene loss – a streamlined toolkit designed to target potent cyanogenic glycosides.[15] In other words, genes not involved in the lemurs' defence against cyanide have experienced relaxed selection and a loss of function.

CHAPTER 20

The Tarsier's Story

SPECIATION GENES

A modern archive of evolutionary changes.
Jürgen Schmitz

As dusk descends in the Sumatran rainforest, a tiny and captivating primate stirs in the dense canopy. Small enough to fit into the palm of your hand, this endearing creature emerges from its daytime hideaway to embark on a night of foraging. Its large, immobile eyes, each bigger than its brain, soon spy a grasshopper crawling along a branch. Although its eyes cannot move, it compensates with a unique bony adaptation: highly modified cervical vertebrae that allow it to rotate its head nearly 180 degrees in each direction, much like an owl, giving it an almost complete 360-degree field of vision. Aided by ultrasound tracking, the small carnivore skilfully snatches the unsuspecting insect with its long, slender fingers, equipped with sticky pads, and quickly devours it.

This animal is a tarsier, one of the world's smallest primates (Plate 30). It belongs to a family of around 14 species, whose name comes from their specialised, extremely elongated foot bones, or tarsi. Tarsiers have powerful hind legs with fused tibia and fibula, allowing them to jump up to 6 metres between trees, a distance over 40 times their body length. While nails replace claws, they retain 'grooming' or 'toilet' claws on the second and third toes, which are used for removing ticks, lice and other parasites from their dense fur. Despite their small size, tarsiers have strong jaws and a wide mouth, enabling them to consume relatively large prey. They are also known for their high-pitched squeaks, typically heard at dusk and dawn when they leave and return to their roosting sites. However, recent studies have shown they can also produce and detect ultrasonic calls beyond human hearing. For example, the Philippine tarsier emits sounds at a dominant frequency of 70 kHz and can hear up to 91 kHz, while humans can only hear up to 20 kHz. This exceptional hearing helps them locate prey, as they feed primarily on creatures that communicate through ultrasound, such as insects, geckos and lizards.

But what can these strange, insular mammals from Southeast Asia – Borneo, Sulawesi, Sumatra, Java and the Philippines – reveal about the evolution of early primates?

Until recently, tarsiers caused problems for systematists, and their line of descent has been the subject of intense debate. Their small size and strange body morphology have led these elfin-like mammals to jump from branch to branch on the primate evolutionary tree. Once, they were thought to have diverged well before the emergence of lemurs and anthropoid (higher) primates, a view known as the 'tarsier-first hypothesis'. Other scientists placed them close to the strepsirrhine lineage, as their teeth, jaws, nipples and uterus resemble those of lemurs and lorises – the 'prosimian monophyly hypothesis'. (Prosimian is an obsolete term for a group of primates that include lemurs and lorises, but not monkeys, apes and humans, i.e. simians.) Another view, the 'haplorrhine hypothesis', is that tarsiers lie close to monkeys and apes, given the similarity of their eyes, noses and placentae. To further complicate matters, some biologists, apparently defeated, resorted to placing them in an unresolved trichotomy of lemurs, tarsiers and higher primates.

However, knowledge of the vitamin C requirements of different primates hinted at the answer. In common with simians, tarsiers must obtain vitamin C from their diet due to 'knock-out' mutations in their L-gulonolactone oxidase (*GULO*) gene, which blocks the final step in the vitamin's synthesis. In contrast, strepsirrhines retain the ability to produce vitamin C, as they lack such mutations. The tarsier's *GULO* pseudogene, which evolved around 61 million years ago and lacks seven of the gene's 12 exons, is also present in monkeys, apes and humans, indicating a close evolutionary relationship among them.

Guinea pigs are the only other mammals (aside from a few bat species) that cannot produce the vitamin. However, their mutations differ and evolved independently around 14 million years ago. Vitamin C, a water-soluble antioxidant, protects organisms from oxidative stress and plays a crucial role in collagen synthesis, which is why a deficiency leads to scurvy. The loss of GULO function in some mammals is best explained by neutral selection. Since vitamin C is readily available in the diets of primates and guinea pigs, the inability to produce it does not pose a significant selective disadvantage. The reason that no other genes in the vitamin's synthetic pathway are lost is that they all encode proteins that serve vital roles in multiple biochemical processes.

Readers will not be surprised to learn that the most robust solution to the long-standing phylogenetic problem of tarsiers lies hidden in their DNA. Jürgen Schmitz and colleagues from the University of Münster in Germany sequenced the complete genome of a Philippine tarsier and compared it to the genomes of humans and lemurs.[1] The results were unequivocal: the 'haplorrhine hypothesis' was fully supported. Indeed, it is now universally accepted that tarsiers are closely associated with anthropoids, not with lemurs and bushbabies (see Figure 19.1). The researchers achieved this breakthrough by studying DNA sequences called transposons, often referred to as 'jumping genes'. Unlike retroposons, a form of jumping gene that relocates through an RNA intermediate (Chapter 2), transposons move using a DNA intermediate. Such sequences, which belong to three different types or families, can move from one position in the genome to another, frequently replicating themselves. While all transposons eventually lose their ability to move, newer transposons can insert themselves into older ones, although the reverse doesn't occur. By examining which transposon families were located within others, the researchers could determine when specific transposon families lost their ability to jump, allowing them to date the emergence of the various transposon insertions. After sifting through millions of transposons, Schmitz's team found over 100 identical insertions in tarsiers and humans that didn't occur in lemurs or bushbabies. Tarsiers are the sister group to anthropoids and belong firmly within the haplorrhine clade, providing a link between basal primates and humans. According to Schmitz, given their phylogenetic placement amongst primates, 'the tarsier genome is a modern archive of evolutionary changes that led to humans'.[2]

Later, by analysing gene sequences from tarsiers and comparing them with those of other primates, the Münster team identified 192 genes undergoing positive selection, meaning they are evolving at different rates than those in other primates.[3] Unsurprisingly, many of these genes are linked to the tarsiers' visual and motor systems, which have adapted for activities like leaping and catching prey. For example, the *FGF* gene family and *IL1A* are thought to contribute to their distinctive traits of disproportionately large extremities and increased muscle mass. Additionally, eight of the identified genes are associated with vision. The researchers then compared the highlighted genes with those implicated in human disorders and found links to 47 diseases, approximately half related to visual and musculoskeletal problems. These associations may be of clinical importance. If specific base changes are linked to the unique musculature or vision of tarsiers, the altered amino acids in the protein are likely crucial for its function and warrant investigation in human disease.

Polymorphic trichromacy

The tarsier's visual system has given primatologists a valuable window into our past. Unlike lorises and lemurs, tarsiers lack a tapetum lucidum, a specialised, mirror-like structure at the back of the eye that enhances night vision

(Chapter 19). When illuminated by a light source, this reflective layer gives animals like cats their characteristic night-time 'eye-shine'. Indeed, before thermal imaging cameras became widely available, nocturnal mammals were best located by detecting their eye-shine using powerful torches.

The absence of the tapetum lucidum implies that the ancestors of haplorrhines (tarsiers and higher primates), unlike the earliest primates, were diurnal and lost their redundant ocular reflector. Such a view means that tarsiers must have returned to a nocturnal lifestyle but, for some reason, could not re-evolve the tapetum. Instead, tarsiers compensated by evolving huge eyes relative to their head size to maximise their light-capturing efficiency. Their relative eye size is unparalleled among living vertebrates, comparable to humans having eyes the size of grapefruits. Larger eyes also increase visual acuity by spreading the visual image across more photoreceptors so that tarsiers have both high resolution and high sensitivity. They also have a greater density of rod receptors, about double that of humans. As a result, the tarsier's primary visual cortex, which contains a complete map of the visual fields covered by the eyes, is proportionally larger than in any other primate. Since tarsiers are tiny, with relatively small brains, this nocturnal adaptation occurred at the expense of other cortical areas, resulting in tiny posterior parietal and prefrontal cortical areas.

Ancestral tarsiers, like living New World monkeys (except howler and owl monkeys) had polymorphic trichromatic vision: that is, they possessed one photoceptor or opsin gene located on an autosomal chromosome sensitive to short wavelengths (*SWS1*) and one of several possible alleles on the X chromosome, the so-called long-to-middle wavelength (L/M) genes. Each different gene produces a protein sensitive to a specific wavelength of light. The allelic variation at the X-linked locus causes heterozygous females to have three types of cones, hence trichromatic vision. In contrast, all males and homozygous females possessed the ancestral mammal condition of dichromacy, or two-colour vision, possessing only two cone types (we will explore the genetics of colour vision further in Chapter 21). Polymorphic trichromacy likely evolved after ancestral primates switched from a nocturnal to a diurnal lifestyle. Such a view is supported by the presence of additional traits associated with enhanced visual acuity under diurnal conditions: forward-facing eyes, smaller corneas relative to eye size, a post-orbital bony plate to prevent the chewing muscles from disrupting the eye position, a fovea (an area of high concentrations of retinal cones), and an increased visual cortex.

The most parsimonious scenario is that polymorphic trichromacy arose in the last common ancestor of tarsiers and New World monkeys, that is in an ancestral haplorrhine. However, the evolution of different cone photoreceptors may have been easier than one might imagine. In tarsiers, the gene polymorphism results from a single base-pair change at a critical spectral tuning site. In other words, it only takes a single mutation for a polymorphism to appear, so trichromacy could easily have evolved independently in different lineages. Indeed, mutations affecting such critical tuning sites are found in lemurs, including Verreaux's sifaka, as well as some bats, findings that support the evolutionary plasticity of opsins.

An interesting question is why ancestral tarsiers possessed polymorphic trichromatic vision, suggestive of diurnal activity, and large eyes, implying nocturnal activity (tarsier fossils from the Middle Eocene had hyperenlarged orbits). Amanda Melin, from the University of Calgary, suggests that these traits coexisted as adaptations to dim or mesopic light levels such as twilight or bright moonlight.[4] Supporting this idea, tarsiers are 'lunar philic', undergoing more foraging and travelling greater distances, with less rest, during full moons. Such conditions would be dark enough to favour large eyes but still be bright enough to select for polymorphic trichromatic vision. Such visual acuity is advantageous for finding green insects in dim forest light, such as the grasshoppers and crickets that form the tarsier's primary food source. In addition, tarsiers are vulnerable to predation, especially from twilight-active cats and civets, and their vision would have helped detect these russet-coloured predators.

However, during the last 5 million years, the tarsiers' L/M polymorphic genes became fixed, so that modern tarsiers are dichromatic. As a result, Philippine and Sulawesi tarsiers only possess an L opsin gene, while the Bornean tarsiers have a single M opsin gene. Why modern tarsiers became independently released from the selective advantage of trichromatic vision remains unclear. It's possible that the allelic fixation resulted from a chance or stochastic event, such as a genetic bottleneck. More likely, given that the fixation occurred in multiple species, is that an environmental change was responsible. Melin and her colleagues speculate that dichromacy might be associated with reduced rainfall across insular Southeast Asia during the Pleistocene. Drier conditions would favour a greater commitment to nocturnality and a reversion to hearing as the primary sensory modality.

Genes and reproductive isolation

Hybrid sterility is a common phenomenon in nature: it prevents the exchange of genes between two species, thus maintaining their reproductive isolation. For reasons not wholly understood, hybrid incompatibility is typically restricted to the heterogametic sex – that is, individuals with non-matching sex chromosomes. In mammals, this is the male, since they possess one X and one Y chromosome, while females have two X chromosomes.

(The opposite is true in moths, butterflies and birds, where the females are the heterogametic sex.) As a result, mammalian XY hybrids are usually infertile. For example, male offspring between tigers and lions ('ligers' and 'tigons') are sterile, while females tend to be more fertile.

Over the last two decades, molecular geneticists have investigated the underlying mechanisms of hybrid sterility and speciation. However, only one so-called speciation gene has been identified, *PRDM9*, which is involved in meiotic recombination in mammals by controlling the position of double-strand breaks in DNA.[5] The protein, a histone methyltransferase, is expressed in germ cells and specifies recombination sites between paternal and maternal chromosomes, controlling how genes are shuffled and passed down through a species. These genetic rearrangements help explain the evolutionary advantage of sexual reproduction: offspring inherit novel combinations of traits distinct from their parents, which can then be refined by natural selection to yield improved adaptations. The finding that alterations in the gene may lead to infertility in human males and hybrid sterility in mice suggests that *PRDM9* could have a role in preventing interbreeding between genetically divergent populations. Such a situation, termed post-zygotic reproductive isolation, prevents gene flow between populations, allowing them to diverge further genetically until they become distinct species.

In 2016, Ben Davies and colleagues at the Wellcome Trust Centre for Human Genetics, University of Oxford, confirmed the involvement of the protein PRDM9 in hybrid infertility and, by extension, speciation. They demonstrated that sterility could be restored in otherwise infertile hybrids of two mouse subspecies by modifying the highly conserved zinc-finger domain of PRDM9.[6] The engineered proteins were designed to bind DNA sequences that are more closely matched between the subspecies. The result was the restoration of symmetry in DNA double-strand breaks, with improved chromosome pairing and spatial alignment during meiosis, and this ultimately enabled the hybrids to produce mature, functional sperm capable of generating offspring.[7] Interestingly, PRDM9-induced hybrid sterility in the wild is not a binary phenomenon. Hybrids tend not to be entirely fertile or completely sterile but rather exhibit a gradual decrease in fertility with increasing divergence of homologues. In other words, not all *PRDM9* alleles are equal.[8]

In effect, evolutionary changes in the DNA-binding domain of PRDM9 likely create incompatible alleles, driving speciation. The reason for labouring this point is that recent research has linked mutations in *PRDM9* to the divergence of tarsiers and their subsequent speciation. Sacha Heerschop and her colleagues from the Institute of Anthropology at the University of Mainz discovered a unique insertion–deletion mutation in the gene of all tarsiers. They proposed that this genetic event underpinned the emergence of tarsiers during the Eocene.[9] Primates evolved 60–80 million years ago and survived two significant climate changes and mass extinctions that created ideal conditions for adaptive radiation. Proto-tarsiers likely endured these evolutionary bottlenecks because *PRDM9* generated new allele combinations better suited to the shifting environmental conditions and newly available ecological niches. A later bottleneck during the Eocene–Oligocene cooling further restricted primate diversity. This climatic event and ongoing tectonic activity caused a geographical shift in tarsier distribution from the mainland to insular Southeast Asia, giving rise to three major clades – Western, Philippine and Sulawesi tarsiers. Such genetic isolation triggered allopatric speciation, resulting in today's taxa. Each tarsier species exhibits distinct *PRDM9* alleles that likely developed through selection and genetic drift, further strengthening the role of *PRDM9* in speciation.

Although *PRDM9* is the only known speciation gene, many other proteins interact to enable its encoded product to bind to recombination sites, break and splice DNA, and correctly align chromosomes. These genes and others yet to be identified could also be key players in generating allele incompatibility. As research continues, our understanding of the genetics of speciation will undoubtedly grow more complex. We will explore another underlying speciation mechanism, reinforcement, in the next chapter.

CHAPTER 21

The Howler Monkey's Story

TRADE-OFFS, REINFORCEMENT AND DUPLICATIONS

Once you eliminate the impossible, whatever remains, no matter how improbable, must be the truth.
Arthur Conan Doyle (1859–1930)

What do uakaris, sakis, titis, howler monkeys, spider monkeys, woolly monkeys, capuchins, squirrel monkeys, owl or night monkeys, marmosets and tamarins all have in common? The surprising answer is that all these New World monkeys, approximately 170 species, evolved from a single ancestor that crossed from Africa to South America on a raft of vegetation during the Eocene, some 40 million years ago (Plate 31). This seemingly unlikely scenario is supported by recent evidence from two distinct disciplines, palaeontology and genetics.

In 2015, a team of palaeontologists led by Mariano Bond of the National University of La Plata described the fossilised teeth of South America's oldest known primate. About the size of a squirrel but with a longer tail and weighing only 250 grams, *Perupithecus* dates from the Eocene, around 40 million years ago.[1] Although bearing little resemblance to any extinct or living South American primate, this diminutive species from western Peru was remarkably similar to *Talahpithecus*, an extinct African primate from the late Eocene. Both species exhibit nearly identical morphological characteristics in their upper molars, indicating that they are likely sister taxa.

Although generally assumed, it was not until the application of molecular genetics that the monophyly of New World monkeys was confirmed. By analysing Alu elements, a form of 'jumping gene', Hans Zischler's team from the German Primate Centre in Göttingen proved that New World monkeys are all genetically closer to each other than to any Old World primate, a finding compatible with their origin from a single species that crossed to the New World.[2] Alternative theories, including vicariance due to the creation of the Atlantic Ocean and colonisation from North America, have been ruled out by molecular dating.

The small size of the last common ancestor of all New World primates would have increased their chance of surviving a long voyage on a drifting raft. Furthermore, their lower resource demands, including a diet of invertebrates and being better able to avoid dehydration, would have favoured their survival compared to larger, heavier animals. Also, South America and Africa were not as far apart as they are now, and sea levels were lower, perhaps even revealing a chain of island stepping-stones. Today, most scientists accept that the ancestor of the New World's primates first evolved in Africa and then crossed the Atlantic Ocean in a single, random dispersal event sometime in the Eocene and before their African cousins went extinct.

Oceanic dispersals, such as those of tenrecs, lemurs and neotropical primates, demonstrate that evolution is shaped not only by standard, routine processes but also by extraordinarily improbable 'black swan' events that are likely never to be repeated. Once you eliminate the impossible, whatever remains, no matter how improbable, must be the truth.

A diverse radiation

The scientific term for the New World primates is Platyrrhini, derived from the Greek for 'broad-nosed', to differentiate them from the apes and Old World monkeys they left behind, the Catarrhini ('down-nosed'). The Platyrrhini–Catarrhini split occurred somewhere on the African continent after the divergence of the tarsier lineage (see Figures 19.1 and 21.1). Once ashore in South America, the new arrivals slowly dispersed and adapted to life in the Amazon Basin. This early phase of their evolution is little known as the humid, heavily forested environment was not conducive to fossil preservation, and carcasses would have rapidly decomposed. Nevertheless, Amazonia's large, forested area functioned as a cauldron of primate biodiversity. By the late Oligocene, after sea levels had fallen and climatic conditions had improved, newly evolved species began to spread out in all directions. Between 20 and 15 million years ago, during the Middle Miocene Climatic Optimum, primates reached their maximum range. Patagonia, the southernmost area that non-human primates have ever inhabited, was colonised via a corridor through eastern South America across a retreating Paranaense Sea, a vast expanse of water that once covered the continent's entire southeastern lowlands. To the north, primates reached the Greater Antilles, including Cuba, Hispaniola and Jamaica, after at least two sweepstake dispersals during the Neogene. The early platyrrhines also increased in weight, reaching body sizes of over 2 kilograms around 20 million years ago.

Such geographical expansions, coupled with marked speciation, especially of larger taxa, were favoured by the warm and humid climate and

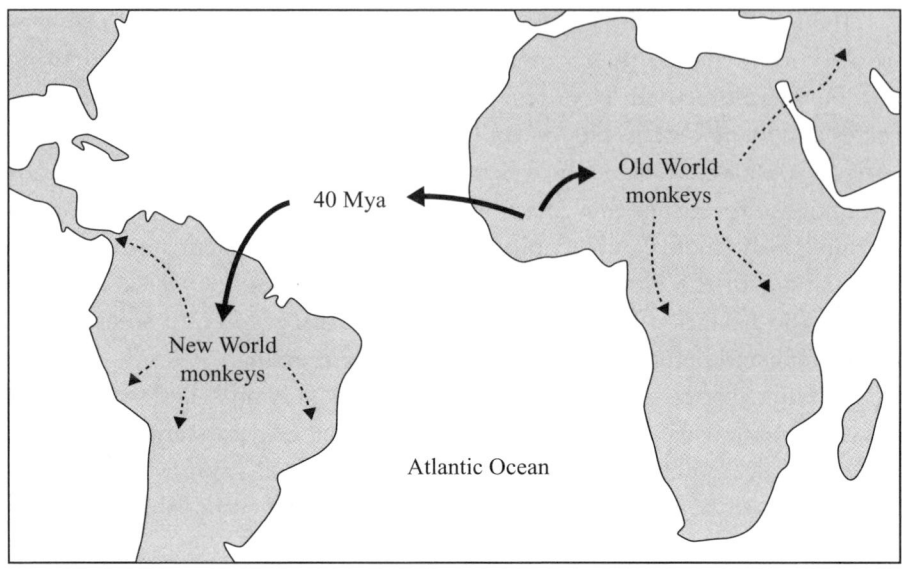

Figure 21.1 A single trans-Atlantic sweepstake dispersal from Africa around 40 million years ago gave rise to all of the New World monkeys. The 170 species include uakaris, sakis, titis, howler monkeys, spider monkeys, woolly monkeys, capuchins, squirrel monkeys, owl monkeys, tamarins and marmosets. Mya, million years ago.

the lower elevation of the Andes. However, global cooling and aridification after the Middle Miocene triggered a geographic contraction and an increased extinction rate.[3] The largest species died out, including some that had reached over 20 kilograms, as did all the lineages that evolved outside Amazonia. For example, after Patagonia's climate deteriorated, its primate population went extinct and left no descendants. Those species that had reached the Greater Antilles fared better and survived until approximately 6,000 years ago when the arrival of humans and increased sea levels led to their extinction.

The five extant families evolved rapidly within a 10-million-year window during the Late Miocene: Pitheciidae (uakaris, sakis and titis), Atelidae (spider, woolly and howler monkeys), Cebidae (capuchins and squirrel monkeys), Aotidae (owl monkeys) and Callitrichidae (marmosets and tamarins). However, their speed of diversification has led to difficulties in establishing a robust phylogeny, with incomplete lineage sorting and short branch lengths compounding the problem.[4] Today, the Platyrrhini are the major diurnal mammals in the Neotropics, being found not just in rainforests but also in habitats that are too hot and dry to support many other types of mammals. They exhibit a striking phenotypic diversity of body and brain size, skeletal morphology and fur colouration, and possess a wide range of social and mating systems.

The eight uakari species, for example, are unusual primates in that their tail length is considerably less than their head and body length, and their bodies are covered with long, loose hair while their heads are bald. Their faces appear almost skull-like, as they possess virtually no subcutaneous fat, and, like their closest relatives, the sakis, they have projecting lower incisors. After splitting from the saki lineage around 5 million years ago, the last common ancestor of uakaris was a lowland floodplain taxon that dispersed to different areas of western Amazonia. Speciation followed due to vicariance, driven by isolation and limited gene flow caused by alterations in ancient river systems and shifts in flooded forest regions.[5] Since Alfred Wallace's influential work, *A Narrative of Travels on the Amazon and Rio Negro* (1853), which inspired the riverine barrier hypothesis, biologists have recognised rivers as significant barriers to gene flow, particularly for smaller platyrrhines such as titis, tamarins and marmosets.

Howler monkeys diverged from their closest relatives, the spider and woolly monkeys, roughly 16 million years ago, with Central and South American species splitting around 9 million years later. Known for their loud, guttural calls, all 15 howler monkey species use these vocalisations to keep their groups together and deter potential threats. Male howler monkeys detect pheromones by smelling and tasting the urine of females to determine if they are in oestrus, a behaviour common among other New World monkeys but rare in Old World monkeys and apes. Like all platyrrhines, howler monkeys have retained a functional *TPR2* gene, vital for pheromone detection by the vomeronasal organ. In contrast, catarrhine *TPR2* became a non-functional pseudogene.

New World monkeys are the only primates with prehensile tails, an adaptation that evolved independently in two groups: howler, spider and woolly monkeys, as well as capuchins. In all these genera, the tail is strong enough to support the animal's entire body weight, enabling access to new feeding opportunities in the treetops and exemplifying how habitat can shape physical traits. A key transitional phase likely involved tail-assisted hindlimb suspension to improve the efficiency of moving through the forest canopy. Then, structural adaptations enhanced the tail's strength and flexibility, including more robust vertebrae, shorter distal vertebrae, an increased number of joints, and larger, more curved articular surfaces. Additionally, expanded muscle attachments enabled greater load-bearing, while the howler monkey lineage also evolved a naked tactile pad to improve their grip. The unique combination of skeletal and muscular modifications makes the platyrrhine's prehensile tail sturdy and dexterous, allowing howler monkeys and capuchins to navigate the precarious high canopy in search of fruit and leaves.

While anthropoids (monkeys, apes and humans) shifted from a nocturnal to a diurnal lifestyle, the owl monkeys re-adapted to nocturnality approximately 15–20 million years ago. To do so, they evolved huge eyes relative to their skull size (hence the name 'owl monkeys'), which, together with large corneas, enable optimal light gathering under dim conditions. They also have more retinal rods and fewer cone photoreceptors than diurnal monkeys, thus exchanging colour vision for high visual sensitivity in low light levels. Unlike most nocturnal mammals, they do not have a tapetum lucidum, suggesting that nocturnality is a recent adaptation. Additionally, their sense of smell is well developed, a sense that aids the location of food and enables communication through scent marking. Genetic analyses suggest that the owl monkeys' most recent common ancestor arose in the central Amazon Basin in the Early Pliocene. Their subsequent speciation appears far more complicated than previously thought, with intense chromosomal rearrangements occurring within a short timescale, driven partly by the formation of the Andes and changes in the Amazonian landscape. Nine species are recognised, distributed in two main clades on either side of the Amazon–Solimões River.[6]

The capuchin monkeys, consisting of around 28 species, are named for the dark fur that forms a cap extending as 'sideburns' – resembling capuchin monks' cowls. They are the most intelligent and adaptable of the New World monkeys, perhaps as smart as chimpanzees. Black-striped capuchins are known for their ability to fashion and use tools, especially to extract embedded foods, a trait that was once attributed solely to humans. They have also been known to take nuts, place them on a stone anvil, and use another stone to crack them open. Because of their ability to learn and remember, capuchins were once trained as 'organ grinder' monkeys. They are often used in movies, and have even been trained to assist disabled people with routine household tasks.

According to genetic studies, the last common ancestor of capuchins lived throughout tropical South America until approximately 6 million years ago, when they diverged into two clades or genera – the 'gracile' or untufted capuchins (*Cebus*) that evolved in Amazonia, and the 'robust' or tufted capuchins (*Sapajus*) that originated in the Atlantic Forest. The gracile clade is characterised by longer limbs relative to body size and rounder skulls, with teeth and jaws adapted to breaking open hard nuts. As their name suggests, the robust capuchins are stockier and more compact, with males sporting tufts of dark hair on the tops of their heads, as well as beards.

The formation of the Amazon River was once believed to have caused the capuchin divergence. However, it is now considered more likely that climatic and floral changes, along with the formation of the Cerrado, a vast intervening tropical savanna region, were responsible for their vicariant speciation.[7] Gracile capuchins expanded throughout Amazonia and into the Andean foothills and Central America, giving rise to 10 white-fronted capuchin species and the Panamanian white-faced capuchin. Meanwhile, the robust capuchins remained in the Atlantic Forest until relatively recently, only reinvading Amazonia during the last million years from the southwest before spreading to all areas, including multiple expansions into drier savanna-like habitats. Their late dispersal accounts for the morphological similarity of the eight recognised robust species, although two, the tufted and large-headed capuchins, have evolved more distinct skull shapes as a result of competition with sympatric gracile capuchin species, phenotypic changes known as character displacement. The rationale for such modifications stems from the competitive exclusion principle, known as Gause's law, which contends that two competing species must adapt to different ecological niches to coexist in a stable environment. Should this not occur, one species will be eliminated or excluded through competition.

Capuchins' advanced cognitive abilities relate to their large brains and high encephalisation quotients (a measure of brain size relative to body size), second only to humans among primates. Other unusual adaptations include long lifespans, a diverse range of behaviours, and the use of tools – traits typically seen in great apes, particularly humans. A recent comparative genomic study highlighted specific capuchin genes linked to ageing and brain development.[8] For instance, 48 genes were linked to long lives, including *PARP1*, which helps cells recover from DNA damage. Another highlighted gene, *MTOR*, known for its role in longevity in other organisms, appears to be a promising target for anti-ageing therapies. Additionally, at least four genes were related to brain size, while others were associated with cognitive abilities. Given the current rate of scientific advancement, a deeper understanding of the genetic basis for the evolution of primate encephalisation and cognition appears to be an achievable goal.

Calls or balls?

Howler monkeys make the loudest roars of any land animal, using a specialised larynx with an enlarged hyoid bone – a U-shaped structure in the neck supporting the tongue – that creates a sound-amplifying air sac. But their forest reverberations come at a cost. Jacob Dunn at the University of Cambridge and his colleagues found that males who make the deepest and loudest calls have the smallest testes and the lowest sperm count.[9] It seems that when it comes to howler monkey sex, it's either big testes or loud calls. This paradoxical observation only made sense once the researchers examined the size and makeup of the different howler monkey social groups. Those species with larger hyoid bones and smaller testes live in single-male groups with a harem of females. In contrast, monkeys like the mantled howler, with smaller hyoids and huge testes, live in multi-male, multi-female groups.

The explanation for the evolutionary trade-off between 'calls and balls' is that a single male in a group of females has exclusive access to them, and, given the absence of sperm competition, a louder call is more important than larger testes for siring offspring. Larger hyoid bones reduce the frequency spectrum and increase the acoustic impression of male body size – a function analogous to investment in a large body mass or enhanced weaponry. However, for those species that live in groups with multiple males, where females are typically promiscuous, the best way to pass on more genes is to evolve larger testes. In other words, with limited resources, howler monkeys must choose whether to invest their energy in roaring or making sperm. The fact that the trade-off involves bigger hyoids for smaller testes rather than size reductions in other organs, such as the liver, heart, kidneys or feet, is that these organs do not increase the monkey's chance of mating or producing offspring.

Without such trade-offs, an organism could optimise all traits simultaneously and become a 'Darwinian demon', a hypothetical organism that would occur if no evolutionary constraints existed. In reality, the best phenotype is always a compromise, and trade-offs are ubiquitous in the evolution of traits related to fitness and reproduction success.

Reinforcement and reproductive isolation

Speciation is a long, slow, and complicated evolutionary process, and scientists only have a basic understanding of the genetic mechanisms involved. However, a recent study of howler monkeys by the University of Michigan has provided further insight into the process.

The Michigan research team, led by Marcella Baiz, has been monitoring and studying mantled and black howler monkeys in Mexico for over two decades. The two species diverged around 3 million years ago, with each population accumulating DNA differences as they adapted to their environments in isolation, that is, in allopatry. Today, the genes of both groups have altered so much that biologists regard them as distinct species. However, within the last 10,000 years, the two howler monkey species have come into contact again, forming a hybrid zone approximately 15 kilometres wide, indicating that reproductive isolation remains incomplete.

Evolutionary biologists accept that natural selection helps to complete the speciation process by strengthening barriers to gene flow between two species, pushing them to complete reproductive isolation. As discussed in previous chapters, natural selection favours individuals who reproduce over those who don't, so low-fitness hybrids (which usually die before they can mate, or are infertile) are selected out. Indeed, nature reduces the incidence of hybrids by increasing the genetic differences between interbreeding species so that hybridisation is less likely. This process is known as reinforcement and, while a familiar concept, has been challenging to prove. Nevertheless, empirical data, including highly complex computer simulations, suggest that reinforcement enhances and strengthens existing species boundaries, thus contributing to the final stages of speciation. However, demonstrating the effect in the wild has been difficult.

Marcella Baiz and her colleagues tackled the problem by comparing the DNA of black and mantled howler monkeys from within their hybrid zone to those living far away.[10] If reinforcement is acting to strengthen reproductive isolation, DNA differences between the two species in the hybrid zone should be greater than those between the same two species in areas without contact. This prediction is precisely what the team observed after analysing genetic markers near genes linked to reproductive isolation. This research is important, as it suggests that multiple types of selection contributed to the reproductive isolation of howler monkeys. Initially, speciation was driven by isolation, in allopatry, followed by divergent selection in the contact zone, or sympatry. Simply put, howler monkey hybridisation plays a direct role in completing the process of speciation by inducing further genetic differences between the two parent species.

Cones and rods

The pathways for evolving colour vision in primates differ from those seen in most other eutherian mammals. The common ancestor of all vertebrates possessed a tetrachromatic system based on the presence of four spectrally different types of cone photoreceptor, each containing a unique opsin or photopigment. Cones are responsible for colour vision at normal light levels. However, the earliest mammals lost two cone cell types while living nocturnally alongside the dinosaurs, a legacy that persists today. As a result, most placental mammal retinas only support dichromatic vision, with two cone cell types, each characterised by a different light-sensitive protein, or opsin: cones with a long/medium-wave-sensitive opsin (L/M; yellow–green), coded by a gene on the X chromosome, and cones with a short-wave-sensitive opsin (*SWS1*; blue–violet), coded by a gene on an autosome (chromosome 7 in humans). The evolutionary loss of two photoreceptor types explains why dogs and horses see primarily in shades of blue, yellow and grey, and cannot distinguish red and orange.

Table 21.1 Three X-linked L/M opsin alleles – X^1, X^2 and X^3 – produce six visual phenotypes in New World monkeys. Males and homozygous females, who possess only one L/M allele along with the shared *SWS1* gene, are dichromatic. Heterozygous females, carrying two different L/M alleles plus the *SWS1* gene, are trichromatic.

Males	Females
X^1Y	X^1X^1
X^2Y	X^1X^2
X^3Y	X^1X^3
	X^2X^2
	X^2X^3
	X^3X^3

Like early tarsiers (Chapter 20), New World monkeys possess alternatives, or polymorphisms, for the X chromosome's L/M gene, each specifying a cone pigment that responds to light within the green-to-red wavelengths. Since there is a pool of three L/M alleles within the New World monkey population, there are six different photopigment phenotypes, although only heterozygous females are trichromatic (see Table 21.1 for explanation).

One might expect that natural selection would favour a particular allele and that, over time, the alternative alleles would disappear. However, for the L/M genes and many other polymorphisms, this is not happening. How can such stable polymorphisms persist? The two favoured explanations are heterozygote advantage and frequency-dependent selection, and both might be germane to the persistence of the L/M alleles in platyrrhines.

A heterozygous advantage exists when the survival rate for heterozygous individuals exceeds that for those that are homozygous. Textbook examples include two well-understood blood disorders in humans, sickle cell disease and thalassaemia. Individuals inheriting two sickle or thalassaemia genes fail to thrive, while those inheriting only one faulty gene remain well and are also partially protected against malaria infection. Overall, the beneficial effect for the heterozygous state outweighs the negative impact for the few homozygous individuals, resulting in the persistence of faulty genes in populations exposed to malaria. For New World monkeys, the heterozygous advantage experienced by the trichromatic female population, presumably in finding red and green fruit with equal ease, is sufficient to maintain both L/M alleles.

Frequency-dependent selection occurs when the fitness of a phenotype increases as it becomes rarer in the population. Regarding platyrrhines, let us assume that red-seeing monkeys are better at seeing red-coloured fruits, while green-seeing monkeys are better at finding green fruits. In a population dominated by green-seeing monkeys, the forest will eventually become dominated by red fruit. At this point, the red-seeing individuals will have the advantage, and their population will increase at the expense of green-seeing animals. The resultant population oscillations will ensure the persistence of both the green and red alleles.

However, two genera of platyrrhine monkeys have departed from this general polymorphic pattern – owl monkeys and howler monkeys. The owl monkeys, or douroucoulis, comprise 11 species inhabiting Panama and South America. Around 15 million years ago, their last common ancestor reverted to a nocturnal lifestyle, and all species have inherited a suite of visual adaptations, including large eyes, a high retinal rod-to-cone ratio, and a corresponding increase in orbit size.[11] In addition, they have become entirely colour-blind, having acquired deleterious mutations in the autosomal *SWS* opsin gene, with only a single intact L/M opsin gene on the X chromosome. Such loss may result from a relaxation of functional constraints on opsin genes, allowing deleterious mutations to accumulate. Alternatively, it could represent an adaptive loss, reducing the metabolically costly neural architecture required for processing colour signals. Whatever the cause, possessing only one cone produces the same result as having none, as colour vision depends on a phenomenon called opponency. Colour vision is more than merely detecting different wavelengths and, instead, is determined by the proportional excitation of different cone types. The result for owl monkeys is an eye that sacrifices sharpness and colour for increased sensitivity.

Interestingly, the owl monkey's retinal adaptations may have evolved through fewer steps than previously assumed. Michael Dyer from St Jude Children's Research Hospital in Memphis, Tennessee, led a team demonstrating that all retinal cells derive from a small group of stem cells known as retinal progenitor cells or RPCs.[12] As an embryo develops, the RPCs go through cycles of division while still maintaining their pluripotential or 'stemness'. Eventually, cells exit the cycle and differentiate into one of the various types of retinal cells. Cells that leave early become cone and ganglion cells required for diurnal vision, while cells exiting later produce rods and their supporting cells that function mainly at night, a sequence of development that never changes. However, owl monkeys have adjusted the timing of this sequence to enhance their night vision, a process termed heterochrony. By delaying when cells exit the stem-cell cycle, owl monkeys have evolved a retina predominantly composed of rods, making it highly adapted for nocturnal life. Another consequence of heterochrony is the absence of a fovea centralis, an area at the retina's centre that provides high visual acuity due to a surplus of cones and a dearth of rods. Owl monkeys are the only higher primate to lack a fovea centralis. It seems that evolution can produce significant phenotypic changes by simply shifting the timing of development rather than undertaking any large-scale genetic alteration – a ubiquitous evolutionary shortcut that we will come across again when discussing human evolution.

Finally, male and female howler monkeys possess trichromacy, but their three-colour vision evolved by a mechanism different from that of other platyrrhines. Rather than being based on a single polymorphic L/M gene, from which different allele variants encode pigments with varying spectral profiles, howler monkeys gained a second L/M gene. Around 7–16 million years ago, their last common ancestor made an error while copying its DNA during germ-cell division. The result was an extra copy of the L/M gene, located immediately upstream of the original on the X chromosome. As previously discussed, such duplications often occur when cells divide and DNA is replicated. They are mistakes but fortuitous ones, for they provide scope for experimentation without disrupting the function of the original gene. The duplicated L/M gene subsequently became modified under selection pressure to produce a protein sensitive to shorter wavelengths, and thereby provided howler monkeys with trichromatic vision. Remarkably, a similar but independent gene duplication occurred in the catarrhines, although much earlier, around 23 million years ago. This event, which involved differing amounts of flanking DNA than in the Howler monkeys, affected a common ancestor of apes and Old World monkeys and explains why humans have three-colour vision.

The evolution of trichromacy in howler monkeys suggests it provided an evolutionary advantage. Howler monkeys are the most folivorous of the New World monkeys, and fruits comprise a relatively small portion of their diet. Leaves form the largest part, especially young, nutritive and digestible ones that are often red and best detected by a red–green signal. Fieldwork investigating the animals' nutritional preferences supports the view that routine trichromacy was environmentally selected as a benefit in folivorous foraging.

CHAPTER 22

The Gibbon's Story

JUMPING GENES

Biology is the science. Evolution is the concept that makes biology unique.
Jared Diamond

After the New World monkeys crossed the Atlantic around 40 million years ago, the relatives they left behind in Africa evolved to become the most widespread and speciose of the non-human primates. The 138-plus living species of Old World monkeys (family Cercopithecidae) occupy many habitats, including rainforests, deserts, snowy mountains and even some cities. They exhibit a striking diversity of social behaviour and diet, such as fruits, leaves, insects, seeds, roots and occasional vertebrates. Their family includes baboons, langurs, colobus monkeys, macaques, mangabeys, mandrills, vervets and proboscis monkeys. All 24 genera differ from their New World cousins in having downward-facing nostrils, while the presence of tails differentiates them from apes. Furthermore, like apes and humans, females show full menstruation, a trait not seen in any earlier phylogenetic lineage of mammal except for the elephant shrews, a spiny mouse, and a few bat species.[1]

Almost nothing is known about the early evolution of Old World monkeys, as the first 12 million years of the fossil record consists of only a few teeth. Nevertheless, their morphology indicates that the earliest species were fruit-eating. Only later would the common ancestor of all extant species evolve their clade-defining feature: molar teeth with two transverse ridges or crests (bilophodonty) for crushing seeds and grinding leaves.[2] Towards the end of the Oligocene and into the Early and Middle Miocene, a warming trend was associated with the expansion of subtropical forests that facilitated the diversification of Old World monkeys in Africa and the Arabian Peninsula, before they spread to colonise Asia.

However, around 25–30 million years ago, before their diaspora, the Old World monkeys underwent a significant divergence. From a human perspective, few evolutionary events were more consequential, as this split, which likely occurred somewhere in East Africa, gave rise to a lineage of large, tailless primates: gibbons, orangutans, gorillas, chimpanzees, bonobos, and ultimately humans. Scientists regard the timing of the ape divergence as robust, as it is supported by both palaeontological and molecular evidence.

During 2011 and 2012, a team led by Nancy Stevens, a vertebrate palaeontologist from Ohio University, unearthed fossils of two previously unknown primates in Tanzania's Rukwa Rift Basin.[3] The finds, consisting of teeth and partial jaws, belonged to an ape and an Old World monkey that lived 25.2 million years ago. Before this discovery, the earliest known fossils of both lineages were younger, dating from the Early Miocene deposits in Kenya, Uganda and North Africa.

Stevens and her team determined the date of the finds by employing advanced mass spectrometry to analyse two layers of volcanic tuffs that sandwiched the fossils. However, because the two species displayed morphological evidence of divergence, the split between Old World monkeys and apes must have occurred earlier. This conclusion aligns with molecular estimates, which placed the divergence between 30 and 25 million years ago. Interestingly, this 5-million-year period coincided with significant environmental, climatic and tectonic changes in East Africa. The end of the Oligocene brought global warming and the tectonic uplift of the eastern side of the East African Rift System. Additionally, the Oligocene–Miocene boundary was marked by the collision of the Afro-Arabian and European landmasses, triggering intermittent faunal exchanges that replaced many native taxa with immigrant species. According to Stevens, the combined environmental pressures were likely pivotal in driving the divergence of the ape lineage from Old World monkeys.

However, during the same 5-million-year period, a key anatomical modification occurred that would profoundly impact human evolution – the disappearance of the ape's tail (see Figure 19.1).

Primate tail loss

Primates' tails differ widely in form and function, reflecting their adaptations to different habitats. Those with non-prehensile tails use them for balance and weight distribution, while those with prehensile tails also use them for grasping branches as they move through the trees. Some species, such as squirrel and spider monkeys, also wrap their tails around their bodies for thermoregulation and comfort during resting or sleep.

Humans also develop a tail during early embryonic development, which grows briefly between the first and second months of gestation. However, human fetal tails get reabsorbed, leaving behind 3–5 fused vertebrae that form our coccyx. Rarely, the genetic program for reabsorption fails, resulting in the birth of a neonate with a tail. The loss of the primate's tail has long been linked to human evolution, enabling bipedalism, the shift from an arboreal to a terrestrial lifestyle and the freeing of hands, facilitating the development of the technology that defines our species. However, despite its significance, the genetic mechanism driving this evolutionary change has remained unclear until recently.

Over several decades, animal studies have highlighted over 100 genes involved in the regulatory networks controlling tail development. In 2024, an American team led by Bo Xia screened each of these genes to determine if any mutation underpinned primate tail loss. Eventually, a modified gene was found in apes and humans but not Old World monkeys. The change involved an Alu element inserted into *TBXT*, a known tail length-regulating gene. However, this insertion was not in the gene's coding sequence, as expected, but in one of its non-coding introns.[4]

Alu elements are a type of transposable element (or 'jumping gene') that exists only in primates. They are discrete DNA sequences capable of 'jumping' from one location to another within the genome, often inserting copies of themselves into protein-coding genes. Their copy-and-paste strategy begins with transcription into RNA, which then utilises the enzyme reverse transcriptase to convert it back into double-stranded DNA before it can become randomly inserted into the genome. Alu elements likely originated from a modified ribosomal RNA that emerged in primate genomes approximately 65 million years ago. Their numbers continue to increase, and today they account for an astonishing 10 per cent of the human genome, with about 1 million copies per cell. Once considered 'junk' DNA, these short sequences have likely played a crucial role in human evolution. They may have driven protein synthesis, regulated gene expression, or contributed to creating new genes or unique gene combinations. Additionally, Alu elements are essential in shaping primates' neurological networks, or connectomes, and managing biochemical processes across the central nervous system.

Xia's team discovered that because of the precise positioning of the Alu element in the intron, alternative splicing of the *TBXT* gene can remove an entire exon, leading to the co-synthesis of a normal-sized and a truncated protein. Laboratory experiments in mice confirmed that expression of the Alu-containing gene results in a complete loss of the tail or a shortened tail depending on the relative abundance of the smaller protein. However, some mice also exhibited neural-tube closure defects, a condition analogous to spina bifida that impacts roughly 1 in 1,000 human newborns. This observation suggests that the selective advantage of tail loss must have been significant, as it likely came with the evolutionary cost of an increased susceptibility to such defects. In other words, a trade-off made in the Late Oligocene may still influence human health today. It never ceases to amaze me how much humans owe to a selfish snippet of DNA that randomly jumped into one intron 25 million years ago: our bipedalism, our big brains, our cognition, and our technology.

In contrast to this extraordinary finding, most intronic insertions of transposable elements are silent, as they are spliced out of the messenger RNA before the protein is produced. However, as additional mutations accumulate over millions of years, new splice sites can form so that the transposon becomes recognised as another exon, a process known as exonisation. Indeed, humans have thousands of Alu-induced exons, a mechanism responsible for over 60 per cent of all new exons. But how could adding a new exon to a gene not be a harmful event? According to the geneticist Rotem Sorek, the answer lies with alternative splicing (see Chapter 17).[5] Genes with new exons produce different RNAs, with and without the extra exon, leading to the synthesis of several different proteins. Some of these may be harmful and eliminated, while others have a neutral effect and are tolerated. Occasionally, the novel protein provides a survival advantage and is selected alongside any ancestral protein. In other words, the new variant protein can be tested without compromising the original function and provides an evolutionary mechanism for generating novel activity with minimal disturbance to the existing functional repertoire. This astonishing revelation deepens our understanding of evolutionary biology and offers new avenues for genomic investigation, as alternative splicing could underpin the emergence of many novel phenotypes.

Furthermore, a unique 'jumping gene' (LAVA) is implicated in the rapid speciation of gibbons, 20 species of critically endangered small apes that inhabit the forests of Southeast Asia (Plate 32).

Genomic plasticity

Gibbons, great apes and humans all belong to the same superfamily, Hominoidea. Within this group, gibbons are classified under the family Hylobatidae, while great apes and humans fall under Hominidae. Gibbons are thus hominoids (apes), but not hominids (great apes). From a phylogenetic perspective, gibbons diverged after the Old World monkeys but before the great apes around 17 million years ago, making them valuable subjects for studying the origins of hominoid traits (see Figure 19.1). However, due to the lack of fossils, the last common ancestor (LCA) of apes and humans remains a mysterious animal. Was it tailless, and did it share the distinctive anatomy of modern gibbons – such as a broad, flat rib cage? While the answer to these questions remains unknown, at least we now know its size.

In 2017, Mark Grabowski of the Senckenberg Centre for Human Evolution and Palaeoenvironment at the University of Tübingen, in collaboration with William Jungers from Stony Brook University, New York, conducted a study to estimate the size of ancient primates.[6] By collating the average and estimated body masses of a diverse range of living and fossil apes (including humans) and other primates, and using innovative comparative methods, the researchers came up with a surprising result. The LCA of apes was smaller than scientists thought, weighing approximately 5.5 kilograms. This result challenges the prevailing assumption of a chimpanzee-sized, chimpanzee-like ancestor.

In other words, gibbons are not a downsized lineage – a conclusion with an important biological implication. Most biologists traditionally believed that early apes developed suspensory and swinging locomotion, termed brachiation, in response to increasing body size, making walking along branches impractical. However, the research by Grabowski and Jungers suggests a different sequence: hanging and swinging behaviours evolved first, with larger body sizes appearing later. The gibbon's adaptation to brachiation appears linked to positive selection for genes such as *TBX5*, which is crucial for forelimb development, and *COL1A1*, which influences the connective tissues in bones and tendons.[7] Brachiation may have arisen due to an arms race with monkeys for fruit resources, with an increased body mass being a later consequence of this competition.

Gibbons stand out for their remarkably high number of large-scale chromosomal rearrangements relative to the ancestral ape karyotype. Approximately 5 million years ago, rapid radiation led to the emergence of the four modern genera. Despite their relatively recent divergence and speciation, gibbons have undergone an exceptional accumulation of chromosomal rearrangements. Each genus, distributed across a different Southeast Asian region, exhibits a distinct karyotype, with diploid chromosome numbers ranging from 38 to 52 – highlighting an extraordinarily rapid pace of karyotype evolution.

The cause for such genomic plasticity remained unknown until a multinational team sequenced the gibbon genome in 2014.[8] Comparative analyses revealed an unexpected role for a unique 'jumping gene', the LAVA element, exclusive to gibbons. With over 1,000 copies in their genome, some of these transposable elements are inserted in a set of genes that ensure proper chromosome separation during cell division. Intriguingly, many of the LAVA-disrupted genes are also mutated in certain human tumours, offering insights that could advance our understanding of human health.

This unique evolutionary development significantly increases the error rate during cell division, leading to a higher likelihood of chromosomal rearrangements in gibbons. Such insertions must have occurred at a low enough level to be compatible with life yet sufficiently high to increase the frequency of chromosome segregation errors to cause speciation. How so many novel karyotypes could have become fixed so quickly in the various populations remains unclear. However, one explanation relates to geographical isolation or vicariance. The divergence of gibbon genera occurred during the Miocene–Pliocene transition, a time marked by significant environmental shifts. The elevation of the Yunnan plateau in Southwest China and fluctuating sea levels likely triggered cycles of forest fragmentation and merging, resulting in alternating periods of range compression and expansion and the rapid speciation of gibbons.

Orangutan adaptations

Orangutans (Malay for 'people of the forest') are the only great apes in Asia and phylogenetically the most distant from humans, having branched off around 12 million years ago (see Figure 19.1). Zoologists recognise three distinct species that diverged around 1 million years ago: the Sumatran and Tapanuli orangutans from the Indonesian island of Sumatra and the Bornean orangutan from Malaysia. They are the largest arboreal mammals in the world, spending all their time – eating, sleeping and travelling – in the forest canopy. Their bodies are adapted to their lifestyle, having evolved four-limbed locomotion known as quadrumanous scrambling. Given their weight, brachiation requires four grasping limbs – hands and feet with short digits – to move between branches. Orangutans also possess highly mobile shoulder joints and hips with a modified ligament that binds the top of the femur to the pelvis. This unique anatomical structure allows extreme flexibility of the hips and enables their legs to reach the back of their heads, yoga-style.

Historically, orangutans occurred across Southeast Asia and China, although their current range is limited to increasingly fragmented forest areas in Sumatra and Borneo. Since the Pleistocene, orangutans on the two islands have experienced significantly different environmental conditions. In the north of Sumatra, where the island's two species live, habitat conditions have remained relatively stable since their speciation, with productive, mineral-rich volcanic soils, lower cloud cover, and consistent rainfall patterns. In contrast, Borneo experienced more pronounced fluctuations in climate and rainforest coverage during glacial cycles.

Even today, orangutans in northeastern Borneo face significant seasonal variations in fruit availability, characterised by short periods of abundance followed by prolonged scarcity. It is also a region particularly vulnerable to extreme droughts and forest fires triggered by El Niño events, which exacerbate food shortages and threaten their survival. As a result, Bornean orangutans are smaller and have evolved a suite of physiological adaptations suited to the island's challenging conditions.[9] Positive selection has led to an enrichment of genes associated with cardiac activity, increasing the efficient use of their limited energy resources. Similar adaptive changes occur in other mammals that live in extreme environments, including humans in oxygen-deficient high altitudes and polar bears in cold waters. Lipid metabolism and energy storage genes also exhibit positive selection, suggesting that metabolic adaptations may provide a physiological buffer against starvation. Indeed, the Bornean orangutan is more adept at fat storage in adipose tissue than the Sumatran orangutan. Such findings mirror human populations living in tropical rainforests with limited and unstable food supplies, as such groups also demonstrate selection for genes related to lipid metabolism and muscle function.[10]

Genes under positive selection in the Sumatran orangutan primarily relate to carbohydrate metabolism and brain development, including functions like learning and memory. These adaptations likely support the species' greater sociability, complex cultural behaviours, tool use, and enhanced problem-solving abilities. Additionally, higher Sumatran population densities since the Pleistocene likely fostered increased opportunities for social learning compared to their Bornean cousins. Although direct evidence is lacking, these observations suggest a fascinating connection between Pleistocene demographic history and the cognitive and cultural evolution of orangutans. In summary, comparative genomics has helped explain how ecological factors can shape the morphology, behaviour and social organisation of closely related allopatric species.

CHAPTER 23

The Gorilla's Story

GHOST ADMIXTURES

Extinction is the rule, survival is the exception.
Carl Sagan (1934–1996)

Today, gorillas, especially the critically endangered mountain subspecies of the eastern gorilla, are among the most iconic and beloved of all primates. Yet, at the time of Darwin's seminal publication *The Origin of Species*, gorillas provoked considerable controversy. Detractors of evolution baulked with 'visceral horror' at the thought that such brutish creatures could be our cousins, and went to extraordinary lengths to promote our differences. Debates about human origins raged, and the Great Ape, which had just been scientifically described (a western gorilla from equatorial Africa), lay at the heart of the clash of minds. The gorilla became a pawn, symbolising the views of the opposing camps: Darwin and his formidable ally, Thomas Huxley, versus Richard Owen, the famed director of the Natural History Museum and a vehement opponent of the theory of evolution.

Owen's riposte to Darwin's heretical ideas was to search for some detail that allowed humans to be classified apart, and he found it in the brain. He announced that humans possessed a unique cerebral area, the hippocampus minor (wrongly, as Huxley would later show), and that our cerebral hemispheres are larger than those of other mammals. Owen concluded that a human was as different from a gorilla as the ape from a platypus, and stressed the impossibility of gorillas standing erect and being counted as human. In Owen's opinion, the brute could not transmute; 'man' was safe, his dignity assured. The irony is that Owen's subsequent anatomical monograph, *Memoir on the Gorilla* (1865), provided crucial evidence for the evolutionary connection between humans and apes.[1] However, Owen's entrenched scepticism never faltered, as evidenced by the word 'savage' in the publication's subtitle (*Troglodytes gorilla*, Savage) and the incorporation of a coloured plate depicting a ferocious, teeth-baring beast.

St George Jackson Mivart, a zoological sophisticate and staunch Roman Catholic, grew increasingly disenchanted with Darwin's ideas, seizing every chance to express his objections. Although initially favouring the concept of evolution, Mivart's ideas later mirrored those of many of society's intelligentsia, challenging the theory's relevance to the human intellect. He argued:

> *Whatever the similarity of man's dead body to a gorilla's, our intellectual, moral and religious nature set us farther from an Anthropoid Ape than such an Ape differs from a lump of granite.*[2]

Darwin, however, recognised that humans did not derive from gorillas as the gutter press liked to imply, typically in prurient and salacious detail; instead, he believed that both species evolved from a distant common primate ancestor, with human and gorilla having separate lines of descent. Darwin was famously hesitant to publish this view, yet it gained broad acceptance among scientists within two decades, laying the groundwork for modern biology. While popular tabloids continued to entertain the public with lurid tales of human descent from apes, more reputable publications reflected the new understanding. In 1888, for instance, the eminent British zoologist Sir Edwin Ray Lankester incorporated Darwin's views of human origins in his contribution on 'Vertebrata' for the highly respected *Encyclopaedia Britannica*:

> *A little reflexion suffices to show that any given living form, such as the gorilla, cannot possibly be the ancestral form from which man was derived, since* ex-hypothesi *that ancestral form underwent modification and development, and in so doing, ceased to exist.*

Darwin eventually published the result of his deliberations in *The Descent of Man* (1871), writing:

> *It is probable that Africa was formerly inhabited by extinct apes closely allied to the gorilla and chimpanzee; and that these two species are now man's nearest allies, it is somewhat more probable that our early progenitors lived on the African continent than elsewhere. But it is useless to speculate on this subject.*[3]

Darwin's prescient statement received scientific support in 2007, when fossils of a primitive gorilla, *Chororapithecus abyssinicus*, were discovered in Ethiopia's

Chorora Formation in the desert scrubland of the Afar region, hence its name.[4] The teeth, from at least three individuals, date from around 8 million years ago and are the same size and shape as those of modern gorillas, even possessing microscopic features indicating a similar adaptation to a fibrous diet. The Japanese and Ethiopian team of palaeontologists concluded that their treasure trove of dental fossils – eight molars and one canine – supports an African origin for the gorilla–human split, likely around 10 million years ago, and not in Eurasia as some scientists had argued.

Living gorillas belong to two species, each divided into two subspecies. The western gorilla consists of the western lowland gorilla, occupying a large continuous range from the Democratic Republic of the Congo (DRC) to Cameroon, and the Cross River population, restricted to the forested hills and mountains of the Cameroon–Nigeria border. The eastern gorilla is split into the eastern lowland gorilla, or Grauer's gorilla, living entirely within the DRC, and the isolated mountain gorilla (Plate 33), found in the Virunga Mountains spanning Rwanda, Uganda and the DRC, and in the Bwindi Impenetrable Forest in Uganda.

Gorillas are the largest primates alive today, with adult males reaching weights of up to 250 kilograms and arm spans extending to 2.6 metres. Eastern gorillas are the larger species, characterised by darker fur, and the mountain subspecies is the darkest. Western gorillas have lighter, often reddish or greyish fur, with mature males sporting notable chestnut colouring on their heads and necks. The mountain gorillas are the smallest, with longer, thicker fur suited to higher altitudes and colder climates. As male gorillas mature, they develop a white or silvery 'saddle' on their backs, earning their title of 'silverbacks'. Once fully mature, usually between the ages of 10 and 12 years, a silverback will dominate and lead a stable and cohesive family group, ranging from four to eight members for western gorillas and up to 30 for eastern gorillas. These males defend their troop of females and young by making intimidating displays including charging and chest-beating, though they are generally calm, gentle and reserved creatures.

Like all great apes apart from humans, gorillas have longer arms than legs and usually move by knuckle-walking, placing weight on the backs of their fingers instead of their palms. They also have opposable thumbs and big toes and can grasp objects with their hands and feet. Adults build nests for day and night use, with daytime nests being simple collections of branches and leaves typically arranged on the ground. Night nests are more complex; they are often built on the ground, but may be constructed in trees when the troop senses danger or feels threatened. The gorilla's large body size and fibre-rich diet led to the evolution of their distinctive digestive anatomy and physiology. Because of their large capacity for microbial fermentation in a sizeable, distended colon, the animals' gastrointestinal tract allows for maximal energy extraction by absorbing volatile fatty acids and microbial components. The longer, dilated intestines account for the animal's more convex abdominal shape than humans.

Gorillas share many human-like behaviours and emotions, displaying traits such as laughter, sadness, and even grief for lost family members. Body posture and facial expressions indicate a gorilla's mood. Gorillas also possess well-developed communication skills, with over 25 distinct vocalisations that they use to express distress, contentment and aggression, as well as for keeping in touch with troop members.

Like most traditional human societies, gorillas predominantly live in reproductive groups, which, in the case of western gorillas, consist of a silverback and one or more females with dependent offspring. Upon reaching sexual maturity, both sexes disperse from their immediate family: females transfer to another social group, while males can spend many years alone before attracting their own females and forming another stable group. A study published in 2019 incorporated genetic analyses to investigate two large gorilla communities in detail.[5] The results revealed that western gorillas live in a more complex, kindred-based, social structure than previously appreciated. In scientific terms, they exhibit hierarchical social modularity (HSM) in which lower-order social units, or families, are nested inside increasingly larger units. This discovery suggests that the basis of human social systems may stretch back to the common ancestor of gorillas and humans, around 10 million years ago, rather than arising from the 'social brain' that evolved after the divergence of hominins from other primates.

Biogeography, speciation and adaptation

The four subspecies of gorilla evolved from a single taxon that inhabited a widespread area of tropical Africa during the Pliocene, around 4 million years ago. Between 1.6 and 0.9 million years ago, this ancestral population became isolated on either side of the Congo River basin, a vicariant event that likely gave rise to the western and eastern species. Although the two species now live far apart, they still look and behave similarly, due partly to gene flow in both directions until at least 78,000 years ago.

Before the Last Glacial Maximum, between 50,000 and 26,000 years ago, the eastern gorillas were more widespread, distributed throughout Eastern Africa and mirroring the present range of eastern chimpanzees. However, a significant climatic shift between 26,000 and 20,000 years ago, with sharp declines in temperature, rainfall and humidity, led to the contraction of rainforest habitats and their replacement by montane and savanna-like ecosystems. During this period, major African lakes – including Lake Albert, Lake Victoria and Lake Edward – dried out, with Lake Tanganyika also shrinking considerably. This arid phase ultimately separated the Virunga and Bwindi gorilla populations, confining them to isolated forest refugia.

Around 14,500 years ago, with the onset of a humid period, lakes started to reform. Forests regrew and created an interconnected ecosystem between the lakes, allowing forest-dependent species to spread and leading to a marked increase in biodiversity. The result was a westward spread of a small population of gorillas from the Virunga region into the lowlands of the DRC. At the same time, a population of Cross River gorillas expanded eastwards from western Africa, where they interbred with the westward-dispersing Virunga gorillas to produce the Grauer's gorillas. About 5,500 years ago, the environment became drier, and the forest habitat declined, with the Sahara becoming increasingly barren and replaced by a sandy desert. As a result, the eastern gorillas – the Grauer's, Virunga and Bwindi populations – became wholly isolated and restricted to their present-day ranges.[6]

Meanwhile, in West Africa, the ancestors of the Cross River gorillas became restricted to the forests near the headwaters of the Cross River and in the Cameroon highlands, although some interbreeding with western lowland gorillas continued until as recently as 500 years ago. This prolonged period of limited genetic exchange was followed by a bottleneck around 320 years ago that caused a 60-fold decrease in the Cross River population, with a marked loss of genetic variation, especially over the last century.[7]

After the climatic oscillations split and isolated the various gorilla populations, their numbers fell, with all subspecies suffering further declines from recent anthropogenic pressures. Indeed, the first human settlers, the Bantu agriculturalists, arrived while forests were contracting, as long as 2,500 years ago. Later, hunting pressurised all the subspecies, and as the human population increased in the twentieth century, so did the threats from poaching, logging, commercial hunting and fossil-fuel exploration. Today, there are only around 316,000 western gorillas and 5,000 eastern gorillas, all classified as critically endangered by the IUCN, except for the mountain gorilla, which is classified as endangered.

While gorillas share many elements of their anatomy and physiology with humans, biologists have resorted to molecular genetics to enable our functional differences to be determined. However, the task has not been an easy one. In 2012, after toiling for five years, 60 scientists announced the genome sequence of Kamilah, a female lowland gorilla and a representative of the last great ape species to have its DNA mapped.[8] The results were surprising, given that humans diverged from gorillas 10 million years ago and chimpanzees 6 million years ago. Although most of the human genome appears closer to chimpanzees, as expected, around 15 per cent is closer to gorillas, while another 15 per cent of the gorilla's DNA best matches that of chimpanzees. According to the study's lead, Aylwyn Scally, from the Wellcome Trust Sanger Institute in England, 'the passage of ancestry across the three genomes changes from position to position'.[9] Overall, 98 per cent of gorilla and human genes are identical, compared to 99 per cent between humans and chimpanzees.

To gain further insight into our differences, Scally's team concentrated on the relatively few areas that appeared to have evolved more rapidly in gorillas. The result was the identification of genes involved in the developmental processes of the ear, hair follicles, gonads and brain, as well as in sound perception. Interestingly, the most rapidly evolving gene, *EVPL*, codes for the protein envoplakin found in the cornified envelope of skin cells or keratinocytes. The researchers suggest that this evolutionary change may relate to the increased cornification of the gorillas' knuckle pads, possibly to support their characteristic walking behaviour, a genetic variant absent in humans.

Another curious observation was the similar rate of evolution of genes related to hearing in gorillas and humans but not chimpanzees. This finding calls into question the reported link between the evolution of auditory genes and the development of human language. Ear morphology is one of the few external traits in which humans are more like gorillas than chimpanzees. However, hearing genes might have undergone parallel evolution for different reasons: human hearing genes to enable speech, and those of gorillas for a different, unknown reason.

Finally, genes involved in sperm production were found to be inactive or scaled back in gorillas. This finding may relate to the animal's polygynous social system, in which the highest-ranking male has exclusive access to females and sires most of the troop's offspring. Competition between males for reproductive success occurs before copulation rather than between different male's sperm within the female reproductive tract. This mating strategy has led to the evolution of the male's large mass and behaviours to protect reproductive access rather than traits that enhance post-copulatory sperm competition. Consequently, silverbacks have small testes, low sperm counts, low sperm swimming speeds, and a high proportion of abnormal forms. Furthermore, the dramatically relaxed selection pressure on sperm-related genes has resulted in many possessing multiple deleterious mutations, with some defects known to be associated with human male infertility.[10] This observation has led to speculation that male gorillas may be at the lowest limit of reproductive function that can be maintained by natural selection, at least in mammals or vertebrates.

'Spooky gene flow'

In 2023, a team led by Martin Kuhlwilm from the Department of Evolutionary Anthropology at the University of Vienna sequenced the genomes of all the gorilla subspecies and obtained a surprising result.[11] Three per cent of eastern gorilla DNA was derived from an archaic 'ghost' lineage that diverged more than 3 million years ago from the common gorilla ancestor. This introgression, which western gorillas do not share, occurred around 40,000 years ago, just before the mountain and eastern lowland gorillas diverged. In other words, an unknown gorilla lineage existed from the Pliocene to the Late Pleistocene, although where it lived is unclear. The evolutionary biologist John Hawks suggests that the extinct population may have inhabited areas where today's eastern gorillas occur, or they may have interacted with the eastern gorillas as they sought a refugium during Late Pleistocene climate oscillations.[12]

Kuhlwilm's research is important, as it sheds light on the existence of an unknown species and how genes from ghost lineages have contributed to evolution. For example, the acquired genetic material included a modified *TAS2R14* gene that encodes a receptor that detects bitter tastes. Eastern gorillas may have benefited from the acquisition, permitting a move into the mountains and a predominantly herbaceous diet. In contrast, western gorillas consume mainly fruit. Other introgressed genes, located in different parts of the genome, also exhibited positive selection, although their adaptive role remains unclear.

Interestingly, archaic DNA in gorillas is mainly absent from the X chromosome, where 'introgression deserts' appear, contrasting with the more random scatter-gun insertions across autosomal chromosomes. A possible explanation is that ancient introgressions on the X chromosome could have reinforced reproductive isolation in hybrids, particularly among males, leading to the rapid purging of archaic DNA from sex chromosomes through selection.

Another mystery is why the eastern gorilla's introgressed DNA represents only a tiny fraction of its genome, occurring at levels similar to that observed in other primates, including bonobos, chimpanzees and humans. There are two possible explanations. One is that DNA from highly divergent populations may have been disadvantageous to descendants, resulting in natural selection reducing its presence over time. Another possibility is that the 'ghost species' had a smaller population size and less genetic diversity, so that as modern primate populations grew the genetic contribution from these ghosts became diluted. These ideas aren't mutually exclusive, and some evolutionary biologists believe that natural selection and population dynamics were likely cofactors.

Primate introgressions were initially detected in DNA extracted from fossilised bones. As we will discuss in Chapter 25, this approach has revealed that humans contain genetic material from at least two extinct hominin species, Neanderthals and Denisovans. However, the search for archaic genetic material in African apes is more challenging because of the paucity of the fossil record, which has likely led to an underestimation of its importance. However, advances in high-quality genome sequencing and improved introgression-detection technologies could compensate for the lack of fossil evidence. Furthermore, such powerful tools may enable evolutionary geneticists to address some intriguing questions: Are ghost introgressions a universal feature of mammalian evolution and speciation? How often are the acquired DNA sequences beneficial and provide survival advantages? Can these ancient introgressions offer a window to determine the traits and adaptations of species that left no fossils?

According to Jente Ottenburghs, a researcher at Wageningen University in the Netherlands, archaic DNA insertions might provide the raw material for rapid evolution.[13] Instead of waiting for novel, random mutations, species could harness the potential of genetic variation by hybridisation with other taxa. Such introgressed genes would have already been honed by natural selection for a specific ecological niche and could speed up the recipient's rate of adaptation and speciation. Furthermore, Ottenburghs muses whether 'spooky gene flow', as he calls it, could explain patterns of rapid evolution in the fossil record. For example, cetaceans evolved from a terrestrial ancestor to a fully aquatic lifestyle in approximately 8–10 million years (Chapter 11). Significantly, their genomes reveal a reticulated evolutionary history with high levels of introgression, raising the possibility that ghost introgressions speeded up their aquatic adaptations. In addition, archaic introgressions might explain the fast diversification rate in other mammalian groups, including horses, elephants and cats.

CHAPTER 24

The Bonobo's Story

VICARIANCE, NEOTENY AND GENETIC FOSSILS

The fundamental difference between our two closest relatives is that one resolves sexual issues with power, while the other resolves power issues with sex.
Frans de Waal (1948–2024)

The bonobo, our closest living relative and the last of the great apes to be discovered, was not identified in some remote, humid rainforest in central Africa but in a collection of skulls stored at the Royal Museum of the Belgian Congo in Tervuren. Henri Schouteden, a museum employee and passionate collector of natural history specimens, persuaded Belgian colonials and Congo's local inhabitants to send him deceased animals for scientific study. After sending out hundreds of letters requesting help, Schouteden received several shipments of specimens, including one in September 1927 that contained the skin and skull of a small 'chimpanzee' killed south of the Congo River. Recognising that the skull was unusually small for the animal's size, he sought the expertise of his friend, primatologist Ernst Schwartz, who noted that it lacked prominent supraorbital ridges and possessed unfused sutures, indicating that it belonged to a young individual. Eventually, Schwartz classified the specimen as belonging to a subspecies of chimpanzee that became known as the 'pygmy chimpanzee'.

In 1933, American zoologist Harold Coolidge, after re-examining the skull and several newly acquired specimens, concluded that the so-called pygmy chimpanzee was in fact a distinct species. He named it *Pan paniscus* ('little Pan') to differentiate it from the common chimpanzee (*Pan troglodytes*). The modern name bonobo emerged later, possibly due to a misspelling of Bolobo, a town along the Congo River where specimens were shipped. Predictably, given how rare it was to discover a new primate species, the identification of the bonobo was celebrated as one of the most important zoological events of the twentieth century.[1] Any attempt to rank such discoveries is inherently subjective, but the bonobo's identification was undeniably significant, leading us to a deeper understanding of human origins and our early evolution and adaptation.

Around 6 million years ago, and 4 million years after the gorilla lineage diverged, the *Pan* and *Homo* lineages parted company (see Figure 19.1). Later, approximately 2 million years ago, the *Pan* lineage split, leading to the separate evolutionary trajectories of chimpanzees and bonobos, vicariant speciation events brought about by the formidable Congo River.

Although they are widely distributed across western and central Africa, chimpanzees have never crossed the Congo River, the deepest in the world. Four subspecies are recognised (Figure 24.1), and their genetic distinctions have arisen following geographical isolation caused by major river systems.[2] For example, the Sanaga River separates the western and Nigeria–Cameroon subspecies from the central and eastern subspecies. Furthermore, the Ubangi River divides the central and eastern groups, while the Dahomey Gap, a vast dry forest–savanna mosaic, and the Niger River serve as barriers between the western and Nigeria–Cameroon subspecies. However, despite the subspecies' adaptations to their local environments, speciation has yet to occur, possibly due to their recent separations, between 500,000 and 100,000 years ago, or

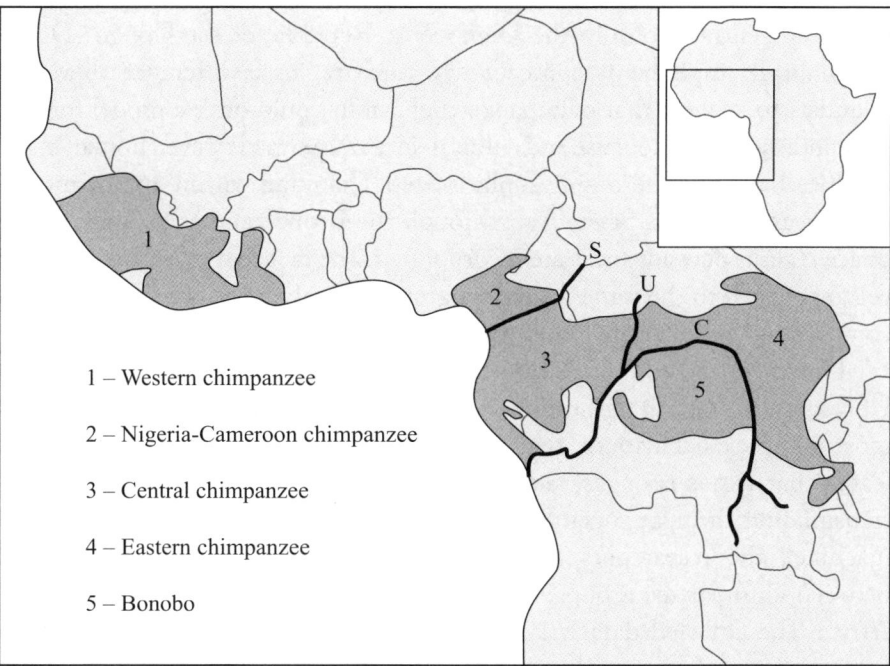

Figure 24.1 A map of equatorial Africa showing the distribution of the four subspecies of chimpanzee (1–4) and the bonobo (5). Note the split of chimpanzee subspecies by the Sanaga River (S) and the Ubangi River (U), while the bonobo is separated from chimpanzees by the Congo River (C).

perhaps because the geographical barriers have not been absolute, allowing some gene flow between groups.

Nevertheless, the genetic evidence that populations on either side of the Sanaga River are monophyletic clades, each containing two distinct subspecies, is supported by immune and behavioural differences. Unlike those to the west, chimpanzees east of the Sanaga River have been under strong selection pressure to reduce the pathogenicity of the simian immunodeficiency virus (SIV), a retrovirus in the same family as the human immunodeficiency virus (HIV).[3] As a result, eastern chimpanzees have altered immune genes, particularly those involved in the production and maturation of T-helper lymphocytes, a type of white blood cell that determines the severity and hence the outcome of AIDS in humans. Genetic adaptations have also involved genes that code for proteins used by the virus to infect, control and kill host cells.

Similarly, cultural behaviours vary amongst subspecies, including tool use, vocal dialects, non-verbal communications (e.g. gestures) and feeding strategies. For example, chimpanzees west of the Sanaga River use stone tools to break open oil palm nuts, while termite fishing is most prevalent to the east. Even the termite-fishing chimpanzees use different probe techniques depending on subspecies: the eastern group in Tanzania use single probes, while the central group in the Democratic Republic of the Congo (DRC) use multiple implements consecutively. Such regional differences have led scientists to suggest that chimpanzee behaviour could offer a model for understanding human tool use and cultural development. However, human technologies have become highly sophisticated, changing within approximately 3,000 years from the Stone Age, through the Bronze and Iron Ages, to the Space Age, as new advances are incorporated. The fact that chimpanzees have yet to progress to the same degree suggests that only humans can build more complex and sophisticated cultures over time.

However, a new multidisciplinary study led by Andrea Migliano at the University of Zurich disputes this conclusion, showing that chimpanzees' most complex behaviours, the use of stone hammers and anvils, chewed leaves that act as sponges, and flexible termite probes, evolved by cultural transmission through social learning.[4] Using what they call a 'genetic time machine', the researchers analysed thousands of years of gene transfer between chimpanzee groups, reflecting their migrations throughout central Africa. The amassed data was then correlated with 15 foraging behaviours from the four subspecies, classified into three levels of complexity: behaviours without tools, simple tool use (e.g. chewed leaves), and complex toolsets (e.g. stones and probes). The analysis revealed a strong link between advanced tool use and populations connected through genetic exchange over the past

5,000–15,000 years, indicating that such behaviours spread through social interactions. Regions where three subspecies overlapped exhibited the most sophisticated tool use, highlighting how cross-group connections foster cultural development. In contrast, simpler behaviours like foraging without tools likely developed independently in various regions.

It is the sexually mature female chimpanzees that migrate to new communities, an evolutionary adaptation that reduces inbreeding and enlarges the gene pool of otherwise isolated populations. Migliano's team found that the movement of females also helped spread new tools and skills to groups lacking them, suggesting that complex tools evolved by building on simpler versions. Indeed, the more females in a group, the greater the number of cultural traits. Furthermore, infant chimpanzees spend up to eight years near their mothers, suggesting that much of their learned behaviour results from the mother's influence.

In summary, sophisticated behaviours don't arise randomly but are transmitted via females between different groups, and have gradually expanded over many generations. However, chimpanzees have yet to evolve human-scale culture, probably due, in part, to their lack of advanced language.

The 'hippie' ape

Unlike chimpanzees, which adapted to relatively open and dry habitats, bonobos never left the protection of trees, being restricted to the humid forests south of the Congo River, a natural barrier maintained by the apes' inability to cross deep water (Plate 34). Chimpanzees avoid water entirely, but bonobos often wade in the shallows to forage, though they cannot swim and never venture beyond waist deep. Such observations raise an intriguing question: how did the ancestral *Pan* population become separated by the river, leading to the allopatric speciation of chimpanzees and bonobos?

A widely accepted explanation was that the Congo River divided a population of ancestral apes when it first formed, an event thought to have occurred around 2 million years ago. However, Japanese researchers at the universities of Kyoto and Nagoya have recently proposed a different hypothesis.[5] By analysing submarine Congo River sediments, the team found that the river formed much earlier than previously thought, around 34 million years ago, and long before the ancestral bonobos inhabited their current range. Crucially, the sediment studies also revealed the river's depth and flow reduced significantly during the Pleistocene when chimpanzees and bonobos diverged. According to the Japanese study, the transient drop in water levels would have allowed one or more founder populations to move into their present territory south of the river, in the DRC, where they have remained ever since. While there are traces of limited admixture, the Congo River has served as an effective barrier between the two species.

Following the ancestral river crossing, chimpanzees and bonobos have evolved along different paths, resulting in unique physical and behavioural traits. Bonobo vocalisations are higher in frequency, consisting of tonal sounds such as peeps and peep-yelps, whereas chimpanzees emit noisier vocalisations that resemble grunts and barks. Physically, bonobos are about 15 per cent smaller than chimpanzees and less robust in build, with long, lanky arms and a small head atop narrow shoulders. They typically have dark faces with pink lips and hair that appears parted down the middle. Bonobos also have a slightly different mode of locomotion. Unlike chimpanzees, who are primarily quadrupedal and rely on all four limbs, bonobos are more bipedal, often standing and walking on their hind legs. In fact, the bonobo's body proportions have been compared to those of australopithecines, a form of prehuman. According to Frans de Waal, Professor of Primate Behaviour at Emory University, 'when apes [bonobos] stand or walk upright, they look as if they stepped straight out of an artist's impression of early hominids.'[6]

Recent research has strengthened the idea that bonobos may most closely resemble the primogenitor of the *Pan* and *Homo* genera. In 2017, a study led by anatomist Rui Diogo from Howard University College of Medicine examined the differences in musculature between chimpanzees and bonobos.[7] Dissecting seven bonobo cadavers from Antwerp Zoo, which houses the largest captive population, the researchers compared the results to those of common chimpanzees, modern humans and other primates. It turns out that humans share more similarities with bonobos (13 major differences) than chimpanzees (20 major differences). This finding suggests bonobos have undergone fewer muscular changes since diverging from their common ancestor, making them a better living model for understanding the last common ancestor of chimpanzees, bonobos and humans. Furthermore, Diogo's findings align with earlier molecular studies that identify bonobos as our closest living relatives.

However, the most profound difference is not their vocalisation or physique but their behaviour. Bonobos are more peaceful and egalitarian than their cousins to the north of the river, maintaining a matriarchal society moulded by cooperation, alliances and, most notably, recreational sex. As a result, bonobos have earned the moniker 'the hippy apes', in contrast to their bellicose relatives across the water.

Bonobos separate sex from reproduction, like humans, but unlike chimpanzees and most other animals. They seem to view sex as a source of pleasure and use it as a social tool to build or mend relationships. According to Frans de Waal, early humans likely exhibited similar behaviour. However, as we developed family systems, the use of recreational sex became more restricted, primarily taking place within family structures. Furthermore, bonobos are one of the few species in which all adult females engage in habitual same-sex sexual interactions, often at a frequency greater than with the opposite sex. Females regularly practice so-called genital–genital rubbing with many females in their group. In contrast, males rarely indulge in same-sex activity. Theories to explain such activity include reduction of social tension, prevention of aggression, or encouragement for social bonding – though none of these ideas explains why same-sex behaviour occurs so frequently and mainly amongst females. However, a study led by Tobias Deschner at the Max Planck Institute for Evolutionary Anthropology suggests it relates to oxytocin.[8] The research team found that females release more oxytocin after same-sex activity than after sex with males. Oxytocin is a hormone synthesised in the hypothalamus and released from the pituitary gland, with many functions, including promoting cooperation, trust and parent–infant bonding. Indeed, after same-sex sexual activity, females remain closer to each other than they do with males after sex. In effect, the greater motivation for cooperation among females, mediated physiologically by oxytocin, is critical to understanding how females attain high dominance ranks in bonobo society.

In the 1970s, Jane Goodall's pioneering field studies in Gombe National Park in Tanzania revealed that chimpanzees, especially males, are aggressive primates that will attack other adults, kill infants, and rape females. Indeed, male chimpanzees will often organise themselves into warring gangs that raid other group's territories, leading to severe injuries and deaths. The cause of such primate aggression has been the subject of intense debate. Most primatologists regarded such war-like behaviour as evolutionarily adaptive. In other words, violence and fighting skills improve a male's chance of survival and fathering more offspring, traits selected for over 2 million years of evolution. However, some anthropologists baulk at this explanation, believing that pressures from human encroachment and habitat destruction led to chimpanzees living closer together, increasing their competitive behaviour. While the debate continues, a monumental collaborative study in 2014 failed to support the human-impact hypothesis, concluding that chimpanzees' lethal aggression has an evolutionary origin.[9]

Infanticide is virtually non-existent among bonobos, as their promiscuous behaviour obscures paternity, removing the motivation for males to harm infants. In addition, marked female bonding provides alliances for mutual support and protection and leaves little opportunity for infanticide. The prevention of male aggression towards infants provides a significant evolutionary advantage for female bonobos, as it increases the survival rate of their offspring. An obvious question, therefore, is why chimpanzees never evolved a similarly beneficial social structure. The answer, it seems, may relate to primate competition.

As mentioned above, the ancestral bonobos crossed the Congo River around 2 million years ago when water levels fell, and occupied the fertile lands to the south. Life was easy as there were no gorillas and few other species to compete with. As a result, the ancestral bonobos had extensive tracts of forest to themselves, with round-the-year food sources that enabled them to travel in larger, more stable groups and form stronger social bonds. In contrast, ancestral chimpanzees in the north faced competition from gorillas for scattered resources such as fruit, fibre-rich vegetation and occasional meat. Female chimpanzees must forage widely, taking their infants in tow to gather enough food, leaving limited opportunities to build strong social bonds. In other words, the environment forced chimpanzees down a different evolutionary path, resulting in a species prone to violence – one characterised by infanticide and male coalitions to defend territories and compete with rivals.

The 'neotenous' ape

Scientists have attributed many of the morphological and behavioural differences between bonobos and chimpanzees to neoteny or paedomorphism, a process where juvenile traits persist into adulthood. For instance, bonobos' skulls maintain their infant-like proportions, growing larger as they age, resulting in narrower faces and smaller teeth than chimpanzees. They also lack the pronounced eyebrow ridges seen in all other adult apes and maintain a youthful, slender build with lighter bones, compared to adult chimpanzees' robust and muscular physique. Unlike chimpanzees, which lose their white tail-tufts after weaning, bonobos retain theirs throughout life, and their voices remain high-pitched. Neoteny also influences their arboreal locomotion, which mirrors the suspensory and quadrupedal behaviours typical of infant chimpanzees. As primatologist and ethologist Isabel Behncke puts it, bonobos embody 'the spirit of Peter Pan's never-ending youth.'[10]

The delay in their developmental timing may also explain the bonobo's unusually high levels of playfulness – rough and tumbling, somersaulting, pirouetting, tickling, and especially play-fighting – perhaps the most of any non-human animal. As chimpanzees grow older, the extent of such activities decreases, while bonobos love to play throughout their lives. Such behaviour represents one of the most complex social interactions in primates and helps to promote creativity and resilience. Playfulness also provides a means to assess the potential of others as competitors or social allies. It enables the evaluation of a partner's commitment to a relationship while simultaneously emphasising readiness to avoid confrontation. Bonobos often undertake facial mimicry during play, which allows the playful mood to be shared between individuals and increases their familiarity and affiliation. Indeed, Behncke believes that when you observe bonobos, 'you see the evolutionary roots of human laughter, dance and ritual.'

Neoteny is involved in the evolution of many other species, from insects and birds to humans. For example, birds evolved from dinosaurs by a block in maturation, resulting in their retention of the large brain, big eyes, absence of teeth, and short faces of infantile dinosaurs. Domesticated animals, including dogs, rabbits and cats, are considered neotenous versions of their wild counterparts. Compared to wolves, dogs retain many anatomical and behavioural features characteristic of puppies: floppy ears, large eyes, playfulness and affection. Neoteny also explains many human traits that distinguish us from other primates, including our globular-shaped skull, large brain, broad face, and small nose and jaw, as well as our body's lack of hair, presence of a hymen and forward-facing vagina. Indeed, maturational block may also underpin the emergence of human-specific cognitive abilities through an extended period of high neuronal plasticity.

Neoteny, therefore, provides life with an efficient and rapid evolutionary route that works by taking something already available and modifying it rather than developing a whole new raft of genetic instructions. The process may also have enabled humans to evolve more rapidly despite sharing most of the same genes as other higher primates.

The 'self-domesticated' ape

After separating from chimpanzees, bonobos evolved into peaceful, socially tolerant, playful, sexual and empathic primates, traits enabled by the relaxed competition for resources. Richard Hare and colleagues have likened these behavioural modifications and the species' neotenous developmental patterns to a syndrome of changes observed in domesticated animals (Chapter 12).[11] How the bonobo's 'self-domestication' occurred is a field of active study that offers insights into the evolution of many of our own behavioural and anatomical features.

One approach has been to use cerebral imaging to detect structural and functional differences in primate brains. Researchers at the Laboratory for Darwinian Neuroscience at Emory University employed non-invasive neuroimaging to compare grey and white matter distribution and their connections in bonobos and chimpanzees.[12] Their findings revealed that bonobos have more grey matter in the amygdala, an almond-shaped mass of neurons in the temporal lobe associated with emotions and empathy in humans. In addition, the connection between the amygdala and the prefrontal cortex is larger in bonobos, a pathway implicated both in top-down control of aggressive impulses and in bottom-up biases against harming others. For example, when our amygdala senses that our actions are causing distress to others, our prefrontal connections modify our behaviour in a more caring, prosocial direction. In effect, the bonobo's modified nervous system enhanced their empathic sensitivity. It also enabled their characteristic behaviours, including sex and play, which helped alleviate tension and kept distress and anxiety at levels that enabled their highly social interactions to evolve.

Another approach is to search for the regulatory genes responsible for the bonobo's evolutionary trajectory against aggression. Sarah Kovalaskas, an anthropologist at Emory University, undertook such a study after spending nine months observing the social development of juvenile bonobos in the DRC.[13] Her research used data from over 70 whole genomes from bonobos and all four subspecies of chimpanzees. A comparative genomic analysis revealed that bonobos had undergone genetic selection in pathways associated with human-like social behaviours. Notably, genes relating to oxytocin, serotonin, gonadotropin and vasopressin showed a strong selection. High oxytocin levels reduce xenophobia and enhance social bonding, while serotonin suppresses reactive aggression and aids in avoiding danger, and gonadotropins may influence sexual behaviours. Combined selection in the vasopressin and gonadotropin pathways could drive increased motivation, reward and social benefits during same-sex sexual interactions. Kovalaskas also identified selection in thyroid-related genes, previously linked to the development of traits observed in domesticated mammals. Lastly, bonobos exhibit significant differences from chimpanzees in genes linked to pancreatic amylase production, an enzyme critical for starch digestion. This observation highlights the importance of diet and resource availability in bonobo evolution.

The bonobo and chimpanzee genes are 99.6 per cent identical, indicating that their physical and behavioural differences likely stem from evolutionary changes outside the coding gene sequences. Indeed, most genetic differences between closely related species probably occur in regulatory or non-protein-coding regions of the genome, with only a tiny portion affecting amino acid sequences in proteins. The search for evolutionary changes in regulatory sequences has been challenging, although several novel strategies are under development (further discussed in Chapter 25). One such technique determines points of contact resulting from genome looping and has already highlighted potential genes of interest.

When stretched out, a primate genome measures about 2 metres in length, but it must fit into a nucleus one-hundredth of a millimetre in diameter.

However, the chromosomes are not randomly squashed within the nuclear space but folded around histone proteins to produce a three-dimensional structure. Furthermore, the chromosomes assume distinct territories, so they have little contact, with each one organised into several compartments that allow distant cis-acting enhancers to loop and contact genes to regulate their activity. Alterations in the 3D arrangement of enhancers and their genes determine whether genes are activated and, if so, where, when and for how long. Such control of gene expression likely underpins most phenotypic differences between closely related species. In 2024, a team from the University of California adapted a machine-learning algorithm that can predict 3D genome contacts from human DNA to look for regulatory differences between chimpanzees and bonobos.[14] The novel approach highlighted several areas of interest. For example, chimpanzees exhibit additional chromatin contact and increased gene expression compared to bonobos at a chromosome 5 locus that overlaps *MYO10*, an essential developmental gene. Additionally, bonobos lack chromatin contact at a chromosome 7 locus compared to chimpanzees and exhibit increased gene expression of *ZNF804B*, a gene highly expressed in human thyroid tissue, raising the intriguing possibility of a role in the primate's 'self-domestication syndrome'.

The 'ghost' ape

After studying archaic DNA from early hominins, Martin Kuhlwilm joined the Comparative Genomics group at the Institute of Evolutionary Biology in Barcelona, led by Tomàs Marquès-Bonet. Without a designated research area, Kuhlwilm recalls sitting with his boss outside the institute, overlooking the Mediterranean, brainstorming potential projects. Their conversation soon led to the idea of searching for 'genomic fossils' within the DNA of chimpanzees. The idea was to determine whether the introgressions reported in gorillas and humans were unusual or whether hybridisation with extinct species was widespread and underpins primate evolution. Despite the logistical challenges, the timing was ideal, as several chimpanzee genomes had already been sequenced, and bonobo DNA was accessible from Lola Ya Bonobo, the world's only bonobo sanctuary in the DRC.

In 2019, having sifted through billions of bases from the genomes of 59 chimpanzees and 10 wild bonobos, Kuhlwilm's team announced their findings.[15] Contrary to their expectations, evidence of 'genetic fossils' popped up in the DNA of bonobos, but not in chimpanzees. The smoking gun was the presence of long stretches of novel DNA absent from chimpanzees. While all species have DNA differences, these usually arise through the gradual accumulation of random mutations that typically appear in short, discrete sequences. In contrast, longer sequences are more likely indicative of introgression through hybridisation with another lineage. But the admixture didn't match the DNA of any known primate. At first, Kuhlwilm's team doubted the significance of their findings. However, after multiple computer simulations and discussions with others in the field, the researchers concluded that the DNA could only have come from an unknown lineage. In other words, the DNA must have originated from an extinct or 'ghost' species.

The bonobo's introgressed DNA makes up approximately 3 per cent of its genome. However, unlike the story of eastern gorillas, the introgressed material was acquired a long time ago, around 400,000 years ago. The unknown archaic primate likely diverged before the Pleistocene, around 3.5 million years ago, and lived in central Africa until at least half a million years ago. Once the ancestral bonobos crossed the Congo River, they would have encountered the 'ghost' species and bred with them to produce fertile hybrids – as *Homo sapiens* did with Neanderthals and Denisovans (Chapter 25) – with the only clues that such liaisons occurred being hidden within the bonobo's genome. Given the seemingly ubiquitous nature of primate introgressive hybridisation – in gorillas, bonobos and humans – it would not be surprising if future studies reveal further historical exchanges in other species, even chimpanzees.

Some of the acquired DNA may have contributed to the bonobo's evolutionary success, boosting their immune response to pathogens and improving their adaptation to different food sources. However, attributing a function is complex, and additional research is needed to confirm such speculation. Furthermore, the small amount of acquired DNA makes learning about the ghost species' phenotype challenging. Nevertheless, these genomic fossils have already informed scientists of roughly when and where the unknown primate lived and possibly, with some imaginative thinking, some of its characteristics. Unfortunately, the chance of finding physical remains and extracting ancient DNA is negligible, as the rapid decomposition of carcasses in the rich, acidic soils of the Congo rainforest would have prevented any fossil formation.

During the Middle Miocene, apes thrived across Africa and Eurasia in what could be considered their golden era. However, their evolutionary trajectory has since been characterised by fragmentation and extinction. Today, great apes, aside from humans, exist as endangered populations confined to scattered equatorial forest refugia. As we will discuss, even the genomes of *Homo sapiens*, now widespread across the globe, reveal evidence of ancient population bottlenecks and extinctions. The human evolutionary biologist Alywyn Scally suggests that the current precarious state of great apes 'may echo aspects of our own ancestral past before the last 100,000 years and perhaps reflect a state experienced repeatedly throughout millions of years of evolution.'[16] Gorillas, chimpanzees and bonobos continued to exchange genes even after diverging from one another. This gene flow, much as in *Homo sapiens* (Chapter 25), may have aided their survival during periods of decline. While the future of our closest living relatives remains uncertain, Scally rightly notes that 'studying great apes deepens our understanding of human evolution, reconnecting us with a fragile past and underscoring the urgent need to protect these remarkable species.'

CHAPTER 25

The Human Story

PALAEOGENOMICS AND ADAPTIVE INTROGRESSIONS

Humans are not the end result of predictable evolutionary progress, but rather a fortuitous cosmic afterthought, a tiny little twig on the enormously arborescent bush of life.
Stephen Jay Gould (1941–2002)

Most readers will be familiar with the traditional 'Out of Africa' model, which argues that modern humans descended from the trees and migrated from Africa to settle in nearly every corner of the planet. However, this straightforward narrative has been significantly refined over the last decade, thanks to breakthroughs in multidisciplinary research, advanced DNA analysis, and improved fossil identification techniques. Our evolving story now includes multiple waves of migration out of Africa, followed by interbreeding with other hominin species, and the development of diverse adaptations to local environments – all contributing to a rapid and unparalleled population expansion.

Emergence of *Homo sapiens*

Let us explore how humans achieved this global dominance while our closest relatives, orangutans, gorillas, chimpanzees and bonobos, remain endangered and restricted to small areas in equatorial Africa and Southeast Asia. After the chimpanzee and bonobo lineages split off, early hominins gradually acquired multiple genetic changes that ultimately gave rise to who we are today – *Homo sapiens*. The precise order of events is unknown but includes the acquisition of various defining traits: bipedalism, high metabolic rate, increased height, large brains, high reproductive rates and exceptional longevity.

Yet our story nearly ended before it began.[1] Between 930,000 and 813,000 years ago, our ancestors faced a severe population bottleneck, with numbers plummeting by 99 per cent to approximately 1,280 reproductive individuals. This crisis, which lasted over 100,000 years, was likely triggered by increased glaciation, leading to lower ocean and land temperatures. Expanding ice sheets absorbed much of the Earth's moisture, making large areas of Africa and Eurasia arid and drought-prone. Eventually, conditions improved. As the climate warmed and early humans mastered fire, populations began to recover rapidly.

One of the early adaptations of hominins was the development of striding bipedalism: a distinctive form of walking in which each leg swings forward in turn, in a coordinated manner. This form of locomotion did not arise from a single mutation but emerged gradually over millions of years of evolutionary experimentation. Early ancestors likely led a mixed lifestyle, relying on trees for safety and food, while gradually spending more time on the ground. This shift coincided with environmental changes during the Late Miocene to Pliocene epochs. Research by Rhianna Drummond-Clarke suggests that bipedalism may have first developed in an arboreal setting, where moving along flexible terminal branches to access food would have favoured upright motion.[2] As hominins adapted to the more open savanna–woodland environments, new selection pressures were encountered – sparse tree cover, increased seasonality and heightened predation risks. These novel conditions provided further evolutionary pressure, leading to more efficient terrestrial locomotion. Over time, the development of bipedalism provided the various *Homo* species with greater mobility, coupled with energy-efficient long-distance endurance. By freeing their arms, early humans were able to use tools, carry food and offspring, hunt and gather more efficiently, and engage in social gestures and communication with their hands.

Bipedalism, however, necessitated many structural changes to our bones, muscles and joints. For example, our skull became positioned directly above the spinal column. This arrangement required the foramen magnum, the opening at the skull's base for the spinal cord, to be horizontally aligned and moved toward the front of the skull. Our spine became curved, and our pelvis widened, shortened and became cup-shaped, with large hip joints and short iliac blades to support our vertical body. The ischial spines became more prominent and shifted towards the middle of the pelvis to provide a greater surface area for ligament attachment to help support our abdominal contents. This unique bony arrangement enabled efficient weight transfer to the legs during various movements – standing, walking, climbing and running. Furthermore, two recent studies have pinpointed the window in human embryonic development during which the pelvis begins to resemble a human form and not an ape-like one, along with hundreds of genes and regulatory DNA regions that drive this transformation.[3] In effect, the growth plate of the ilium bone turned 90 degrees sideways. At the same time, ossification (bone formation) was delayed, allowing the pelvis to flare out and enabling our muscles to support upright walking. These findings add to the growing literature showing that evolution often produces new physical features by acting on genetic switches that affect early embryonic development rather than relying on the emergence of a new suite of genes.

Our legs grew thicker, more angled, and relatively longer than our arms, with fully extendable knee joints that lock for greater stability. The human foot became highly specialised, with a large heel to bear the increased weight, a robust ankle joint and a medial arch that acts as a shock absorber. But organisms cannot optimise every trait simultaneously, and bipedalism inevitably led to evolutionary trade-offs – the loss of our ancestral climbing adaptations, including an opposable big toe, curved fingers and toes, and shoulder-joint adaptations for arboreal movement. Bipedalism also had other downsides that affect us today. The spine and knee joints, which bear greater weight in a bipedal stance, are particularly vulnerable to wear and tear, while stiff necks and lower back pain are the leading causes of time off work in the Western world today. Joint degeneration and eventual destruction due to osteoarthritis have been a health issue since hominins adopted bipedalism, with evidence of the condition in the vertebrae of prehistoric hunter-gatherers. In other words, enhanced joint stability without compromising locomotion efficiency remains an evolutionary constraint.

Bipedalism requires just a quarter of the energy chimpanzees expend while knuckle-walking on all fours, and, in other ways as well, humans are remarkably distinct from other primates in their energy use. A recent study by Harvard researchers, led by Andrew Yegian, found that we have the highest metabolic rate of any mammal.[4] Non-human primates allocate approximately 30–50 per cent more calories to their resting metabolic rates than other similar-sized mammals. Humans take this trend to the extreme, expending about 60 per cent more calories than most mammals. Notably, a regulatory mutation that enhances the expression of the *ACSF3* gene – through increased enhancer activity – appears linked to elevated basal metabolic rate and increased stature in modern human populations. This mutation, which is absent in non-human primates, likely contributed to shifts in energy expenditure and physical growth in response to dietary changes involving increased meat consumption.[5]

Generally, a higher metabolic rate generates more body heat, making it challenging to stay cool in tropical climates. As a result, animals like chimpanzees – which have relatively high resting metabolisms – tend to conserve energy by limiting physical activity, often spending much of the day sitting and eating. Humans, however, have evolved a unique solution to this challenge: the ability to efficiently dissipate heat through sweating. According to Yegian, this adaptation makes us energetically distinct as a species. By breaking the usual link between resting and active metabolic rates, our ancestors were able to sustain extended periods of physical activity, such as long-distance hunting, while still supporting the energy demands of growing brains, higher reproductive rates and longer lifespans.

These uniquely human traits would not have been possible without our ability to sweat efficiently – our primary method of cooling the body. Humans can sweat up to 1 litre per hour, a rate 4–10 times higher than that of chimpanzees. This crucial adaptation stems from two key evolutionary changes: the loss of body fur, which improves airflow and facilitates evaporation, and the development of a high concentration of eccrine sweat glands, approximately 10 times more numerous than in other primates. The vital role of these skin glands is underscored by the danger of hyperthermia faced by rare individuals born with too few of them, or with malfunctional ones.

The development of eccrine glands is controlled by a gene regulated by an enhancer – a short DNA sequence that serves as a binding site for proteins to boost the gene's expression. Researchers at the Perelman School of Medicine in Philadelphia discovered that the human version of this enhancer is significantly more active than its counterparts in apes and monkeys.[6] Throughout human evolution, the enhancer accumulated multiple point mutations, collectively increasing the activity of the associated gene and resulting in a higher density of eccrine glands in human skin. This study also highlights how sequential mutations in a single enhancer can progressively alter gene activity, driving key adaptations such as the enhanced thermoregulation that distinguishes humans from other primates.

A recent analysis of mammalian DNA has offered new insights into how humans became the only 'naked ape' among roughly 500 primate species. In 2022, researchers led by Nathan Clark from the University of Utah's Department of Human Genetics carried out a comparative genomic study across more than 60 mammalian species, encompassing both hairy and hairless animals.[7] Using a computational tool designed to identify DNA regions evolving at varying rates, the team analysed nearly 20,000 genes and 350,000 regulatory elements. Their findings revealed that, unlike our closest relatives, the chimpanzees and bonobos, humans have accumulated mutations in numerous genes and regulatory regions associated with keratin production and other hair-related proteins. Interestingly, similar genetic patterns were found in other hairless mammals: armadillos, manatees, whales, porpoises, walruses, elephants, rhinoceroses and naked mole-rats. Because these species are only distantly related, the study suggests that hairlessness evolved independently at least nine times in separate mammalian lineages.

The stage was now set for the emergence of one of humanity's most remarkable features: our large and complex brain, often considered among the most intricate physical structures in the known universe. Over millions of years, primate evolution has driven a dramatic increase in both the size and complexity of the brain. Since our lineage split from *Pan* around 6 million years ago, the human brain has more than tripled in volume, and it is now about eight times larger than that of New World monkeys, comprising over 80 billion neurons. This expansion, however, did not occur uniformly. Different brain regions developed at varying rates, resulting in changes in their relative proportions. The cerebral cortex, in particular, has undergone substantial growth, especially the prefrontal cortex, which is critical for executive functions such as decision-making, personality expression and social behaviour. Just as vital, though less apparent, are the internal rewiring and physiological changes that have refined how neurons communicate within the brain and with the rest of the body.

A recent study suggests that human brain size increased gradually within each ancient species rather than through sudden leaps between species.[8] This conclusion, which challenges traditional thinking, was deduced after researchers compiled the largest dataset of ancient hominin fossils, covering 7 million years, and using advanced computational and statistical methods to cover gaps in the fossil record. As expected, the genetic events that drove this enlargement are complex, ranging from single nucleotide substitutions to significant structural modifications to the genome. Such changes resulted in a diverse set of functional outcomes – alterations in protein sequences, variations in cis-regulatory elements, the emergence of new genes and the loss of existing ones.[9]

Unexpectedly, a Chinese group has added the involvement of 'junk' or non-coding DNA to the growing list of underlying mechanisms. In 2023, a team from Peking University reported that several genes that underpin our brain's large lobes and complex information networks emerged initially from junk DNA that acquired the ability to code for proteins.[10] Most novel genes are produced from errors during cell division that lead to duplications from which new functions can slowly arise after random mutations. Indeed, some duplicated genes are implicated in human encephalisation. However, de novo genes can derive from junk DNA that acquires the ability to make proteins, often an all-or-nothing transformation to functionality. While coding genes produce messenger RNAs, which exit the nucleus and travel to ribosomes to make proteins, RNA transcribed from non-coding DNA cannot cross the nuclear membrane. However, the research carried out at Peking University has shown that some junk-DNA-derived RNA, so-called long non-coding RNA (lncRNA), is modified and can reach the cytoplasm and produce proteins.

After comparing the genomes of humans, chimpanzees and macaques, the researchers identified 74 examples of junk DNA that had transformed into protein-coding DNA. Humans and chimpanzees share 29 of these de novo genes, indicating that they emerged after humans and chimpanzees diverged from the ancestor they shared with macaques. The remaining 45 de novo genes are unique to humans and have only appeared during the last 6 million years. Crucially, nine of these are active in the human brain and have been shown to cause altered cell kinetics in cerebral organoids (artificial conglomerations of brain cells), and enlarged brains with cortical expansion and human-like ridges and grooves in genetically modified mice.

After approximately 600,000 years ago, the increase in hominin brain size was no longer linked to total body mass; brains enlarged independently by around 30 per cent. This dramatic increase enabled the evolution of our broad cognitive and cultural abilities. However, there were downsides to this evolutionary trajectory. The increasingly large-brained fetuses had to be delivered through narrow pelvises that were already adapted to bipedalism, making birth increasingly problematic. Given their wider pelvises, primates are precocial, giving birth to relatively large, mature and mobile offspring. Humans are an exception: our newborns are altricial, an evolutionary trade-off to facilitate childbirth. For example, human full-term babies possess around 25 per cent of the adult brain mass, compared to 40–70 per cent in our closest relatives. But many mammals, including bears, dogs and cats, show more pronounced altriciality – so why don't humans give birth to less developed babies that could easily pass through the birth canal?

Human neonates born at an earlier developmental stage have many disadvantages. Prematurity, for example, is associated with impaired cognitive function in later life, as well as sudden infant death syndrome, asthma and feeding difficulties. Similarly, accelerating postnatal brain growth is not an option, as fast growth is associated with autism spectrum disorders and intellectual disability. Indeed, the human brain is unique among mammals due to its exceptionally slow development as the connections between neurons in the cerebral cortex, essential for cognitive processes, take years to mature fully. In contrast, brain maturation is complete within months in species like mice or macaques. Our slow cerebral development, known as neoteny, is controlled by several unique genes that, if non-functional, lead to neurodevelopmental disorders.[11] Longer gestation is also not possible, because it would make excessive metabolic demands on the mother. Finally, our pelvis dimensions are evolutionarily constrained, and the birth canal cannot become more spacious, as broader pelvises lead to impaired walking, backache, and structural instability of the pelvic floor.

Overall, modern human pregnancy length and fetal brain development have evolved as the result of trade-offs between easier childbirth, increased metabolic demands, health risks and cognitive disadvantages. Such multifactorial selection pressures have led to a 'compromise phenotype distribution' that maximises the mean population fitness despite the risk of severe impairment or death for any outlier.[12]

By around 200,000 years ago, *Homo sapiens* had lost the distinctive heavy, robust prognathism (protrusion of the lower jaw) of apes and archaic humans. The distinctive facial plan of primates before *Homo sapiens* resulted from increased osteoblast activity, cells that in our ancestors continuously deposited skull bone until adulthood. In contrast, the modern human's face grew with a counterbalanced action of osteoclasts, cells that, in our case, dissolved and removed bone matrix, especially in the lower part of the skull.[13] This evolutionary shift resulted in a smaller, flatter face, with smaller upper and lower jaws, reduced brow ridges, a smaller nose, and a distinctive pointed chin. In addition, modern humans exhibit many other neotenic features, including a rounded skull shape, thinness of cranial bones, larger eyes and smaller teeth. Our dental changes mean that many individuals today do not have enough space in their mouths for their third molars or wisdom teeth.

Human newborns are uniquely helpless among primates, requiring prolonged care and support from family or community members for survival, protection and learning. This dependency necessitated the development of the tight family bonds and social networks, which in turn paved the way for our complex cultural systems and remarkable technological achievements. *Homo sapiens* had arrived: we were ready to leave Africa.

Out of Africa

Darwin's idea of an African origin for all humanity was proved beyond doubt in 1987 with a publication in the prestigious scientific journal *Nature*.[14] A research team from the University of California analysed mitochondrial genomes from different modern human populations and found that the highest genetic diversity occurred in Africa, decreasing progressively in populations further from the continent. This pattern is typical of a 'serial founder effect', where genetic variation is lost as small groups of individuals branch off from a much larger population to form new ones. Even within Africa, *Homo sapiens* was far from a homogeneous population, consisting of many widely dispersed groups influenced by climate-driven migrations: movements that facilitated gene flow and the exchange of tool technologies. Our ancestral genetic diversity may have been even greater, enriched by interbreeding with other archaic human species that likely coexisted with our ancestors.

While Africa is the birthplace of *Homo sapiens*, scientists are less certain when, how or why our ancestors left. We were not even the first *Homo* species to leave the continent. *Homo erectus*, a species like us but with much smaller brains, ventured into Eurasia around 1.6 million years ago to reach island Southeast Asia before becoming extinct as early as 100,000 years ago. Another species related to *Homo habilis* may even have left at an earlier date. Indeed, Africa is viewed as a melting pot of *Homo* species, possibly interbreeding, with multiple dispersals from Africa during favourable climatic conditions. They included dispersals across southern Asia and into western Europe 900,000 years ago. Movements back to Africa may have also occurred. One of these early migrations from Africa gave rise to the Neanderthals in Eurasia and the Denisovans in Asia, although, given the lack of fossils, their precise link to the continent remains unclear.

Modern humans began to leave Africa in two, maybe more, migrations between 160,000 and 50,000 years ago. The earliest dispersals colonised the Levant, but these attempts ended in extinction. Later, climatic fluctuations led to periods of greening in the Middle East, with an associated increase in animals to hunt, which encouraged further travels via the northern Sinai Peninsula and the Bab al-Mandab strait at the southern tip of Arabia. Which route was taken was determined by climatic conditions and sea levels. Around 50,000 years ago, there were at least three human species: *Homo sapiens* in Africa and the Middle East, Neanderthals in Eurasia, and Denisovans across Asia.

Once established in Arabia, *Homo sapiens* began their extraordinary migrations, travelling further than any earlier species of the genus *Homo*. During these journeys, they interbred with Neanderthals, Denisovans and other archaic humans that existed at the time (explored further below), indicating that human evolution underwent at least three resets in the past 50,000 years. By around 45,000 years ago, they had reached New Guinea and Australia.

Between 30,000 and 15,000 years ago, humans crossed the Bering Strait into North America during an interglacial period, reaching the tip of South America approximately 12,000 years ago. This 20,000-kilometre migration, the longest undertaken by humans, spanned countless generations and thousands of years. Once in South America, the early migrants split into four major groups: the Amazonians, the Andeans, the Chaco Amerindians and the Patagonians. Due to founder effects, these small, isolated groups retained only a fraction of their ancestors' genetic diversity, particularly in relation to genes involved in the immune system.[15] This loss of variation may have limited their ability to resist new diseases, helping to explain why some

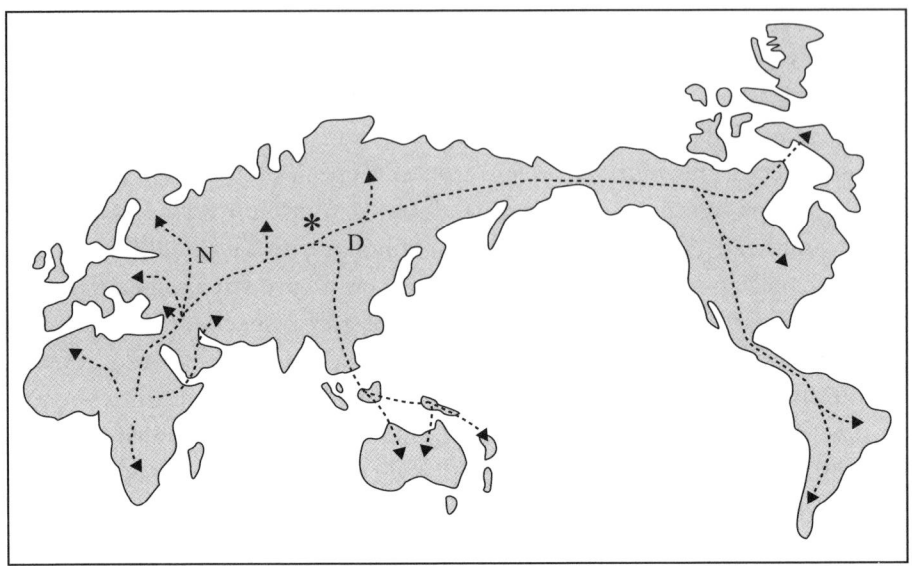

Figure 25.1 Origin and dispersal of *Homo sapiens*. N, Neanderthals; D, Denisovans; *, Denisova Cave in the Altai Mountains, Siberia.

indigenous communities were especially vulnerable to illnesses introduced by later arrivals, such as European colonists.

Unlike the overland migrations into the Americas, the expansion into remote Oceania – including regions as distant as Tonga and Samoa – required advanced maritime technology and did not occur until about 3,000 years ago. New Zealand was among the last landmasses to be inhabited, with settlers from East Polynesia arriving between 1250 and 1300 CE (Figure 25.1). However, why modern humans were so successful and spread across the globe, while archaic humans became extinct 40,000 years ago, remains a mystery.

Adaptive introgressions

In 2010, the results of a four-year effort to sequence the Neanderthal genome were published.[16] Geneticists, led by Svante Pääbo from the Max Planck Institute for Evolutionary Anthropology in Leipzig (Plate 35), were able to extract sufficient DNA from fossils to compare it with modern human DNA. The results shocked the scientific community. It turned out that between 1 and 4 per cent of the DNA from modern European and Middle Eastern individuals is derived from Neanderthals, the first conclusive evidence of interbreeding. However, Neanderthal DNA found in one person may not be the same as that found in another person, and it is now estimated that, overall, around 20 per cent of Neandertal DNA may survive today.[17]

Several months after the Neanderthal paper, Pääbo's laboratory stunned the world yet again. Two young researchers, Johannes Krause and Qiaomei Fu, sequenced the mitochondrial DNA from a tiny finger bone unearthed from Denisova Cave in southern Siberia.[18] The DNA didn't match either Neanderthals or modern humans as expected, but came from a new and unsuspected archaic hominin – now called Denisovan, after the cave where the fossil was found. The pea-sized bone also provided high-quality nuclear DNA that allowed the sequencing of almost 70 per cent of the Denisovan genome – a groundbreaking achievement that contributed to Pääbo being awarded the 2022 Nobel Prize in Physiology or Medicine.

The first surprise was that Denisovans were more closely related to Neanderthals than either species was to modern humans. The genetic differences suggest that the two archaic species diverged between 470,000 and 380,000 years ago, and that both separated from their common ancestor with humans between 770,000 and 550,000 years ago. The second surprise came when Pääbo's team compared the Denisovan genome to various modern human populations to see if any are more closely related to the archaic humans than others. While no introgressions were found in African populations, 4.8 per cent of Melanesian genomes – people from New Guinea and Bougainville – derive from Denisovans. The interbreeding likely occurred when our ancestors dispersed outwards from Africa and moved to eastern Eurasia after 50,000 years ago. Subsequent research showed that Aboriginal Australians and Filipino Negritos also possess similar levels of Denisovan DNA, suggesting that Denisovans were widespread across Asia. Indeed, there is evidence that at least three distinct Denisovan groups may have existed, with one population interbreeding with the easternmost Neanderthals, given that approximately 17 per cent of their genome derives from Neanderthals. Intriguingly, about 4 per cent of the Denisovan genome originates from a 'ghost species', an unidentified archaic human population that diverged from modern humans over a million years ago. Our once simple ancestral story is becoming increasingly complicated and messy, with evidence of interbreeding between at least three *Homo* species in Asia before 40,000 years ago.

Ancestors of Neanderthals and Denisovans expanded into Eurasia tens of thousands of years before the arrival of modern humans. Indeed, when *Homo sapiens* reached Eurasia, the various archaic populations had already adapted to a wide range of environments, including cold temperatures, low and high ultraviolet levels, high altitude, lowlands and perhaps even humid tropical regions.[19] Therefore, interbreeding provided an opportunity to introduce beneficial genetic variants that helped accelerate human adaptations.[20]

Our Neanderthal genes, acquired between 50,500 and 43,500 years ago, were rapidly subject to natural selection – positive and negative – so that only around 1–4 per cent remain today in any individual.[21] These include genes related to skin and hair pigmentation, metabolism and immunity. One haplotype, present in 70 per cent of Europeans, encompasses the *BNC2* gene expressed in skin cells and is associated with skin pigmentation and freckling. Other skin-related genes control the proliferation and differentiation of epidermal cells, modify the cellular response to ultraviolet radiation, and are associated with blue iris pigmentation and blond and red hair colour. Many genes are related to immunity and protect against various infections, especially viral diseases, including HIV, and bacterial diseases, including *Helicobacter pylori* infections. These observations are not surprising, as archaic humans in Eurasia would have encountered novel pathogens that induced adaptations to their immune systems: genomic modifications that benefited *Homo sapiens*.

A Denisovan introgression in Inuit people increases their tolerance to cold. In almost 100 per cent of Greenlandic Inuit and several other populations, the acquired DNA includes two genes, *WARS2* and *TBX15*, linked to heat generation from a specific type of body fat called brown fat. Neonates possess a lot of brown fat, especially behind their shoulder blades, an essential heat source before they can shiver. Brown fat breaks down blood sugar and fat molecules to create heat that maintains body temperatures when triggered by cold conditions. The introgressed sequence is absent in African populations but appears at low-to-moderate frequencies across Eurasia, becoming nearly universal in the Inuit. Remarkably, these genes are highly pleiotropic, influencing other traits, including stature, ear shape, hair pigmentation and facial morphology. Researchers suggest that this archaic genetic adaptation gave modern humans a survival advantage as they traversed Siberia and Beringia and entered the Americas.[22]

The Tibetans, like the Inuit, also benefited from Denisovan genes. Over 80 per cent of the present-day population of Tibet possess a segment of DNA on chromosome 2 that encodes EPAS1, a transcription factor induced by hypoxia that aids survival at altitude and low oxygen levels. This gene, absent or very rare in other Asian populations, remained selectively neutral for a long time and only became positively selected when modern Tibetans lived permanently on the Tibetan plateau after the Last Glacial Maximum. The fact that Melanesians lack this introgression suggests that genetically different Denisovan populations must have inhabited Asia.

Local adaptations

Since acquiring genes from Neanderthals and Denisovans, human populations have been exposed to new environmental selection pressures, including novel pathogen exposures and cultural innovations, such as the spread of farming and dietary shifts. As a response, many local populations have evolved a higher-than-average fitness due to genomic and phenotypic adaptations to their habitat and lifestyle.

For example, malaria has threatened human existence for millennia and has acted as the most potent selective force in recent human evolution, given its propensity to infect children and cause death before reproductive age. Around 20,000 years ago, a spontaneous mutation arose in a haemoglobin gene in Africa that was protective against *Plasmodium falciparum*, the organism responsible for the most aggressive form of malaria. While heterozygotes have a lower risk and reduced severity from the infection, homozygotes suffer from sickle cell disease, with high comorbidities and mortality. Since the gene has persisted in malarial areas, the benefit for those with one copy of the gene must outweigh the increased mortality for those with two gene copies. In other parts of the world, different inherited blood disorders provide similar levels of malarial resistance, including thalassaemias, G6PD deficiency and ovalocytosis.

Over the past 12,000 years, many human populations have increased their ability to digest carbohydrates after shifting to a starch-rich diet. According to a recent multinational study, the average number of genes coding for amylases (starch-digesting enzymes) has increased sevenfold in human genomes compared to those of chimpanzees, bonobos and Neanderthals.[23] Amylases in saliva and pancreatic secretions break down starch into sugar, providing essential energy for the body. The selection for multiple amylase genes correlates with the spread of agriculture from the Middle East into Europe, as farmers introduced high-carbohydrate cereals like wheat and other grains. Interestingly, similar amylase increases have occurred in agricultural populations on different continents, irrespective of the type of starch-containing crop.

Another local adaptation involved the enzyme lactase. To digest lactose, the sugar found in milk, humans require the enzyme lactase, produced by the cells lining the small intestine. Nearly all infants produce lactase, but in most individuals its production decreases significantly after weaning and through adolescence. However, a genetic adaptation known as lactase persistence independently evolved multiple times over the past 10,000 years due to mutations in the lactase gene's upstream controlling element. This trait spread among milk-drinking populations in Eurasia and Africa, so that today approximately one-third of adults are lactase-persistent.

The Bajau people, often called Sea Nomads, are marine hunter-gatherers, travelling the seas of Southeast Asia on houseboats for over 1,000 years. Their existence depends on the food they collect through free diving, and they are renowned for their extraordinary breath-holding abilities, reaching depths of over 70 metres with only a set of weights and wooden goggles. Due to the mammalian dive reflex, the spleen contracts when diving and releases oxygen-carrying blood cells. Over the past millennium, the Bajau have undergone positive selection for a genetic variant of the *PDE10A* gene that increases their spleen size and acts via the thyroid to increase red cells, providing them with a larger reservoir of oxygen-carrying blood cells.[24]

Other examples of diving adaptations are seen in the Haenyeo, the all-female divers from Jeju Island in Korea, celebrated for their extraordinary ability to dive in cold waters. Indeed, breath-hold diving is so deeply embedded in Jeju's culture that even the linguistic tendency to shorten words is thought to have arisen from the need for quick communication at the water's surface. To survive and thrive in such a harsh environment, the Haenyeo have evolved genetic adaptations linked to cold resistance, higher pain thresholds and lower diastolic blood pressure. Over the past 10,000 years, natural selection has enabled the female divers to repeatedly forage along the sea floor for up to 4–5 hours each day in conditions that would challenge most humans.[25]

Although these recent genetic adaptations, driven by local selection pressures, have helped specific groups of modern humans to thrive, they do not fully explain our global success. As more genomes from diverse populations are analysed, Pääbo's team anticipates that very few, if any, genetic differences will be found that distinguish all modern humans from Neanderthals and Denisovans. Instead, they suggest that the essence of being a modern human lies in a unique combination of genetic traits, with none necessarily shared by every individual alive today. The challenge for future research is to identify these genomic changes.[26]

All alone

Approximately 40,000 years ago, *Homo sapiens* was suddenly alone, the only member of the *Homo* genus left on Earth. Our many archaic brothers and sisters had turned to dust, and their disappearance raises one of the biggest questions in the human story: where did they all go? It's a hotly debated topic, with many proffered solutions: climate change, environmental catastrophe, genocide, genetic assimilation, disease, or competitive exclusion through culture and language. Whatever the answer, the survival advantage of *Homo sapiens*, no matter how small, cannot be in doubt, given our global dominance – currently 8.2 billion individuals and likely to rise in the years ahead.

Our archaic cousins, of course, are not entirely lost: small fragments of their DNA lie scattered throughout our genomes, actively functioning alongside our own and contributing to our success. However, we are arguably the most baffling species of all the mammals, not because we represent any apogee of evolution, but because we alone can question how evolution happens from the strange standpoint of being one of them. Ever since Darwin's insight, we have sought to understand how *Homo sapiens* came to occupy a 'tiny little twig on the enormously arborescent bush of life'. Yet, despite 160 years of inquiry, many questions remain unanswered. Who were our immediate ancestors? What drove our exodus from Africa? How many archaic human species were there? How did our human language, culture, and sexuality evolve? Despite the many gaps in our knowledge, we have made remarkable progress. Indeed, if recent discoveries are any indication, the story told in this chapter may soon need to be modified, if not completely rewritten.

EPILOGUE

The descent of mammals

Ninety-nine per cent of all mammals that have ever existed have vanished, gone forever. Their demise is a natural and continuous feature of life's history, neatly encapsulated in Carl Sagan's pithy words: 'extinction is the rule, survival is the exception'. Added to this, occasional mass extinctions punctuate the background rate when most of the planet's species disappear in a geological blink of an eye. Indeed, Earth's history has recorded five such events, prompting the question: when is the sixth one due? As many readers will be aware, most scientists believe we may be in the middle of it already, an extinction event that, for the first time, is caused by just one species, *Homo sapiens*. Around 100,000 years ago, extinction rates began to rise as humans expanded from Africa to other continents, leading to the decline of large mammals, a trend that accelerated around 12,000 years ago. It is a process arcanely referred to as Anthropocene Defaunation or, more prosaically, the Sixth Mass Extinction. It results from a litany of human activities: over-exploitation of resources and other species, habitat destruction, introduction of non-native species, and greenhouse gas emissions.

Scientists estimate that during the last 10,000 years, over 255 species of mammal have been lost, one-third of that total in the last 500 years. Despite our awareness, mammal extinctions continue: the Yangtze River dolphin or baiji was considered extinct in 2007, and the Christmas Island pipistrelle in 2009. The kouprey, also known as the grey or forest ox, which inhabits parts of Cambodia and Laos, may already be extinct by the time you read this book. Other species, such as the vaquita, a porpoise endemic to the Gulf of California, and the Hainan gibbon, are now reduced to such tiny remnant populations, comprising only a handful of individuals, that their future looks bleak. In 2019, the Australian government officially declared the Bramble Cay mosaic-tailed rat extinct: the first mammal species lost from human-induced climate change. Its habitat, a small low island or cay composed of coral and vegetation at the northern tip of the Great Barrier Reef, was inundated by storm surges as sea levels continue to rise. Climate change will likely drive further extinctions through additional mechanisms, including increasing global temperatures – such as the extreme heatwave in 2018 that wiped out one-third of Australia's spectacled flying foxes, and the country's 2019–2020 bushfires that impacted an estimated 140 million mammals.

Humans are not immune to the dire effects of biodiversity loss and the degradation of ecosystems. Intact environments provide key functions – such as carbon removal, nutrient recycling and pollination – that underpin global economies, livelihoods and health. Billions of people depend directly on natural systems, including oceans, rivers, forests, grasslands and mangroves, for essential resources like food, drinking water, medicine and protection against extreme environmental events. For instance, over 400 million people and around 84 mammal species rely on the Chinese Yangtze River basin, now one of Earth's most damaged ecosystems. This vast area is affected today by unprecedented industrial and plastic pollution, pesticides from agricultural runoff, overfishing, construction of dams, sediment pollution and the loss of lakes and wetlands.[1]

Sadly, according to a study by Tobias Andermann and colleagues from the University of Gothenburg, the outlook for mammals will only worsen.[2] The researchers collected data on all the mammal species that became extinct since the beginning of the Late Pleistocene and correlated the findings to changes in the human population size over the same time scale. The conclusions were stark: the number of humans on the planet directly predicted past extinctions, while climate change had minimal effect. Based on mathematical modelling, the Gothenburg team also predicted that by 2100 all areas of the world will have experienced a second wave of extinctions, several orders of magnitude greater than what is happening at present. Indeed, it seems that Australia and the Caribbean have already entered such a phase. A 2019 study conducted by researchers at the University of Southampton suggests that mammals could shrink in size by up to 25 per cent over the next 100 years. In the future, small, insect-eating mammals with rapid reproduction, short lifespans and the ability to adapt to diverse habitats – such as rodents – are expected to dominate. Conversely, larger mammals with slower lifecycles and more specialised habitat needs are at a higher risk of extinction.[3]

All life on Earth will ultimately face extinction in a runaway greenhouse scenario when absorbed solar radiation surpasses the planet's ability to emit thermal radiation: a fate projected to occur several billion years in the future. However, the Earth may become uninhabitable for mammals much earlier than this. A new supercomputer simulation by a team from the University of Bristol has predicted that mammals will become extinct in approximately 250 million years when the planet's continents are expected to merge, forming Earth's next supercontinent.[4]

The Earth's foundations are dynamic, composed of solid continental and oceanic plates that rest above the upper mantle, or asthenosphere. This layer, though solid, is pliable and flows slowly like a fluid due to heat convection forces from the outer core lying below. Over the last 2 billion years, this process has shaped Earth's surface, pulling apart the plates to form oceans and continents, only to bring them back together into a supercontinent approximately every 600 million years. Scientists project that in about 250 million years, the next supercontinent, Pangaea Ultima, will form as Earth's landmasses merge, likely near the equator. The formation of the predicted landmass will drive more volcanic activity, resulting in increased levels of carbon dioxide, which will not be sequestered as it is today. Most land will be far from the ocean, and the typical mechanism of gas-trapping within rocks near water will be severely limited. Furthermore, the sun will shine 2.5 per cent more brightly, with more intense solar radiation bombarding the Earth. The result will be a single continent with exceedingly hot surface temperatures, transforming much of the landmass into a vast, lifeless desert.

The history of life on Earth shows that when confronted with inhospitable conditions, creatures experience one of three fates: they can leave, adapt or die. But, as we have seen, humans will experience such overwhelming conditions in the future that adaptation will be impossible. We will have to either leave the Earth or perish. However, unlike all other life forms, we are, to some extent, masters of our destiny. Indeed, we are already on the verge of colonising the Moon and our nearest planet, Mars. If we survive well into the future, and this is by no means certain, then in the view of Professor Michio Kaku, we have a chance of making our home amongst the stars, and perhaps even achieving immortality.[5]

Glossary

Adaptive radiation – the rapid evolution from a common ancestor of several species that occupy different ecological niches.
Admixture – the assimilation of genes from one population into another.
Aestivation – a state of animal dormancy during hot and dry periods.
Afrotheria – a newly identified monophyletic clade of species that originated in Africa, notable for their diverse forms and ecological roles (elephants, sirenians, hyraxes, aardvarks, sengis, golden moles and tenrecs).
Allantois – a placental membrane lying between the inner amnion and the outer chorion.
Alleles – alternative forms of a gene at a particular locus.
Allometry – the study of relationships between of body size and shape, anatomy, physiology and behaviour.
Altricial – having young that are hatched or born in a very immature and helpless condition, requiring parental care for some time.
Amniotes – a clade of terrestrial vertebrates in which the embryo develops within a set of protective extra-embryonic membranes – the amnion, chorion and allantois.
Amnion – an outer membrane forming a fluid-filled cavity (the amnion sac) that encloses the embryo.
Angiosperms – flowering plants that produce seeds enclosed within a fruit.
Anuric – no urine, or without urine.
Apex predator – a predator residing at the top of a food chain, on which no other creatures prey.
Apoptosis – a process of programmed cell death that occurs in multicellular organisms.
Artiodactyl – an even-toed ungulate; a member of the order Artiodactyla.
Autapomorphy – a specialised character or trait that is unique to a given taxon.
Autosome – any chromosome that is not a sex chromosome.
Back-crossing – the crossing of a hybrid species with one of its original parent genotypes or a similar genetic line.
Base – one of the nitrogen-containing molecules that are building blocks of the genetic code.
Base pair – a unit of two bases in a molecule of DNA or RNA. In DNA, adenine always pairs with thymine (A-T), and guanine always pairs with cytosine (G-C).
Beringia – a series of landforms that once existed periodically and in various configurations between northeastern Asia and northwestern North America and that were associated with periods of worldwide glaciation and subsequent lowering of sea levels.
Biological constraint – a factor which makes populations resistant to evolutionary change.

Bilophodonty – having molar teeth in which the four cusps are joined by two transverse ridges.
Boreoeutheria – a magnorder of placental mammals that evolved in the northern hemisphere and diverged to produce the Laurasiatheria and Euarchontoglires. With few exceptions, male boreoeutherians have a scrotum, an ancestral feature of the clade.
Brachydonty – having teeth with short crowns and well-developed roots (as in humans).
Carrier's constraint – the observation that air-breathing vertebrates that flex their bodies sideways while walking find it difficult to move and breathe at the same time.
Cenozoic – the latest era of geologic time that includes the Palaeogene, Neogene and Quaternary periods, spanning from 66 million years ago to the present age, characterised by the formation of modern continents, glaciation, and the diversification of plants, birds and animals.
Chorion – the double-layered membrane that forms the fetal layer of the placenta.
Chromatin – a highly organised condensed structure consisting of protein, RNA and DNA that forms the chromosomes in the nucleus of eukaryotic cells.
Cingulata – a clade that includes extinct and living armadillo-like mammals. The name refers to the girdle-like shell of living armadillos.
Clade – a group consisting of an ancestor and all its descendants, a single 'branch' on the tree of life.
Class – a taxonomic rank below phylum and above order. Groups such as Mammalia and Reptilia are classes.
Cloaca – a single outlet for the genital, urinary and digestive tracts.
Comparative genomics – a field of biological research in which genomic features of different organisms are compared.
Contingency – in the Gouldian sense, a particular sequence of events critically determines the course of history: A leads to B, B to C, C to D, and so on.
Convergent evolution – the process whereby unrelated species independently evolve similar traits, using different genetic mechanisms, because of having to adapt to similar environments or ecological niches. See also *parallel evolution*.
Cope's rule – the tendency for organisms in evolving lineages to increase in size with time.
Cretaceous – a geological period extending from 145 to 66 million years ago.
Crown group – a group of living species and their ancestors back to the most recent common ancestor. See also *stem species*.
Cynodonts – a late clade of synapsids that gave rise to true mammals.
Digitigrade – a foot posture in which the digits are in contact with the ground, but not the sole or heel of the foot.
Diphyodonty – having two successive sets of teeth (deciduous and permanent), one succeeding the other.
Diploid – the presence of two sets of chromosomes in somatic cells.
Dispersal – the spread of species to a new area, a complex process that involves emigration and establishment.
DNA – Deoxyribonucleic acid, a double-stranded molecule held together by weak bonds between pairs of nucleotides, which encodes hereditary information.
DNA polymerase – an enzyme that creates new DNA molecules by assembling nucleotides, the building blocks of DNA.
DNA sequence – the linear order of the base pairs along a DNA molecule.

Dollo's law – states that an organism never returns exactly to a former state, even if it finds itself placed in conditions identical to those in which it has previously lived.

Ecological release – a population increase or explosion occurring when a species is freed from limiting factors in its environment.

Edentulous – lacking teeth.

Encephalisation – an evolutionary increase in the complexity or relative size of the brain, involving a shift of function from non-cortical parts of the brain to the cortex.

Enhancer – a regulatory DNA sequence that, when bound by specific proteins called transcription factors, enhances the transcription of an associated gene.

Epigenetics – external modifications to DNA that, although not altering its sequence, modify the expression of genes.

Eocene – the second epoch of the Palaeogene, occurring from 56 to 33.9 million years ago.

Euarchontoglires – a superorder of placental mammals that includes rodents, rabbits, treeshrews, colugos and primates.

Eukaryote – an organism whose cells contain DNA within a nucleus.

Eulipotyphla – an order of placental mammals that includes the solenodons, hedgehogs and moles.

Eusocial – showing an advanced level of social organisation, in which a single female produces offspring and non-reproductive individuals help care for the young.

Eutherians – placental mammals, including all extinct sideshoots from their Mesozoic days.

Exaptation – a shift in the function of a trait during evolution.

Evo-devo – shorthand for evolutionary developmental biology; concerned with how changes in embryonic development during a single generation relate to the evolutionary changes that occur between generations.

Evolution vortex – the result of an increase in the impact of genetic drift on a population, due to the population's decreased size.

Fitness – in evolutionary biology refers to an organism's ability to survive and successfully reproduce, thereby passing on its genetic material to the next generation.

Folivorous – feeding on leaves.

Fossorial – (of an animal) burrowing.

Founder effect – the loss of genetic variation that occurs when a new population is established by a very small number of individuals from a larger population.

Frugivorous – feeding on fruits and seeds.

Gause's law – states that two species with identical niches, and which compete for a single resource, cannot coexist indefinitely.

Gene – The fundamental unit of heredity that encodes for a specific protein or carries out a specific function.

Gene splicing – the process by which non-coding regions (introns) are excised from pre-coding RNA and the coding sequences (exons) joined together to form a mature mRNA molecule.

Genetic drift – the statistical drift over time of gene frequencies in a population, due to random sampling effects in the formation of successive generations.

Genotype – the genetic makeup, as distinct from the physical appearance, of an organism. See also *phenotype*.

Genus (plural *genera*) – a taxonomic rank above species and below family.
Haploid – Having a single set of chromosomes, as in a gamete (sperm or egg).
Heterochromatin – a tightly packed or condensed form of DNA that is inaccessible to transcription factors, such that its genes are usually rendered silent.
Heterochrony – a genetic shift in timing of the development of a tissue or anatomical part, or in the onset of a physiological process, relative to an ancestor.
Heterogametic sex – the sex of a species in which an individual's gametes have non-matching sex chromosomes.
Heterothermic – a physiological term for animals that vary between self-regulating their body temperature and allowing the surrounding environment to affect it.
Heterozygous – having two different versions or alleles of the same gene (one inherited from the father and one inherited from the mother).
Holarctic – a biogeographical area that comprises the majority of habitats found across the continents of the Northern Hemisphere.
Holocene – the current geological epoch that began 11,700 years ago, after the last major ice age.
Hominids – a family of primates (Hominidae) including orangutans, gorillas, chimpanzees, bonobos and humans.
Hominins – a tribe (Homini, a subgroup of the family Hominidae) comprising chimpanzees, bonobos and humans.
Hominoids – a superfamily of primates (Hominoidea) that includes the gibbons and the hominids.
Homology – the similarity between biological structures or genes in different organisms that is due to shared ancestry.
Homoplasy – the development of organs or other bodily structures within different species, which resemble each other and have the same functions, but did not have a common ancestral origin.
Homozygous – the state of having two identical versions or alleles of a particular gene.
Hybridisation – the process of a species breeding with another species or variety.
Hypsodonty – having teeth with high crowns and short roots that provide a lot of material for wear (as in herbivores).
Incomplete lineage sorting (ILS) – when alleles within a species' gene pool fail to fully sort into distinct lineages during speciation events, leading to conflicting gene trees compared to the species tree.
Introgression – the transfer of genetic information from one species to another because of hybridisation between them and repeat back-crossing.
Infraclass – a taxonomic group that is below subclass and above superorder.
Insectivorous – feeding on insects, worms and other invertebrates.
Intraordinal – within a taxonomic order.
Jurassic – a geologic period from the end of the Triassic period, 201 million years ago, to the beginning of the Cretaceous period, 145 million years ago.
K–Pg boundary – the geological division marking the transition between the Cretaceous (K) and Palaeogene (Pg) periods, approximately 66 million years ago, notable for a mass extinction event that wiped out many species, including non-avian dinosaurs.
Karyotype – an individual's complete set of chromosomes.
Keystone species – a species which has a disproportionately large effect on its natural environment relative to its abundance.

Lagerstätte – a site of exceptional fossilisation, often with faunal and floral diversity (from the German for 'storage space').
Last common ancestor (LCA) – the most recent individual from which a set of organisms are inferred to have descended.
Laurasiatheria – a superorder of placental mammals that groups together true insectivores, bats, carnivores, pangolins, even-toed and odd-toed ungulates, and all their extinct relatives.
Macropodidae – a family of 67 herbivorous marsupials, that includes kangaroos, wallabies, rock-wallabies, tree-kangaroos, wallaroos, pademelons and quokkas. The name comes from the Greek for 'large feet'.
Magnorder – the highest taxonomic ranking below class, one rank above superorder.
Marsupial – a type of mammal from Australasia or South or Central America that is not completely developed when it is born and is carried around on the mother's body, where it is fed and protected until it is completely developed.
Mesaxonic – having an enlarged middle digit that forms the axis of the foot. See also *paraxonic*.
Mesic – an environment or habitat containing a moderate amount of rain.
Mesozoic – the geological era extending from 252 to 66 million years ago – the 'Age of Reptiles'.
Messenger RNA (mRNA) – a single-stranded molecule of RNA that corresponds to the genetic sequence of a gene, and is read by a ribosome in the process of synthesising a protein.
Metatherians – marsupials and all the extinct mammal groups most closely related to them.
Miocene – the first geological epoch of the Neogene period, extending from 23 to 5.3 million years ago.
Modern synthesis – the fusion of Mendelian genetics and Darwinian evolution that resulted in a unified theory of evolution. The term was coined by Julian Huxley in his 1942 book, *Evolution: The Modern Synthesis*.
Monodactyly – having one finger or toe.
Monophyly – a group of species is monophyletic if they all share the same common ancestor and share it with no other species.
Molecular clock – the concept that the difference between two homologous DNA sequences (expressed as the number of nucleotide substitutions separating them) is proportional to the amount of time since they last shared an ancestor.
Müllerian mimicry – a form of mimicry in which two or more harmful or unpalatable species evolve similar appearances, enhancing their collective protection by reinforcing predators' tendency to avoid them.
Mutualism – a positive interaction between species.
Multituberculate – any member of an extinct group of small, rat-like mammals that existed from about 178 to 50 million years ago, when they were the most common mammals.
Mya – million years ago.
Mysticeti – a clade that includes all living baleen whales.
Natural selection – the process in which organisms better adapted to their environment tend to survive and reproduce.
Nectarivorous – feeding on nectar.
Negative selection – a natural process by which deleterious genes are selectively purged from the population.

Neogene – the second period in the Cenozoic era, extending from 23 to 2.5 million years ago.
Neolithic – relating to the latest period of the Stone Age.
Neutral evolution – a state in which genetic variations in a population result from mutations and genetic drift and not by natural selection.
Neutral mutations – gene changes that are not affected by natural selection.
*Notoryctida*e – a family consisting of two species of marsupial mole from Australia's interior.
Odontoceti – a clade that includes all living toothed whales.
Oripulation – grasping with the mouth rather than the hand.
Orogeny – deformation imposed during mountain building.
Paedomorphosis – a genetic and phenotypic change in which adults of a species retain traits that were previously only seen in juveniles.
Palaeogene – the first period in the Cenozoic era, extending from 66 to 23 million years ago.
Palynivorous – feeding on pollen.
Pangaea – a supercontinent that that existed approximately 300–200 million years ago, incorporating almost all the landmasses on Earth.
Panthalassic Ocean – the vast super-ocean that encompassed the Earth and surrounded the supercontinent Pangaea.
Parallel evolution – when a particular phenotype evolves by similar genetic mechanisms in different species. See also *convergent evolution*.
Paraphyly – a group of organisms that includes an ancestor but not all of its descendants.
Paraxonic – the condition in which the axis of the animal's weight passes between the third and fourth digit. See also *mesaxonic*.
Patagium – a membrane or fold of skin between the forelimbs and hindlimbs on each side of a bat or gliding mammal.
PCR – polymerase chain reaction, a widely used technique to amplify specific stretches of DNA.
Perissodactyl – an odd-toed ungulate; a member of the order Perissodactyla.
Pelycosaurs – the earliest of the synapsids.
Phanerozoic – the current and latest of the four geological aeons of the Earth, covering the period from 538.8 million years ago to the present. Includes the Palaeozoic, Mesozoic and Cenozoic eras.
Phenotype – the observable physical or biochemical characteristics of an organism. See also *genotype*.
Phenotypic plasticity – the ability of an organism to change in response to stimuli or inputs from the environment.
Phorusrhacids – an extinct clade of large, flightless, carnivorous birds that were the apex predators in South America throughout the Cenozoic. Colloquially known as 'terror birds', they fed on mammals and reptiles.
Plantigrade – walking on the full length of the foot: toes, sole and heel.
Platyrrhines – the five families of New World monkeys, named after their flat noses.
Pilosa – a clade of xenarthrans that includes anteaters and sloths. The name comes from the Latin word for 'hairy'.
Pleiotropy – the production by a single gene of two or more seemingly unrelated phenotypic traits.

Pleistocene – a geological epoch, the first of the Quaternary period, extending from 2.5 million years ago to 11,700 years ago.
Plesiomorphic – the ancestral character state for a particular clade.
Pliocene – the most recent epoch of the Neogene period, extending from 5.3 to 2.5 million years ago.
Polyembryony – a condition in which two or more embryos develop from a single fertilised egg.
Polygyny – a mating system in which males mate with several females during a breeding season.
Polytomy – multiple simultaneous speciation events in a phylogeny.
Positive allometry – when one part of the body grows faster than another part.
Precocial – a state in which the young are relatively mature and mobile from the moment of birth.
Pseudogene – a segment of DNA that structurally resembles a gene but is not capable of coding for a protein.
Prosimian – an obsolete term for any of a suborder (Prosimii) of lower primates such as lemurs and lorises.
Purifying selection – see *negative selection*.
Quaternary – the most recent geological period, extending from 2.5 million years ago to the present day.
Relaxed selection – when environmental changes eliminate or weaken a source of selection that was formerly important for the maintenance of a specific trait.
Retrogene – a processed copy of a gene that arises from the reverse transcription of its messenger RNA (mRNA) and subsequent integration into the genome.
Retroposon – a type of transposable element (or 'jumping gene') that moves within the genome via an RNA intermediate. See also *transposon*.
Retrovirus – a family of viruses that have an RNA-based genome. To insert into the DNA-based genome of their mammalian hosts, retroviruses use their own enzyme, reverse transcriptase, to convert their RNA genome to its equivalent DNA.
Reverse transcriptase – see *retrovirus*.
Ribosome – the cellular machinery responsible for making proteins.
Ruminant – an even-toed ungulate mammal (an artiodactyl) that regurgitates and re-chews its food from the stomach (rumen), including cattle, sheep, antelopes, deer and giraffes.
Sanguivory – blood-sucking.
Scansorial – adapted or specialised for climbing.
Semelparous – reproducing or breeding only once in a lifetime.
Simian – any of a suborder (Anthropoidea) of primates that includes monkeys, apes and humans.
Sirenia – a mammalian order of fully herbivorous and aquatic mammals that includes the dugong and three species of manatee (also known as sea cows or sirenians).
Sparassodonta – an extinct order of carnivorous metatherian mammals native to South America.
Stem species – species that comprise a set of extinct taxa that are not in the crown group but are more closely related to the crown groups than to any other. See also *crown group*.

Sweepstake dispersal – an unexpected and unpredictable dispersal across a substantial barrier.
Synapomorphy – a characteristic present in an ancestral species and shared exclusively by its evolutionary descendants.
Synapsids – a clade that includes pelycosaurs, therapsids, cynodonts and true mammals.
Tapetum lucidum – a layer of tissue in the eyes of many mammals that reflects light back through the retina.
Taxon (plural *taxa*) – a taxonomic group of any rank, such as a species, family or class.
Taxonomy – the science of defining species into evolutionary lineages and categories.
Testicondy – the condition of having testes located within the abdomen.
Tetrapod – any four-limbed vertebrate.
Therapsids – a clade that gave rise to cynodonts and true mammals.
Therians – a subclass including marsupials and placentals but excluding monotremes.
Transcription – the process whereby one strand of a DNA molecule is used as a template for synthesis of a complementary RNA by RNA polymerase.
Transcription factor – a protein that controls the rate of gene expression by binding to specific DNA sequences.
Transgenic – an organism or cell whose genome has been artificially altered by the introduction of a DNA sequence from another species.
Translation – the process by which cells make protein using the genetic information encoded in messenger RNA (mRNA).
Transposon – a type of transposable element (or 'jumping gene') that moves via a DNA intermediate. See also *retroposon*.
Triassic – a geological period from the end of the Permian period, 251 million years ago, to the beginning of the Jurassic period, 201 million years ago.
Tridactyly – having three fingers or toes.
Trophic cascade – an ecological phenomenon in which the removal of an apex predator causes dramatic changes in the ecosystem structure and nutrient recycling.
Tuff – a light, porous rock formed by the consolidation of volcanic ash.
Ungulate – any animal with hooves.
Unguligrade – walking on the tiptoe with the weight supported by hooves.
Vermilingua – a suborder of Xenarthra comprising the American anteaters.
Vicariance – the geographical separation and isolation of a subpopulation of a species by a geographical barrier, such as a mountain or a body of water, resulting in the emergence of a new species.
Vomeronasal organ – a chemosensitive system located within the nasal chamber involved in perceiving and processing stimuli related to social and reproductive behaviour.
Xenarthra – a basal clade of placental mammals, originating in South America, characterised by unique vertebral articulations (xenarthrous processes), containing armadillos, sloths and anteaters.
Zalambdodonty – having teeth with two ridges that meet at an angle, resembling the letter lambda.

Dramatis Personae

Below is a list of species mentioned in the text. The vernacular names are given as they appear in the book, together with their scientific names. The names of families, such as echidnas and bears, are not included. Species are arranged in alphabetical order.

Aardvark *Orycteropus afer*
African buffalo *Syncerus caffer*
African crested rat *Lophiomys imhausi*
African forest elephant *Loxodonta cyclotis*
African manatee *Trichechus senegalensis*
African savanna elephant *Loxodonta africana*
African wild ass *Equus africanus*
Alpaca *Lama pacos*
Amami spiny rat *Tokudaia osimensis*
Amazon manatee *Trichechus inunguis*
American bison *Bison bison*
American black bear *Ursus americanus*
American lion *Panthera atrox*
American mastodon *Mammut americanum*
Andean white-eared opossum *Didelphis pernigra*
Antilopine kangaroo *Osphranter antilopinus*
Arizona bark scorpion *Centruroides sculpturatus*
Asian black bear *Ursus thibetanus*
Asian elephant *Elephas maximus*
Asian golden cat *Catapuma temminckii*
Asian water buffalo *Bubalus bubalis*
Asiatic mouflon *Ovis gmelini*
Aurochs *Bos primigenius*
Aye-aye *Daubentonia madagascariensis*
Bactrian camel *Camelus bactrianus*

Baird's tapir *Tapirus bairdii*
Beluga *Delphinapterus leucas*
Benin tree hyrax *Dendrohyrax interfluvialis*
Bennett's tree-kangaroo *Dendrolagus bennettianus*
Bezoar ibex *Capra aegagrus*
Black howler monkey *Alouatta caraya*
Black rat *Rattus rattus*
Black rhinoceros *Diceros bicornis*
Black-and-white ruffed lemur *Varecia variegata*
Black-and-white tree-kangaroo *Dendrolagus mbaiso*
Black-striped capuchin *Sapajus libidinosus*
Blue antelope *Hippotragus leucophaeus*
Blue whale *Balaenoptera musculus*
Blue wildebeest *Connochaetes taurinus*
Bonobo *Pan paniscus*
Bornean orangutan *Pongo pygmaeus*
Bowhead whale *Balaena mysticetus*
Bramble Cay mosaic-tailed rat *Melomys rubicola*
Brown bear *Ursus arctos*
Brown four-eyed opossum *Metachirus nudicaudatus*
Brown long-eared bat *Plecotus auritus*
Brown rat *Rattus norvegicus*
Brown-throated sloth *Bradypus variegata*

Bronze quoll *Dasyurus spartacus*
Bumblebee bat *Craseonycteris thonglongyai*
Cactus mouse *Peromyscus eremicus*
Cape grysbok *Raphicerus melanotis*
Capybara *Hydrochoerus hydrochaeris*
Cave bear *Ursus spelaeus*
Cave lion *Panthera spelaea*
Chacoan pygmy opossum *Chacodelphys formosa*
Cheetah *Acinonyx jubatus*
Chimpanzee *Pan troglodytes*
Christmas Island pipistrelle *Pipistrellus murrayi*
Columbian mammoth *Mammuthus columbi*
Common bottlenose dolphin *Tursiops truncatus*
Common hippopotamus *Hippopotamus amphibius*
Common minke whale *Balaenoptera acutorostrata*
Common noctule *Nyctalus noctula*
Common spiny mouse *Acomys cahirinus*
Common wombat *Vombatus ursinus*
Coypu *Myocastor coypus*
Crabeater seal *Lobodon carcinophaga*
Cuban solenodon *Atopogale cubanus*
Damaraland mole-rat *Fukomys damarensis*
Dassie rat *Petromus typicus*
Denisovan *Homo denisova*
Domestic donkey *Equus asinus*
Domestic horse *Equus caballus*
Dromedary (Arabian camel) *Camelus dromedarius*
Dugong *Dugong dugon*
Eastern gorilla *Gorilla beringei*
Eastern grey kangaroo *Macropus giganteus*
Eastern long-beaked echidna *Zaglossus bartoni*
Eastern quoll *Dasyurus viverrinus*

Egyptian fruit bat *Rousettus aegyptiacus*
European rabbit *Oryctolagus cuniculus*
Fat-tailed dunnart *Sminthopsis crassicaudata*
Feathertail glider (Pygmy gliding possum) *Acrobates pygmaeus*
Fishing cat *Prionailurus viverrinus*
Flat-headed cat *Prionailurus planiceps*
Four-horned antelope *Tetracerus quadricornis*
Gaur *Bos gaurus*
Giant anteater *Myrmecophaga tridactyla*
Giant armadillo *Priodontes maximus*
Giant panda *Ailuropoda melanoleuca*
Giant short-faced bear *Arctodus simus*
Gilbert's potoroo *Potorous gilbertii*
Golden bamboo lemur *Hapalemur aureus*
Golden-headed lion tamarin *Leontopithecus chrysomelas*
Goodfellow's tree-kangaroo *Dendrolagus goodfellowi*
Grant's golden mole *Eremitalpa granti*
Gray mouse opossum *Tlacuatzin canescens*
Great flying fox *Pteropus neohibernicus*
Greater bamboo lemur *Hapalemur simus*
Greater fairy armadillo *Calyptophractus retusus*
Greater hedgehog tenrec *Setifer setsus*
Green iguana *Iguana iguana*
Grévy's zebra *Equus grevyi*
Grey bamboo lemur *Hapalemur griseus*
Grey whale *Eschrichtius robustus*
Grey wolf *Canis lupus*
Groundhog *Marmota monax*
Guiana dolphin *Sotalia guianensis*
Guanaco *Lama guanicoe*
Guinea pig *Cavia porcellus*
Haast's eagle *Hieraaetus moorei*
Hainan gibbon *Nomascus hainanus*
Highland streaked tenrec *Hemicentetes nigriceps*
Highveld mole-rat *Cryptomys pretoriae*

Hispaniolan hutia *Plagiodonta aedium*
Hispaniolan solenodon *Solenodon paradoxus*
Hoatzin *Opisthocomus hoazin*
Honey possum *Tarsipes rostratus*
Iberian lynx *Lynx pardinus*
Impala *Aepyceros melampus*
Indian rhinoceros *Rhinoceros unicornis*
Indri *Indri indri*
Jaguar *Panthera onca*
Jaguarundi *Herpailurus yagouaroundi*
Java mouse-deer *Tragulus javanicus*
Javan rhinoceros *Rhinoceros sondaicus*
Koala *Phascolarctos cinereus*
Kouprey *Bos sauveli*
Laotian rock rat *Laonastes aenigmamus*
Lar gibbon *Hylobates lar*
Large-headed capuchin *Sapajus macrocephalous*
Leadbeater's possum *Gymnobelideus leadbeateri*
Leopard *Panthera pardus*
Leopard seal *Hydrurga leptonyx*
Lemuroid ringtail possum *Hemibelideus lemoroides*
Leopard cat *Prionailurus bengalensis*
Lesser hedgehog tenrec *Echinops telfairi*
Lion *Panthera leo*
Little brown bat *Myotis lucifugus*
Little red kaluta *Dasykaluta rosamondae*
Llama *Lama glama*
Long-nosed bandicoot *Perameles nasuta*
Long-nosed potoroo *Potorous tridactylus*
Lowland streaked tenrec *Hemicentetes semispinosus*
Lowland tapir *Tapirus terrestris*
Lumholtz's tree-kangaroo *Dendrolagus lumholtzi*
Mahogany glider *Petaurus gracilis*
Mainland clouded leopard *Neofelis nebulosa*
Malayan tapir *Tapirus indicus*
Maned rat *Lophiomys imhausi*

Mantled howler monkey *Alouatta palliata*
Marbled cat *Pardofelis marmorata*
Masai giraffe *Giraffa tippelskirchi*
Minckley's cichlid *Herichthys minckleyi*
Monito del monte *Dromiciops gliroides*
Mountain pygmy possum *Burramys parvus*
Mountain tapir *Tapirus pinchaque*
Mountain zebra *Equus zebra*
Muskox *Ovibus muschatus*
Nancy Ma's night monkey *Aotus nancymaae*
Naked mole-rat *Heterocephalus glaber*
Narwhal *Monodon monoceros*
Natal droptail ant *Myrmicaria natalensis*
Neanderthal *Homo neanderthalensis*
Nile crocodile *Crocodylus niloticus*
Nilgai *Boselaphus tragocamelus*
Nine-banded armadillo *Dasypus novemcinctus*
Northern brown bandicoot *Isoodon macrourus*
Northern common cuscus *Phalanger orientalis*
Northern giraffe *Giraffa camelopardalis*
Northern marsupial mole *Notoryctes caurinus*
Northern naked-tailed armadillo *Cabassous centralis*
Northern quoll *Dasyurus hallucatus*
Northern three-striped opossum *Monodelphis americana*
Northern tamandua *Tamandua mexicana*
Numbat (Banded anteater) *Myrmecobius fasciatus*
Nyala *Tragelaphus angasii*
Okapi *Okapia johnstoni*
Onager *Equus hemionus*
Pacarana *Dinomys branickii*
Pallas's cat *Otocolobus manul*
Pampas cat *Leopardus colocola*

Panamanian white-faced capuchin *Cebus imitator*
Pancho's monito del monte *Dromiciops bozinovici*
Pen-tailed treeshrew *Ptilocercus lowii*
Philippine colugo *Cynocephalus volans*
Philippine tarsier *Carlito syrichta*
Pink fairy armadillo *Chlamyphorus truncatus*
Plains zebra *Equus quagga*
Platypus *Ornithorhynchus anatinus*
Polar bear *Ursus maritimus*
Pronghorn *Antilocapra americana*
Przewalski's horse *Equus przewalskii*
Puma *Puma concolor*
Pygmy hippopotamus *Choeropsis liberiensis*
Pyrenean ibex *Capra pyrenaica*
Red kangaroo *Osphranter rufus*
Red-necked wallaby *Notamacropus rufogriseus*
Red panda *Ailurus fulgens*
Reticulated giraffe *Giraffa reticulata*
Ring-tailed lemur *Lemur catta*
Ring-tailed possum *Pseudocheirus peregrinus*
Rock hyrax *Procavia capensis*
Royal antelope *Neotragus pygmaeus*
Rusty-spotted cat *Prionailurus rubiginosus*
Saola *Pseudoryx nghetinhensis*
Seba's short-tailed bat *Carollia perspicillata*
Serval *Leptailurus serval*
Sharpe's grysbok *Raphicerus sharpei*
Short-beaked echidna *Tachyglossus aculeatus*
Silky anteater *Cyclopes didactylus*
Sir David's long-beaked echidna *Zaglossus attenboroughi*
Sloth bear *Melursus ursinus*
Snow leopard *Panthera uncia*
Soprano pipistrelle *Pipistrellus pygmaeus*
Southern giraffe *Giraffa giraffa*
Southern grasshopper mouse *Onychomys torridus*
Southern greater glider *Petauroides volans*
Southern marsupial mole *Notoryctes typhlops*
Southern opossum *Didelphis marsupialis*
Southern tamandua *Tamandua tetradactyl*
Spectacled bear *Tremarctos ornatus*
Spectacled flying fox *Pteropus conspicillatus*
Sperm whale *Physeter macrocephalus*
Spotted-tailed quoll *Dasyurus maculatus*
Squirrel glider *Petaurus norfolcensis*
Steenbok *Raphicerus campestris*
Steller's sea cow *Hydrodamalis gigas*
Straight-tusked elephant *Palaeoloxodon antiquus*
Sugar glider *Petaurus breviceps*
Sulawesi bear cuscus *Ailurops ursinus*
Sulawesi pygmy cuscus *Strigocuscus celebensis*
Sumatran orangutan *Pongo abelii*
Sumatran rhinoceros *Dicerorhinus sumatrensis*
Sun bear *Helarctos malayanus*
Sunda clouded leopard *Neofelis diardi*
Sunda colugo *Galeopterus variegatus*
Tailless tenrec (Common tenrec) *Tenrec ecaudatus*
Tapanuli orangutan *Pongo tapanuliensis*
Tasmanian devil *Sarcophilus harrisii*
Tasmanian pademelon *Thylogale billardierii*
Tasmanian tiger *Thylacinus cyanocephalus*
Tibetan antelope *Pantholops hodgsonii*
Tibetan kiang *Equus kiang*
Tiger *Panthera tigris*
Tiger quoll *Dasyurus maculatus*
Topi *Damaliscus lunatus*
Touan short-tailed opossum *Monodelphis touan*

Tufted capuchin *Sapajus apella*
Vangunu giant rat *Uromys vika*
Vaquita *Phocoena sinus*
Verreaux's sifaka *Propithecus verreauxi*
Vicuña *Lama vicugna*
Virginia opossum *Didelphis virginiana*
Waigeo cuscus *Spilocuscus papuensis*
Walrus *Odobenus rosmarus*
Water chevrotain (African mouse-deer) *Hyemoschus aquaticus*
Water opossum *Chironectes minimus*
Web-footed tenrec *Microgale mergulus*
West Indian manatee *Trichechus manatus*
Western gorilla *Gorilla gorilla*
Western grey kangaroo *Macropus fuliginosus*
Western long-beaked echidna *Zaglossus bruijnii*
Western quoll *Dasyurus geoffroii*
Western tarsier *Cephalopachus bancanus*
White rhinoceros *Ceratotherium simum*
White-eared kob *Kobus leucotis*
White-eared opossum *Didelphis albiventris*
White-tailed deer *Odocoileus virgineanus*
Wild Bactrian camel *Camelus ferus*
Wildcat *Felis silvestris*
Wilson's bird-of-paradise *Diphyllodes respublica*
Woolly mammoth *Mammuthus primigenius*
Woolly rhinoceros *Coelodonta antiquitatis*
Yak *Bos mutus*
Yangtze River dolphin *Lipotes vexillifer*
Yellow-bellied three-toed skink *Saiphos equalis*

Notes

Prologue: Linnaeus's Legacy

1. Mate, B.R., Ilyashenko, V.Y., Bradford, A.L. et al. 2015. Critically endangered western gray whales migrate to the eastern North Pacific. *Biology Letters*, **11**, 20150071. doi.org/10.1098/rsbl.2015.0071.
2. Swartz, S.L. 2014. *Lagoon Time: A Guide to Gray Whales and the Natural History of San Ignacio Lagoon*. The Ocean Foundation.
3. Gregory, W.K. 1910. The Order of Mammals. *Bulletin of the American Museum of Natural History*, **27**, 10–12.
4. While scientists may have modified Linnaeus's categories and changed some species' names, many remain in use today. Examples include the American bison (*Bison bison*), wolf (*Canis lupus*), red fox (*Vulpes vulpes*), European hedgehog (*Erinaceus europaeus*), black rat (*Rattus rattus*), and wood mouse (*Apodemus sylvaticus*).
5. There are several exceptions that Linnaeus was unaware of – such as the Bismarck masked flying fox and the Dayak fruit bat from New Guinea and Malaysia, respectively. But these are physiological oddities, rare examples of male lactation that still seek an evolutionary explanation.
6. Schiebinger, L. 1993. Why mammals are called mammals: gender politics in eighteenth-century natural history. *American Historical Review*, **98**, 382–411.
7. Burgin, C.J., Colella, J.P., Kahn, P.L. et al. 2018. How many species of mammal are there? *Journal of Mammalogy*, **99**, 1–14. doi.org/10.1093/jmammal/gyx147.
8. Morales, A.E. and Carstens, B.C. 2018. Evidence that *Myotis lucifugus* 'subspecies' are five nonsister species, despite gene flow. *Systems Biology*, **67**, 756–769. doi.org/10.1093/sysbio/syy010.
9. Parsons, D.J., Pelletier, T.A., Wieringa, J.G. et al. 2022. Analysis of biodiversity data suggests that mammal species are hidden in predictable places. *Proceedings of the National Academy of Science of the United States of America*, **119**, e2103400119. doi.org/10.1073/pnas.2103400119.
10. For engaging accounts of our less famous but no less fascinating warm-blooded ancestors see Wallace, D.R. 2004. *Beasts of Eden*, University of California Press; Kemp, T. 2005. *Origin and Evolution of Mammals*. Oxford University Press; Rose, K.D. 2006. *The Beginning of the Age of Mammals*, Johns Hopkins University Press; Panciroli, E. 2021. *Beasts Before Us: The Untold Story of Mammal Origins and Evolution*. Bloomsbury.
11. Prothero, D.R. 2017. *Princeton Field Guide to Prehistoric Mammals*, Princeton: Princeton University Press; Brusatte, S. 2022. *The Rise and Reign of the Mammals*, London: Picador.
12. Hotalling, S., Kelley, J.L. and Frandsen, P.B. 2021. Toward a genome sequence for every animal: where are we now? *Proceedings of the National Academy of Science of the United States of America*, **118**, e2109019118. doi.org/10.1073/pnas.2109019118.
13. Zoonomia Consortium. 2020. A comparative genomics multitool for scientific discovery and conservation. *Nature*, **587**, 240–245. The project's ambitious aim is encapsulated in its moniker 'Zoonomia', derived from the Greek words *zoo* meaning animal, and *nomia* meaning governing laws.
14. Watson, J.D. and Crick, F.H.C. 1953. A structure for deoxyribose nucleic acid. *Nature*, **171**, 737–738. doi.org/10.1038/171737a0.

15. Carlin, J.L. 2011. Mutations are the raw materials of evolution. *Nature Education Knowledge* **3**(10), 10. www.nature.com/scitable/knowledge/library/mutations-are-the-raw-materials-of-evolution-17395346.
16. Ho, S.Y.M., Chen, A.X.Y., Lins, L.S.F. *et al*. 2016. The genome as an evolutionary timepiece. *Genome Biology and Evolution*, **8**, 3006–3010. doi.org/10.1093/gbe/evw220.
17. Forest, F. 2009. Calibrating the Tree of Life: fossils, molecules and evolutionary timescales. *Annals of Botany*, **104**, 789–794. doi.org/10.1093/aob/mcp192.
18. Rhie, A., McCarthy, S.A., Fredrigo, O. *et al*. 2021. Towards complete and error-free genome assemblies of all vertebrate species. *Nature*, **592**, 737–746. doi.org/10.1038/s41586-021-03451-0.
19. Cabreira, S.F., Schultz, C. L., da Silva, L.R. *et al*. 2022. Diphyodont tooth replacement of *Brasilodon* – a late Triassic eucynodont that challenges the time of origin of mammals. *Journal of Anatomy*, **241**, 1424–1440. doi.org/10.1111/joa.13756
20. Damas, J., Corbo, M., Kim, J. *et al*. 2022. Evolution of the ancestral mammalian karyotype and syntenic regions. *Proceedings of the National Academy of Science of the United States of America*, **119**, e2209139119. doi.org/10.1073/pnas.2209139119.
21. Gill, P.G., Purnell, M.A., Crumpton, N. *et al*. 2014. Dietary specializations and diversity in feeding ecology of the earliest stem mammals. *Nature*, **512**, 303–305. doi.org/10.1038/nature13622.
22. Rowe, T.B., Macrini, T.E. and Luo, Z-X. 2011. Fossil evidence on origin of the mammalian brain. *Science*, **332**, 955–957. doi.org/10.1126/science.1203117.
23. Interestingly, once the enlarged olfactory cortex had evolved, later mammals recommissioned and re-tooled their nerve cells, or neurons, for colour vision, echolocation and, in the case of the platypus, even electroreception.
24. Newham, E., Gill, P.G., Brewer, P. *et al*. 2020. Reptile-like physiology in early Jurassic stem-mammals. *Nature Communications*, **11**, 5121. doi.org/10.1038/s41467-020-18898-4.
25. Ramírez-Chaves, H.E., Weisbecker, V., Wroe, S. *et al*. 2016. Resolving the evolution of the mammalian middle ear using Bayesian inference. *Frontiers in Zoology*, **13**, 39. doi.org/10.1186/s12983-016-0171-z. The definitive modern mammal ear evolved three times, in the monotremes, the therians and the multituberculates. However, the multituberculates do not feature in this book as they were a mammalian order that became extinct, probably because of competitive exclusion following the evolution of modern rodents.
26. Araújo, R., David, R., Benoit, J. *et al*. 2022. Inner ear biomechanics reveals the Late Triassic origin of mammalian endothermy. *Nature*, **607**, 726–731. doi.org/10.1038/s41586-022-04963-z.

Chapter 1: The Platypus's Story

1. Shaw, G. 1799, The Duck-billed Platypus, *Platypus anatinus*. *The Naturalist's Miscellany*, 10, 385–386. London, Nodder & Co. doi.org/10.5962/p.304567.
2. Home, E. 1802. A Description of the Anatomy of the *Ornithorhynchus paradoxus*. *Philosophical Transactions of the Royal Society of London*, **92**, 67–84.
3. Journal of Lieutenant George Tobin on HMS *Providence* 1791–1793. Mitchell Library, State Library of New South Wales, ML A562, CY 1421.
4. Shaw, G. 1792. The Porcupine Ant-Eater, *Myrmecophaga aculeata*. *The Naturalist's Miscellany*, **3**, London: F.P. Nodder & Co.
5. Mosely, H.N. 1885. On the Ova of Monotremes. In: *Report of the Fifty-fourth Meeting of the British Association for the Advancement of Science held at Montreal in August and September 1884*. London: John Murray.
6. Zhou, Y., Shearwin-Whyatt, L., Li, J. *et al*. 2021. Platypus and echidna genomes and evolution. *Nature*, **592**, 756–762. doi.org/10.1038/s41586-020-03039-0.
7. Fenelon, J.C., Bennetts, A., Anthwal, N. *et al*. 2023. Getting out of a mammalian egg: the egg tooth and caruncle of the echidna. *Developmental Biology*, **495**, 8–18. doi.org/10.1016/j.ydbio.2022.12.005.
8. Gould, S.J. and Vrba, E.S. 1982. Exaptation: a missing term in the science of form. *Paleobiology*, **8**, 4–15.

9. Oftedal, O.T. 2002. The mammary gland and its origins during synapsid evolution. *Journal of Mammary Gland Biology and Neoplasia*, 7, 225–252. doi.org/10.1023/A:1022896515287.
10. Howard, B.A. and Gusterson, B.A. 2000. Human breast development. *Journal of Mammary Gland Biology and Neoplasia*, 5, 119–137. doi.org/10.1023/A:1026487120779.
11. Schep, R., Necsulea, A., Rodriguez-Carballo, E. *et al.* 2016. Control of *Hoxd* gene transcription in the mammary bud by hijacking a preexisting regulatory landscape. *Proceedings of the National Academy of Sciences of the United States of America*, 113, E7720–E7729. doi.org/10.1073/pnas.1617141113.
12. Capuco, A.V. and Akers, R.M. 2009. The origin and evolution of lactation. *Journal of Biology*, 8, 37. doi.org/10.1186/jbiol139.
13. Brawand, D., Wahli, W., and Kaessmann, H. 2008. Loss of egg yolk genes in mammals and the origin of lactation and placentation. *PLoS Biology* 6(30), e63. doi.org/10.1371/journal.pbio.0060063.
14. Kawasaki, K. and Weiss, K.M. 2003. Mineralized tissue and vertebrate evolution: the secretary calcium-binding phosphoprotein gene cluster. *Proceedings of the National Academy of Sciences of the United States of America*, 100, 4060–4065. doi.org/10.1073/pnas.0638023100.
15. Holt, C. and Carver, J.A. 2012. Darwinian transformations of a 'scarcely nutritious fluid' into milk. *Journal of Evolutionary Biology*, 25, 1253–1263. doi.org/10.1111/j.1420-9101.2012.02509-x.
16. Enjapoori, A.K., Grant, T.R., Nicol, S.C. *et al.* 2014. Monotreme lactation protein is highly expressed in monotreme milk and provides antimicrobial protection. *Genome Biology and Evolution*, 6, 2754–2773. doi.org/10.1093/gbe/evu209.
17. Newman, J., Sharp, J.A., Enjapoori, K. *et al.* 2018. Structural characterization of a novel monotreme-specific protein with antimicrobial activity from the milk of the platypus. *Acta Crystallographica Section F Structural Biology*, 74, 39–45. doi.org/10.1107/S2053230X17017708.
18. Kumar, A., Parveen, S., Sharma, I. *et al.* 2019. Structural and mechanistic insights into EchAMP: A antimicrobial protein from Echidna milk. lactation. *Biochimica et Biophysica Acta*, 6, 1260–1270. doi.org/10.1016/j.bbamem.2019.03.020.
19. Ramírez-Chaves, H.E., Weisbecker, V., Wroe, S. *et al.* 2016. Resolving the evolution of the mammalian ear using Bayesian inference. *Frontiers in Zoology*, 13, 39. doi.org/10.1186/s12983-016-0171-z.
20. Anthwal, N., Fenelon, J.C., Johnston, S.D. *et al.* 2020. Transient role of the middle ear as a lower jaw support across mammals. *eLife*, 9, e57860. doi.org/10.7554/eLife.57860.
21. Le Maître, A., Grunstra, N.D.S., Pfaff, C. *et al.* 2020. Evolution of the mammalian ear: an evolvability hypothesis. *Evolutionary Biology*, 47, 187–192. doi.org/10.1007/s11692-020-09502-0.
22. Zhou *et al.* 2021 (see note 6).
23. Itoigawa, A., Hayakawa, T., Zhou, Y. *et al.* 2022. Functional diversity and evolution of bitter taste receptors in egg-laying mammals. *Molecular Biology and Evolution*, 39, msac107. doi.org/10.1093/molbev/msac107.
24. Hurum, J.H., Luo, Z-X., and Kielan-Jaworowska, Z. 2006. Were mammals originally venomous? *Acta Palaeontologica Polonica*, 51, 1–11.
25. Wong, E.S.W., Nicol, S., Warren, W.C. *et al.* 2013. Echidna venom gland transcriptome provides insights into the evolution of monotreme venom. *PLoS One*, 8(11), e79092. doi.org/10.1371/journal.pone.0079092.
26. Whittington, C. M., and Belov, K. 2014. Tracing monotreme venom evolution in the genomics era. *Toxins*, 6, 1260–1273. doi.org/10.3390/toxins6041260.
27. Whittington, C.M., Papenfuss, A.T., Bansal, P. *et al.* 2008. Defensins and the convergent evolution of platypus and reptile venom genes. *Genome Research*, 18, 986–994. doi:10.1101/gr.7149808.
28. Ohno, S. 1970. *Evolution by Gene Duplication*. Berlin: Springer-Verlag.
29. Wagner, A. 2014. *Arrival of the Fittest: Solving Evolution's Greatest Puzzle*. London: Oneworld, p. 4.
30. Zhou *et al.* 2021 (see note 6).
31. Pettigrew, J.D., Manger, P.R., Fine, S.L.B. 1998. The sensory world of the platypus. *Philosophical Transactions of the Royal Society of London: Biological Sciences*, 353, 1199–1210.
32. Pettigrew, J.D. 1999. Electroreception in monotremes. *Journal of Experimental Biology*, 202, 1447–1454.

33. Flannery, T. 1998. *Throwim Way Leg: Tree-Kangaroos, Possums, and Penis Gourds. On the Track of Unknown Mammals in Wildest New Guinea.* New York: Atlantic Monthly Press.
34. Flannery, T.F., Rich, T.H., Vickers-Rich, P. *et al.* 2022. A review of monotreme (Monotremata) evolution. *Alcheringa*, **45**, 1–18. doi.org/10.1080/03115518.2022.2025900.
35. Chimento, N.R., Agnolín, F.L., Manabe, M. *et al.* 2023. First monotreme from the late cretaceous of South America. *Communications Biology*, **6**, 146. doi.org/10.1038/s42003-023-04498-7.
36. Robertson, D.S., Lewis, W.M., Sheehan, P.M. *et al.* 2013. K–Pg extinction patterns in marine and freshwater environments: the impact winter model. *Journal of Geophysical Research. Biogeosciences*, **18**, 1006–1014. doi.org/10.1002/jgrg.20086.
37. Baldwin, S.L., Fitzgerald, P.G. and Webb, L.E. 2012. Tectonics of the New Guinea region. *Annual Review of Earth and Planetary Sciences*, **40**, 495–520. doi.org/10.1146/annurev-earth-040809-152540.
38. Sir David's long-beaked echidna was named after Sir David Attenborough, the famous naturalist and TV presenter. The species may be extinct, and all our knowledge comes from a poorly preserved museum specimen, collected in 1961. However, hope for its survival has been boosted following the recent discovery of its burrows in the Cyclops Mountains in West Papua.
39. The fallacy of monotreme primitivity is the subject of Stephen Jay Gould's essays, 'To be a platypus' and 'Bligh's bounty', published in *Bully for Brontosaurus* (W.W. Norton & Co., 1991). For a more general discourse of the public perception of Australian mammals, see Jack Ashby's *Platypus Matters* (William Collins, 2022).

Chapter 2: The Monito del Monte's Story

1. Simpson, G.G. 1936. Studies of the earliest mammalian dentitions. *Dental Cosmos*, **78**, 791–800, 940–953.
2. Rich, T.H., Vickers-Rich, P., Constantine, A. *et al.* 1997. A tribosphenic mammal from the Mesozoic of Australia. *Science*, **278**, 1438–1442. doi.org/10.1126/science.278.5342.1438.
3. Flynn, J.J., Parrish, J.M., Rakotosamimanana, B. *et al.* 1999. A Middle Jurassic mammal from Madagascar. *Nature*, **401**, 57–60. doi.org/10.1038/43420.
4. Flannery, T.F., Rich, T.H., Vickers-Rich, P. *et al.* 2022. The Gondwanan origin of Tribosphenida (Mammalia). *Alcheringa: An Australian Journal of Palaeontology*, **46**, 277–290. doi.org/10.1080/03115518.2022.2132288.
5. Christidis, L. and Schodde, R. 1991. Relationships of the Australo-Papuan songbirds: protein evidence. *Ibis*, **133**, 277–285.
6. Hochmuth, K., Gohl, K. and Uenzelmann-Neben, G. 2015. Playing jigsaw with Large Igneous Provinces – a plate tectonic reconstruction of Ontong Java Nui, West Pacific. *Geochemistry, Geophysics, Geosystems*, **16**, 3789–3807. doi:10.1002/2105GC006036; Jacobs, L.L., Strganac, C., and Scotese, C. 2011. Plate motions, Gondwana dinosaurs, Noah's arks, beached Viking funeral ships, ghost ships, and landspans. *Anais da Academia Brasileira de Ciências*, **83**, 3–22. doi.org/10.1590/S0001-37652011000100002.
7. Halliday, T. 2022. *Otherlands: A World in the Making.* London: Allen Lane. pp. 115–134.
8. Emerling, C.A., Delsuc, F. and Nachmann, M.W. 2018. Chitinase genes (*CHIAs*) provide genomic footprints of a post-Cretaceous dietary radiation in placental mammals. *Science Advances*, **4**, eaar6478. doi.org/10.1126/sciadv.aar6478.
9. Jacobs, G.H., and Rowe, M.P. 2004. Evolution of vertebrate colour vision. *Clinical and Experimental Optometry*, **87**, 206–216. doi.org/10.1111/j.1444-0938.2004.tb05050.x.
10. Hurum, J.H., Luo, Z-X. and Kielan-Jaworowska, Z. 2006. Were mammals originally venomous? *Acta Palaeontologica Polonica*, **51**, 1–11.
11. Carrier, D. 1987. Lung ventilation during walking and running in four lizard species. *Experimental Biology*, **47**, 33–42. doi.org/10.1017/S0094837300008903.
12. Carrier, D. 1987. The evolution of locomotor stamina in tetrapods: circumventing a mechanical constraint. *Paleobiology*, **13**, 326–341. doi:10.1017/S0094837300008903.
13. Kielan-Jaworowska, Z. and Hurum, J.H. 2006. Limb posture in early mammals: sprawling or parasagittal. *Acta Palaeontologica Polonica*, **51**, 393–406.

14. Wang, H.-B., Hoffmann, S., Wang, D-C. *et al.* 2022. A new mammal from the Lower Cretaceous Jehol Biota and implications for eutherian evolution. *Philosophical Transactions of the Royal Society B*, **377**, 20210042. doi.org/10.1098/rstb.2021.0042.
15. Luo, Z-X., Ruf, I., Schultz, J.A. *et al.* 2011. Fossil evidence on evolution of inner ear cochlea in Jurassic mammals. *Proceedings of the Royal Society B*, **278**, 28–34. doi.org/10.1098/rspb.2010.1148.
16. Hepson, J.A. 1973. Endothermy, small size and origins of mammalian reproduction. *American Naturalist*, **107**, 446–452.
17. Haig, D. 1992. Genomic imprinting and the theory of parental–offspring conflict. *Seminars in Developmental Biology*, **3**, 153–160; Curley, J.P., Barton, S., Surani, A. *et al.* 2004. Coadaptation in mother and infant regulated by a paternally expressed imprinted gene. *Proceedings of the Royal Society B*, **271**, 1303–1309. doi.org/10.1098/rspb.2004.2725.
18. Renfree, M.B., Suzuki, S. and Kaneko-Ishino, T. 2013. The origin and evolution of genomic imprinting and viviparity in mammals. *Philosophical Transactions of the Royal Society B*, **368**, 20120151. doi.org/10.1098/rstb.2012.0151.
19. Gao, W., Sun, Y-B., Zhou, W-W. 2019. Genomic and transcriptome investigations of the evolutionary transition from oviparity to viviparity. *Proceedings of the National Academy of Sciences of the United States of America*, **116**, 3646–3655. doi.org/10.1073/pnas.1816086116.
20. White, H.E., Tucker, A.S., Fernandez, V. *et al.* 2023. Pedomorphosis in the ancestry of marsupial mammals. *Current Biology*, **33**, 2136–2150. doi.org/10.1016/j.cub.2023.04.009.
21. Weaver, L.N., Fulghum, H.Z., Grossnickle, D.M. *et al.* 2022. Multituberculate mammals show evidence of life history strategy similar to that of placentals, not marsupials. *The American Naturalist*, **200**, 383–400. doi.org/10.1086/720410.
22. Shiura, H., Ono, R., Tachibana, S. *et al.* 2021. PEG10 viral aspartic protease domain is essential for the maintenance of fetal capillary structure in the mouse placenta. *Development*, **148**, dev199564. doi.org/10.1242/dev.199564.
23. Hone, D.W.E., Dececchi, T.A., Sullivan, C. *et al.* 2022. Generalised diet of *Microraptor zhaoianus* included mammals. *Journal of Vertebrate Paleontology*, **42**(2). doi.org/10.1080/02724634.2022.2144337.
24. Suárez, R., Paolino, A., Fenlon, L.R. *et al.* 2018. A pan-mammalian map of interhemispheric brain connections predates the evolution of the corpus callosum. *Proceedings of the National Academy of Sciences of the United States of America*, **115**, 9622–9627. doi.org/10.1073/pnas.1808262115.
25. Zuo, Z-X., Ji, Q., Wible, J.R. *et al.* 2003. An Early Cretaceous tribosphenic mammal and metatherian evolution. *Science*, **302**, 1934–1940. doi.org/10.1126/science.1090718.
26. Bi, S., Zheng, X., Wang, X. *et al.* 2018. An early Cretaceous eutherian and the placental–marsupial dichotomy. *Nature*, **558**, 390–395. doi.org/10.1038/s41586-018-0210-3.
27. Ji, Q., Luo, Z-X., Yuan, C-X. *et al.* 2002. The earliest eutherian mammal. *Nature*, **416**, 811–822. doi.org/10.1038/416816a.
28. Luo, Z-X., Yuan, C-X., Meng, Q-J. *et al.* 2011. A Jurassic eutherian mammal and divergence of marsupials and placentals. *Nature*, **476**, 442–445. doi.org/10.1038/nature10291.
29. Wilson, G.P., Ekdale, E.G., Hoganson, J.W. *et al.* 2016. A large carnivorous mammal from the Late Cretaceous and the North American origin of marsupials. *Nature Communications*, **7**, 13734. doi.org/10.1038/ncomms13734.
30. Guernsey, M.W., Chuong, E.B., Cornelis, G. *et al.* 2017. Molecular conservation of marsupial and eutherian placentation and lactation. *eLife*, **6**, e27450. doi.org/10.7554/eLife.27450.
31. Eberle, J.J., Clemens, W.A., McCarthy, P.J. *et al.* 2019. Northernmost record of the Metatheria: a new Late Cretaceous pediomyid from the North Slope of Alaska. *Journal of Systematic Palaeontology*, **17**, 1805–1824. doi.org/10.1080/14772019.2018.1560369.
32. Goin, F.L., Woodburne, M.O., Zimicz, A.N. *et al.* 2016. *A Brief History of South American Metatherians. Evolutionary Contexts and Intercontinental Dispersals*. Dordrecht, Springer.
33. Williamson, T.E., Brusatte, S.L. and Wilson, G.P. 2014. The origin and early evolution of metatherian mammals: the Cretaceous record. *Zookeys*, **465**, 1–76. doi.org/10.3897/Zookeys.465.8178; Bennett, C.V., Upchurch, P., Goin, F.J. *et al.* 2018. Deep time diversity of metatherian animals: implications for evolutionary history and fossil-record quality. *Paleobiology*, **44**, 171–198. doi:10.3897/Zookeys.465.8178.

34. Ladevèze, S., de Muizon, C., Beck, R.M.D. *et al.* 2011. Earliest evidence of mammalian social behaviour in the basal Tertiary of Bolivia. *Nature*, **474**, 83–86. doi.org/10.1038/nature09987.
35. Tarquini, S.D., Ladevèze, S. and Prevosti, F.J. 2022. The multicausal twilight of South American native mammalian predators (Metatheria, Sparassodonta). *Scientific Reports*, **12**, 1224. doi.org/10.1038/s41598-022-05266-z.
36. Feng, S., Bai, M., Rivas-González, I. *et al.* 2022. Incomplete lineage sorting and phenotypic evolution in marsupials. *Cell*, **185**, 1646–1660. doi.org/10.1016/j.cell.2022.03.034.
37. Eldridge, M.D.B., Beck, R.M.D., Croft, D.A. *et al.* 2019. An emerging consensus in the evolution, phylogeny, and systematics of marsupials and their fossil relatives. *Journal of Mammalogy*, **100**, 802–837. doi.org/10.1093/jmammal/gyz018.
38. Pavan, S.E. 2016. Phylogeny, systematics and biogeography of short-tailed opossums (Didelphidae: *Monodelphis*). *CUNY Academic Works*. academicworks.cuny.edu/gc_etds/1472.
39. Castro, M.C., Dahur, M.J. and Ferreira, G.S. 2021. Amazonia as the origin and diversification area of Didelphidae (Mammalia: Metatheria), and a review of the fossil record of the clade. *Journal of Mammalian Evolution*, **28**, 583–598. doi.org/10.1007/s10914-021-09548-7.
40. Lemos, B. and Cerqueira, R. 2002. Morphological differentiation in the white-eared opossum group (Didelphidae: *Didelphis*). *Journal of Mammalogy*, **83**, 354–369. doi:10.1644/1545-1542(2002)083<0354:MDITWE>2.0.CO;2.
41. Voss, R.S., and Jansa, S.A. 2021. *Opossums: An Adaptive Radiation of New World Marsupials*. Baltimore: Johns Hopkins University Press.
42. Nigenda-Morales, S.F., Harrigan, R.J. and Wayne, R.K. 2018. Playing by the rules? Phenotypic adaptation to temperate environments in an American marsupial. *PeerJ*, **6**, e4512. doi.org/10.7717/peerj.4512.
43. Ojala-Barbour, R., Pinto, C.M., Brito M, J. *et al.* 2013. A new species of shrew-opossum (Paucituberculata: Caenolestidae) with a phylogeny of extant Caenolestids. *Journal of Mammalogy*, **94**, 967–982. doi.org/10.1644/13-MAMM-A-018.1.
44. Woodburne M.O. and Zinsmeister, W.J. 1984. Fossil land mammal from Antarctica. *Science*, **218**, 284–286. doi.org/10.1126/science.218.4569.284.
45. Goin, F.J., Case, J.A., Woodburne, M.O. et al. 1999. New discoveries of 'opposum-like' marsupials from Antarctica (Seymour Island, Medial Eocene). *Journal of Mammalian Evolution*, **6**, 335–365. doi.org/10.1023/A:1027357927460.
46. Reguero, M.A., Marenssi, S.A. and Santillana, S.N. 2002. Antarctic Peninsula and South America (Patagonia) Paleogene terrestrial faunas and environments: biogeographic relationships. Palaeogeography, Palaeoclimatology, *Palaeoecology*, **179**, 189–210. doi.org/10.1016/S0031-0182(01)00417-5.
47. Bijl, P.K., Bendle, J.A.P., Bohaty, S.M. *et al.* 2013. Eocene cooling linked to early flow across the Tasmanian gateway. *Proceedings of the National Academy of Sciences of the United States of America*, **110**, 9645–9650. doi.org/10.1073/pnas.1220872110.
48. Beck, R.M.D., Godthelp, H., Weisbecker, V. *et al.* 2008. Australia's oldest marsupial fossils and their biogeographical implications. *PLoS One*, **3**(3), e1858. doi.org/10.1371/journal.pone.0001858.
49. Nilsson, M.A., Churakov, G., Sommer, M. *et al.* 2010. Tracking marsupial evolution using archaic genomic retroposon insertions. *PLoS Biology*, **8**(7), e1000436. doi.org/10.1371/journal.pbio.1000436.
50. Fontúrbel, F.E., Franco, L.M., Bozinovic, F. *et al.* 2022. The ecology and evolution of the monito de monte, a relic species from the southern South America temperate forests. *Ecology and Evolution*, **12**, e8645. doi.org/10.1002/ece3.8645.
51. Feng *et al.* 2022 (see note 36).
52. Quintero-Galvis, J.F., Saenz-Agudelo, P., Celis-Diez, J.L. 2021. The biogeography of *Dromiciops* in southern South America: Middle Miocene transgressions, speciation and association with *Nothofagus*. *Molecular Phylogenetics and Evolution*, **163**, 107234. doi.org/10.1016/j.ympev.2021.107234.
53. Quintero-Galvis, J.F., Saenz-Agudelo, P., Amico, G.C. *et al.* 2022. Genomic diversity and demographic history of the *Dromiciops* genus (Marsupialia: Microbiotheriidae). *Molecular Phylogenetics and Evolution*, **168**, 107405. doi.org/10.1016/j.ympev.2022.107405.

Chapter 3: The Marsupial Mole's Story

1. Gallus, S., Janke, A., Kumar, V. et al. 2015. Disentangling the relationship of the Australian marsupial orders using retrotransposon and evolutionary network analyses. *Genome Biology Evolution*, **7**, 985–992. doi.org/10.1093/gbe/evv052.
2. Duchêne, D.A., Bragg, J.G., Duchêne, S. et al. 2018. Analysis of phylogenomic tree space resolves relationships among marsupial families. *Systematic Biology*, **67**, 400–412. doi.org/10.1093/sysbio/syx076.
3. Lopes, F., Oliveira, L.R., Kessler, A. et al. 2021. Phylogenomic discordance in the eared seals is best explained by incomplete lineage sorting following explosive radiation in the southern hemisphere. *Systematic Biology*, **70**, 786–802. doi.org/10.1093/sysbio/syaa099; Rivas-González, I., Rousselle, M., Li, F. et al. 2023. Pervasive incomplete lineage sorting illuminates speciation and selection in primates. *Science*, **380**, eabn4409. doi.org/10.1126/science.abn4409.
4. Feng, S., Bai, M., Rivas-González, I. et al. 2022. Incomplete lineage sorting and phenotypic evolution in marsupials. *Cell*, **185**, 1646–1660. doi.org/10.1016/j.cell.2022.03.034.
5. Benshemesh, J. and Johnson, K. 2003. Biology and conservation of marsupial moles (*Notoryctes*). In: M. Jones, C. Dickman and M. Archer (eds), *Predators with Pouches: The Biology of Carnivorous Marsupials*. Melbourne: CSIRO, pp. 464–474.
6. Beck, R.M.D. 2008. A dated phylogeny of marsupials using a molecular supermatrix and multiple fossil constraints. *Journal of Mammalogy*, **89**, 175–189. doi.org/10.1644/06-MAMM-A-437.1.
7. Archer, M., Beck, R., Gott, M. et al. 2011. Australia's first marsupial mole (Notoryctemorphia) resolves controversies about their evolution and palaeoenvironmental origins. *Proceedings of the Royal Society B*, **278**, 1478–1506. doi.org/10.1098/rspb.2010.1943.
8. Kear, P.B., Aplin, K.P. and Westman, M. 2016. Bandicoot fossils and DNA elucidate lineage antiquity amongst xeric-adapted Australian marsupials. *Scientific Reports*, **6**, 37537. doi.org/10.1038/srep37537.
9. Freyer, C., Zeller, U. and Renfree, M.B. 2003. The marsupial placenta: a phylogenetic analysis. *Journal of Experimental Zoology A: Ecological and Integrative Physiology*, **299**, 59–77. doi.org/10.1002/jez.a.10291.
10. Fisher, D.O., Dickman, C.R., Jones, M.E. et al. 2013. Sperm competition drives the evolution of suicidal reproduction in mammals. *Proceedings of the National Academy of Sciences of the United States of America*, **110**, 17910–17914. doi.org/10.1073/pnas.1310691110.
11. Tian, R., Geng, Y., Yang, C. et al. 2021. A chromosome-level genome of *Antechinus flavipes* provides a reference for an Australian marsupial genus with male death after mating. *Molecular Ecology Resources*, **22**, 740–754. doi.org/10.1111/1755-0998.13501.
12. Owen, D. and Pemberton, D. 2005. *Tasmanian Devil: a Unique and Threatened Animal*. Crows Nest: Allen & Unwin.
13. Epstein, B., Jones, M., Hamede, R. et al. 2016. Rapid evolutionary response to a transmissible cancer in Tasmanian devils. *Nature Communications*, **7**, 12684. doi.org/10.1038/ncomms12684.
14. Gooley, R.M., Hogg, C.J., Fox, S. et al. 2020. Inbreeding depression in one of the last DFTD-free wild populations of Tasmanian devils. *PeerJ*, **8**, e9220. doi.org/10.7717/peerj.9220.
15. Johnson, R.N., O'Meally, D., Chen, Z. et al. 2018. Adaptation and conservation insights from the koala genome. *Nature Genetics*, **50**, 1102–1111. doi.org/10.1038/s41588-018-0153-5.
16. Shiffman, M.E., Soo, R.M., Dennis, P.G. et al. 2017. Gene and genome-centric analysis of koala and wombat fecal microbiomes point to metabolic specialization for *Eucalyptus* digestion. *PeerJ*, **5**, e4075. doi.org/10.7717/peerj.4075.
17. Henneberg, M.J., Lambert, K.M. and Leigh, C.M. 1997. Fingerprint homoplasy: koalas and humans. *naturalSCIENCE*, **1**(4). hdl.handle.net/2440/5433 (accessed July 2025)
18. Yum, S-M., Baek, I-K., Hong, D. et al. 2020. Fingerprint ridges allow primates to regulate grip. *Proceedings of the National Academy of Sciences of the United States of America*, **117**, 31665–31673. doi.org/10.1073/pnas.2001055117.
19. Hocknull, S.A., Lewis, A., Arnold, L.T. et al. 2020. Extinction of eastern Sahul megafauna coincides with sustained environmental deterioration. *Nature Communications*, **11**, 2250. doi.org/10.1038/s41467-020-15785-w.

20. Yang, P.J., Lee, A.B., Chan, M. *et al.* 2021. Intestines of non-uniform stiffness mold the corners of wombat feces. *Soft Matter*, **17**, 475–488. doi.org/10.1039/d0sm01230k.
21. Couzens, A.M.C. and Prideaux, G.J. 2018. Rapid Pliocene adaptive radiation of modern kangaroos. *Science*, **362**, 72–75. doi.org/10.1126/science.aas8788.
22. Den Boer, W., Campione, N.E. and Kear, B.P. 2019. Climbing adaptations, locomotory disparity and ecological convergence in ancient stem 'kangaroos'. *Royal Society Open Science*, **6**, 181617. doi.org/10.1098/rsos.181617.
23. Janis, C.M., O'Driscoll, A.M. and Kear, B.P. 2023. Myth of the *QANTAS* leap: perspectives on the evolution of kangaroo locomotion. *Alcheringa: An Australian Journal of Palaeontology*, **47**, 671–685. doi.org/10.1080/03115518.2023.2195895.
24. Potter, S., Bragg, J.G. Blom, M.P.K. *et al.* 2017. Chromosomal speciation in the genomic era: disentangling phylogenetic evolution of rock-wallabies. *Frontiers in Genetics*, **8**. doi.org/10.3389/fgene.2017.00010; Potter, S., Bragg, J.G., Turakulov, R. *et al.* 2022. Limited introgression between rock-wallabies with extensive chromosomal rearrangements. *Molecular Biology and Evolution*, **39**, msab333. doi.org/10.1093/molbev/msab333.
25. Eldridge, M. and Couslon, G. 2015. Family Macropodidae (kangaroos and wallabies). In: D.E. Wilson and R.A. Mittermeir (eds), *Handbook of the Mammals of the World*, Vol. 5. Barcelona: Lynx Edicions, pp. 630–735, .
26. Eldridge, M. 2011. Out on a limb: tree-kangaroos. *Explore*, **33**(3), 8–11. australian.museum/learn/animals/mammals/out-on-a-limb (accessed July 2025).
27. Louis Antoine Marie Joseph Dollo (1857–1931) was a French-born Belgian palaeontologist and curator of Belgium's Royal Museum of Natural History, famous for his studies of Iguanodon dinosaurs. Around 1890 he formulated his Law of Irreversibility, which states that an organism cannot return, even partially, to a previous state already realised in its ancestral series. This hypothesis was widely accepted until Michael Whiting's 2003 discovery that certain insects that had lost their wings regained them millions of years later.
28. Frankham, G.J., Handasyde, K.A. and Eldridge, M.D.B. 2015. Evolutionary and contemporary responses to habitat fragmentation detected in a mesic zone marsupial, the long-nosed potoroo (*Potorous tridactylus*) in south-eastern Australia. *Journal of Biogeography*, **43**, 653–665. doi.org/10.1111/jbi.12659.
29. Bradshaw, D. and Bradshaw, F. 2012. The physiology of the honey possum, *Tarsipes rostratus*, a small marsupial with a suite of highly specialised characters: a review. *Journal of Comparative Physiology B*, **182**, 469–489. doi.org/10.1007/s00360-011-0632-9.
30. Feigin, C.V., Moreno, J.A., Ramos, R. *et al.* 2023. Convergent deployment of ancestral functions during the evolution of mammalian flying membranes. *Science Advances*, **9**, 7511. doi.org/10.1126/sciadv.ade7511.
31. Eldridge, M.D.B., Potter, S., Helgen, K.M. *et al.* 2018. Phylogenetic analysis of the tree-kangaroos (*Dendrolagus*) reveals multiple divergent lineages within New Guinea. *Molecular Phylogenetics and Evolution*, **127**, 589–599. doi.org/10.1016/j.ympev.2018.05.030.
32. Malekian, M., Cooper, S.J.B., Norman, J.A. *et al.* 2010. Molecular systematics and evolutionary origins of the genus *Petaurus* (Marsupialia: Petauridae) in Australia and New Guinea. *Molecular Phylogenetics and Evolution*, **54**, 122–135. doi.org/10.1016/j.ympev.2009.07.026.
33. Mitchell, K.J., Pratt, R.C., Watson, L.N. *et al.* 2014. Molecular phylogeny, biogeography, and habitat preferences evolution of marsupials. *Molecular Biology and Evolution*, **31**, 2322–2330. doi.org/10.1093/molbev/msu176.
34. van Ufford, A.Q. and Cloos, M. 2005. Cenozoic tectonics of New Guinea. *American Association of Petroleum Geologists Bulletin*, **89**, 119–140. doi.org/10.1306/08300403073.
35. Ruedas, L.A. and Morales, J.C. 2005. Evolutionary relationships among genera of Phalangeridae (Metatheria: Diprotodontia) inferred from mitochondrial DNA. *Journal of Mammalogy*, **86**, 353–365. doi.org/10.1644/BER-117.1.
36. Raterman, D., Meredith, R.W., Ruedas, L.A. *et al.* 2006. Phylogenetic relationships of the cuscuses and brushtail possums (Marsupialia: Phalangeridae) using the nuclear gene *BRCA1*. *Australian Journal of Zoology*, **54**, 353–361. doi.org/10.1071/ZO05067.
37. Kealy, S., Donnellan, S.C., Mitchell, K.J. *et al.* 2019. Phylogenetic relationships of cuscuses (Diprotodontia: Phalangeridae) of island Southeast Asia and Melanesia based on the mitochondrial *ND2* gene. *Australian Mammalogy*, **42**, 266–276. doi.org/10.1071/AM18050.

38. Leavesley, M. 2005. Prehistoric hunting strategies in New Ireland, Papua New Guinea: the evidence of the cuscus (*Phalanger orientalis*) remains from Buang Merabak cave. *Asian Perspectives*, **44**, 207–218. doi.org/10.1353/asi.2005.0010.
39. For reproductive constraints on morphology, see:
    Shoulder girdle – Seers, K. 2004. Constraints on the morphological evolution of marsupial shoulder girdles. *Evolution*, **58**, 2353–2370. doi.org/10.1111/j.0014-3820.2004.tb01609.x.
    Limbs – Kelly, E.M. and Sears, K.E. 2011. Limb specialization in living marsupials and eutherian mammals: constraints on mammalian limb development. *Journal of Mammalogy*, **92**, 1038–1049. doi.org/10.1644/10-MAMM-A-425.1.
    Jaw – Fabre, A-C., Dowling, C., Miguez, R.P. et al. 2021. Functional constraints during development limit jaw shape evolution in marsupials. *Proceedings of the Royal Society B*, **288**, 20210319. doi.org/10.1098/rspb.2021.0319.
    Skull – Bennett, C.V. and Goswami, A. 2013. Statistical support for the hypothesis of developmental constraint in marsupial skull evolution. *BMC Biology*, **11**, 52. doi.org/10.1186/1741-7007-11-52.
    Mouth – Goswami, A., Randau, M., Polly, P.D. et al. 2016. Do developmental constraints and high integration limit the evolution of the marsupial oral apparatus? *Integrative and Comparative Biology*, **56**, 404–415. doi.org/10.1093/icb/icw039.
40. Conway Morris, S. 2003. *Life's Solutions: Inevitable Humans in a Lonely Universe*. Cambridge: Cambridge University Press; Conway Morris, S. 2014. *The Runes of Evolution: How the Universe Became Self-Aware*. West Conshohocken: Templeton Press.
41. Gould, S.J. 1990. *Wonderful Life: The Burgess Shale and the Nature of History*. London: Hutchinson Radius.
42. Blount, Z.D., Lenski, R.E. and Losos, J.B. 2018. Contingency and determinism in evolution: replaying life's taper. *Science*, **362**, eaam5979. doi.org/10.1126/science.aam5979.
43. Worthy, T.H., Hand, S.J., Archer, M. et al. 2019. Evidence for a giant parrot from the early Miocene of New Zealand. *Biology Letters*, **15**, 20190467. doi.org/10.1098/rsbl.2019.0467; Boast, A.P., Chapman, B., Herrera, M.B. et al. 2019. Mitochondrial genomes from New Zealand. Extinct Adzebills (Aves: Aptornithidae: *Aptornis*) support a sister-taxon relationship with the Afro-Madagascan Sarothruridae. *Diversity*, **11**(2), 24. doi.org/10.3390/d11020024.
44. Xie, V.C., Pu, J., Metzger, B.P.H. et al. 2021. Contingency and chance erase necessity in the experimental evolution of ancestral proteins. *eLife*, **10**, e67336. doi.org/10.7554/eLife.67336.

Chapter 4: The Tasmanian' Tiger's Story

1. Feigin, C.Y., Newton, A.H., Doronina, L. et al. 2018. Genome of the Tasmanian tiger provides insights into the evolution and demography of an extinct marsupial carnivore. *Nature Ecology & Evolution*, **2**, 182–192. doi.org/10.1038/s41559-017-0417-y.
2. Feigin, C.Y., Hewton, A.H. and Pask, A.J. 2019. Widespread cis-regulatory convergence between the extinct Tasmanian tiger and gray wolf. *Genome Research*, **29**, 1648–1658. doi:10.1101/gr.244251.118.
3. Feigin, C. 2019. The shared evolution of the Tasmanian tiger and the wolf. *Pursuit*, pursuit.unimelb.edu.au (accessed July 2025).
4. Peel, E., Silver, L., Brandies, P. et al. 2022. Genome assemblies of the numbat (*Myrmecobius fasciatus*), the only termitivorous marsupial. *Gigabyte*, **2022**, 1–17. doi.org/10.46471/gigabyte.47.
5. Cook, L.E., Feigin, C.Y., Hills, J.D. et al. 2025. Gene regulatory dynamics during craniofacial development in a carnivorous marsupial. *eLife*, **14**, RP103592. doi.org/10.7554/eLife.103592.1.
6. Attard, M.R.G., Chamoli, U., Ferrara, T.L. et al. 2011. Skull mechanics and implications for feeding behaviour in a large marsupial carnivore guild: the thylacine, Tasmanian devil and spotted-tailed quoll. *Journal of Zoology*, **285**, 292–300. doi.org/10.1111/j.1469-7998.2011.00844.x.
7. Beer, M.A., Proft, K.M., Veillet, A. et al. 2024. Disease-driven top predator decline affects mesopredator population genomic structure. *Nature Ecology & Evolution*, **8**, 293–303. doi.org/10.1038/s41559-023-02265-9.
8. Campbell, K.H., McWhir, J., Ritchie, W.A. et al. 1996. Sheep cloned by nuclear transfer from a cultured cell line. *Nature*, **380**, 64–66. doi.org/10.1038/380064a0.

9. Folch, J., Cocero, M.J., Chesné, P. *et al.* 2009. First birth of an animal from an extinct subspecies (*Capra pyrenaica pyrenaica*) by cloning. *Theriogenology*, **71**, 1026–1034. doi.org/10.1016/j.theriogenology.2008.11.005.
10. Feigin, C., Frankenberg, S. and Pask, A. 2022. A chromosome-scale hybrid assembly of the extinct Tasmanian Tiger (*Thylacinus cynocephalus*). *Genome Biology and Evolution*, **14**, evac048. doi.org/10.1093/gbe/evac048.
11. Deakin, J.E. 2018. Chromosome evolution in marsupials. *Genes*, **9**, 72. doi.org/10.3390/genes9020072.
12. Doudna, J. and Sternberg, S.H. 2018. *A Crack in Creation: Gene Editing and the Unthinkable Power to Control Evolution*. London, HarperCollins.
13. Shapiro, B. 2017. Pathways to de-extinction: how close can we get to resurrection of an extinct species? *Functional Ecology*, **31**, 996–1002. doi.org/10.1111/1365-2435.12705.
14. Kelly, E. and Phillips, B.L. 2019. Targeted gene flow and rapid adaptation in an endangered marsupial. *Conservation Biology*, **33**, 112–121. doi.org/10.1111/cobi.13149.
15. Odenbaugh, J. 2023. Philosophy and ethics of de-extinction. *Cambridge Prisms: Extinction*, **1**, e7. doi.org/10.1017/ext.2023.4.
16. Carlquist, S. 1965. *Island Life: A Natural History of the Islands of the World*. New York: Natural History Press, p.166.

Chapter 5: The Aardvark's Story

1. Springer, M.S. 2022. Afrotheria. *Current Biology*, **32**, R197–R212. doi.org/10.1016/j.cub.2022.02.001.
2. Simpson, G.G. 1945. *The Principles of Classification and a Classification of Mammals*. New York: Bulletin of the American Museum of Natural History.
3. Simpson, G.G. 1996. *The Dechronization of Sam Magruder*. New York: St Martin's Press. Afterword by S.J. Gould.
4. Springer, M.S., Cleven, G.C., Madsen, O. *et al.* 1997. Endemic African mammals shake the phylogenetic tree. *Nature*, **388**, 61–63. doi:10.1038/40386.
5. van Dijk, M.A., Madsen, O., Catzeflis, F. *et al.* 2001. Protein sequence signatures support the African clade of mammals. *Proceedings of the National Academy of Sciences of the United States of America*, **98**, 188–193. doi:10.1073/pnas.98.1.188.
6. Nikaido, M., Nishihara, H., Fukumoto, Y. *et al.* 2003. Ancient SINEs from African endemic mammals. *Molecular Biology and Evolution*, **20**, 522–527. doi:10.1093/molbev/msg052.
7. Kellogg, M.E., Burkett, S., Dennis, T.R. *et al.* 2007. Chromosomal painting in the manatee supports Afrotheria and Paenungulata. *BMC Evolutionary Biology*, **7**, 6. doi.org/10.1186/1471-2148-7-6.
8. Mess, R. and Carter, A.M. 2006. Evolutionary transformations of fetal membrane character in Eutheria with special reference to Afrotheria. *Journal of Experimental Zoology, Part B, Molecular Development and Evolution*, **306**, 140–163. doi.org/10.1002/jez.b.21079.
9. Sánchez-Villagra, M.R., Narita, Y. and Kuratami, S. 2007. Thoracolumbar vertebral number: the first skeletal synapomorphy for afrotherian mammals. *Systematics and Biodiversity*, **5**, 1–7. doi.org/10.1017/S1477200006002258.
10. Springer, M.S. 2022. Afrotheria. *Current Biology*, **32**, R197–R212. doi.org/10.1016/j.cub.2022.02.001.
11. Kawasaki, K., Hu, J.C-C. and Simmer, J.P. 2014. Evolution of Klk4 and enamel maturation in eutherians. *Biological Chemistry*, **395**, 1003–1013. doi.org/10.1515/hsz-2014-0122.
12. Gheerbrant, E., Amaghzaz, M., Bouya, B. *et al.* 2014. *Ocepeia* (Middle Paleocene of Morocco): the oldest skull of an Afrotherian mammal. *PLoS One* **9**(2), e89739. doi.org/10.1371/journal.pone.0089739.
13. Isaac, N.J.B., Turvey, S.T., Collen, B. *et al.* 2007. Mammals on the EDGE: conservation priorities based on threat and phylogeny. *PLoS One*, **2**(3), e296. doi.org/10.1371/journal.pone.0000296.
14. Gumbs, R., Scott, O., Bates, R. *et al.* 2024. Global conservation status of the jawed vertebrate Tree of Life. *Nature Communications* **15**, 1101. doi.org/10.1038/s41467-024-45119-z.

15. Despite being derived from an older lineage and being the sole living representative of its family, the platypus gets a lower score than the aardvark because of its divergence with echidnas being more recent than originally thought. The result is that it has a shorter terminal branch than the aardvark and hence an ED score of 50 MY.
16. Kingdon, J. 1997. *The Kingdon Field Guide to African Mammals*. London: Academic Press.
17. Snyder, H.K., Maia, R., D'Alba, L. *et al*. 2012. Iridescent colour production in hairs of blind golden moles (Chrysochloridae). *Biology Letters*, 8, 393–396. doi.org/10.1098/rsbl.2011.1168.
18. Springer, M.S., Emerling, C.A. and Gatesy, J. 2023. Three blind moles: molecular evolutionary insights on the tempo and mode of convergent eye degeneration in *Notoryctes typhlops* (southern marsupial mole) and two Chrysochlorids (golden moles). *Genes*, 14, 2018. doi.org/10.3390/genes14112018.
19. Lewis, E.R., Narins, P.M., Jarvis, J.U. *et al*. 2006. Preliminary evidence for the use of microseismic cues for navigation by the Namib golden mole. *Journal of the Acoustical Society of America*, 119, 1260–1268. doi.org/10.1121/1.2151790. PMID: 16521787.
20. Arnold, P., Hagemann, J., Gilissen, E. *et al*. 2022. Otter shrew mitogenomes (Afrotheria, Potamogalidae) reconstructed from historical museum skins. *Mitochondrial DNA Part B*, 7, 1699–1701. doi.org/10.1080/23802359.2022.2122747.
21. Simpson, G.G. 1940. Mammals and land bridges. *Journal of the Washington Academy of Science*, 30, 137–163.
22. Wallace, A.R. 1880. *Island Life: or, the Phenomena and Causes of Insular Faunas and Floras, including a Revision and Attempted Solution of the Problem of Geological Climates*. London: Macmillan, pp. 71–72.
23. Censky, E.J., Hodge, K. and Dudley, J. 1998. Over-water dispersal of lizards due to hurricanes. *Nature*, 395, 556. doi.org/10.1038/26886.
24. Masters, J.C., Génin, F., Zhang, Y. *et al*. 2022. Geodispersal as a biogeographic mechanism for Cenozoic exchanges between Madagascar and Africa. In: Goodman, S.M. (ed.), *The New Natural History of Madagascar*. Princeton: Princeton University Press; Mazza, P.P.A., Buccianti, A. and Savorelli, A. 2019. Grasping at straws: a re-evaluation of sweepstakes colonisation of islands by mammals. *Biological Reviews*, 94, 1364–1380. doi.org/10.1111/brv.12506; Stankiewicz, J., Thiart, C., Masters, J.C. *et al*. 2006. Did lemurs have sweepstake tickets? An exploration of Simpson's model for the colonization of Madagascar by mammals. *Journal of Biogeography*, 33, 221–235. doi.org/10.1111/j.1365-2699.2005.01381.x.
25. Crowley, B.E., Godfrey, L.R. and Samonds, K.E. 2023. What can hippopotamus isotopes tell us about past distributions of C_4 grassy biomes on Madagascar. *Plants, People, Planet*, 5, 997–1010. doi.org/10.1002/ppp3.10402.
26. Masters, J.C., Génin, F., Zhang, Y. *et al*. 2022. Biogeographic mechanisms involved in the colonization of Madagascar by African vertebrates: rifting, rafting and runways. *Journal of Biogeography*, 48, 492–510. doi.org/10.1111/jbi.14032.
27. Ali, J.R. and Huber, M. 2010. Mammalian biodiversity in Madagascar controlled by ocean currents. *Nature*, 463, 653–656. doi.org/10.1038/nature08706.
28. Everson, K.M., Soarimalala, V., Goodman, S.M. *et al*. 2016. Multiple loci and complete taxonomic sampling resolve the phylogeny and biogeographic history of tenrecs (Mammalia: Tenrecidae) and reveal higher speciation rates in Madagascar's humid forests. *Systematic Biology*, 65, 890–909. doi.org/10.1093/sysbio/syw034.
29. Murphy, W.J., Pringle, T.H., Crider, T.A. *et al*. 2007. Using genomic data to unravel the root of the placental mammal phylogeny. *Genome Research*, 17, 413–421. doi.org/10.1101/gr.5918807.
30. Stankowich, T. and Stensrud, C. 2019. Small but spiny: the evolution of antipredator defences in Madagascar tenrecs. *Journal of Mammalogy*, 100, 13–20. doi.org/10.1093/jmammal/gyz003.

Chapter 6: The Hyrax's Story

1. Seiffert, E.R. 2007. A new estimate of afrotherian phylogeny based on simultaneous analysis of genomic, morphological, and fossil evidence. *BMC Evolutionary Biology*, 7, 224. doi.org/10.1186/1471-2148-7-224.

2. Oates, J.F., Woodman, N., Gaubert, P. *et al.* 2012. A new species of tree hyrax (Procaviidae: *Dendrohyrax*) from West Africa and the significance of the Niger–Volta interfluvium in mammalian biogeography. *Zoological Journal of the Linnean Society*, 192, 527–552. doi.org/10.1093/zoolinnean/zlab029.
3. Amoroso, E.C. and Perry, J.S. 1964. The foetal membranes and placenta of the African elephant (*Loxodonta africana*). *Philosophical Transactions of the Royal Society B*, 248, 1–34. doi.org/10.1098/rstb.1964.0007.
4. Rasmussen, L.E. and Munger, B.L. 1996. The sensorineural specializations of the trunk tip (finger) of the Asian elephant, *Elephas maximus*. *The Anatomical Record*, 246, 127–134.
5. Sarko, D.K., Rice, F.L. and Reep, R.L. 2015. Elaboration and innervation of the vibrissal system in the rock hyrax (*Procavia capensis*). *Brain, Behaviour and Evolution*, 85, 170–188. doi.org/10.1159/000381415.
6. Kershenbaum, A., Ilang, A., Blaustein, L. *et al.* 2012. Syntactic structure and geographical dialects in the songs of male rock hyraxes. *Proceedings of the Royal Society B*, 279, 2974–2981. doi.org/10.1098/rspb.2012.0322.
7. Demartsev, V., Haddas-Sasson, M., Ilany, A. *et al.* 2022. Male rock hyraxes that maintain an isochronous song rhythm achieve higher reproductive success. *Journal of Animal Ecology*, 92, 1520–1531. doi.org/10.1111/1365-2656.13801.
8. Hakeem, A.Y., Hof, P.R., Sherwood, C.C. *et al.* 2005. Brain of the African elephant (*Loxodonta africana*): neuroanatomy from magnetic resonance images. *The Anatomical Record Part A: Discoveries in Molecular, Cellular, and Evolutionary Biology*, 287A, 1117–1127. doi.org/10.1002/ar.a.20255.
9. Demartsev *et al.* 2022 (see note 7).
10. Liu, A.G., Seiffert, E.R. and Simons, E.L. 2008. Stable isotope evidence for an amphibious phase in early proboscidean evolution. *Proceedings of the National Academy of Sciences of the United Sates of America*, 105, 5786–5791. doi.org/10.1073/pnas.0800884105.
11. Mirceta, S., Signore, A.V., Burns, J.M. 2013. Evolution of mammalian diving capacity traced by myoglobin net surface charge. *Science*, 340, 1234192. doi.org/10.1126/science.1234192.
12. Domning, D.P. 2001. The earliest known fully quadrupedal sirenian. *Nature*, 413, 625–627 doi.org/101038/35098072. Benoit, J., Adnet, S., El Mabrouk, E. *et al.* 2013. Cranial remain from Tunisia provides new clues for the origin and evolution of Sirenia (Mammalia, Afrotheria) in Africa. *PLoS One*, 8(1), e54307. doi.org/10.1371/journal.pone.0054307.
13. Heritage, S. and Seiffert, E.R. 2022. Total evidence time-scaled phylogenetic and biographic models for the evolution of sea cows (Sirenia, Afrotheria). *PeerJ*, 10, e13886. doi.org/10.7717/peerj.13886.
14. Shephard, G.E., Müller, R.D., Liu, L. *et al.* 2010. Miocene drainage reversal of the Amazon River driven by plate–mantle interaction. *Nature Geoscience*, 3, 870–875. doi:10.1038/NGEO1017.
15. Tian, R., Zhang, Y., Kang, H. *et al.* 2024. Sirenian genomes illuminate the evolution of fully aquatic species within the mammalian superorder Afrotheria. *Nature Communications*, 15, 5568. doi.org/10.1038/s41467-024-49769-x.
16. Signore, A.V., Paijmans, J.L.A., Hofreiter, M. *et al.* 2019. Emergence of a chimeric globin pseudogene and increased haemoglobin oxygen affinity underlie the evolution of aquatic specializations in Sirenia. *Molecular Biology and Evolution*, 36, 1134–1147. doi.org/10.1093/molbev/msz044.
17. Le Duc, D., Velluva, A., Cassatt-Johnstone, M. *et al.* 2022. Genomic basis for skin phenotype and cold adaption in the extinct Steller's sea cow. *Science Advances*, 8, eabl6496. doi.org/10.1126/sciadv.abl6496.
18. Signore, A.V., Morrison, P.R., Brauner, C.J. *et al.* 2023. Evolution of an extreme hemoglobin phenotype contributed to the sub-Arctic specialization of extinct Steller's sea cow. *eLife*, 12, e85414. doi.org/10.7554/eLife.85414.
19. Bauer, G.B., Reep, R.L. and Marshall, C.D. 2018. The tactile senses of marine mammals. *International Journal of Comparative Psychology*, 31. doi.org/10.46867/ijcp.2018.31.02.01.
20. Bachteler, D. and Dehnhardt, G. 1999. Active touch performance in the Antillean manatee: evidence for a functional difference of the facial tactile hairs. *Zoology*, 102, 61–69.

21. Reep, R.L., Stoll, M.L., Marshall, C.D. *et al.* 2001, Microanatomy of facial vibrissae in the Florida manatee: the basis for specialized sensory function and oripulation. *Brain, Behaviour and Evolution*, **58**, 1–14. doi.org/10.1159/000047257.

Chapter 7: The Elephant's Story

1. Gheerbrant, E. 2009. Paleocene emergence of elephant relatives and the rapid radiation of African ungulates. *Proceedings of the National Academy of Sciences of the United States of America*, **106**, 10717–10721. doi.org/10.1073/pnas.0900251106.
2. Rasmussen, L.E. and Munger, B.L. 1996. The sensorineural specialization of the trunk tip (finger) of the Asian Elephant, *Elephas maximus*. *The Anatomical Record*, **246**, 127–134.
3. Moore, A.M., Hartstone-Rose, A. and Gonzalez-Socoloske, D. 2021. Review of sensory modulation of sirenians and the other extant Paenungulata clade. *The Anatomical Record*, **305**, 715–735. doi.org/10.1002/ar.24741.
4. Deiringer, N., Schneeweiss, U., Kaufmann, L.V. *et al.* 2023. The functional anatomy of elephant trunk whiskers. *Communications Biology*, **6**, 591. doi.org/10.1038/s42003-023-04945-5.
5. Whitney, M.R., Angielczyk, K.D., Peecock, B.R. *et al.* 2021. The evolution of the synapsid tusk: insights from dicynodont therapsid tusk histology. *Proceedings of the Royal Society B*, **288**, 20211670. doi.org/10.1098/rspb.2021.1670.
6. Campbell-Station, S.C., Arnold, B.J., Gonçalves, D. *et al.* 2021. Ivory poaching and the rapid evolution of tusklessness in African elephants. *Science*, **374**, 483–487. doi.org/10.1126/science.abe.7389.
7. Cantalapiedra, J.L., Sanisidro, Ó., Zhang, H. *et al.* 2021. The rise and fall of proboscidean ecological diversity. *Nature Ecology and Evolution*, **5**, 1266–1272. doi:10.1038/s41559-021-01498-w.
8. van der Valk, T., Pečnerová, P., Díez-del-Molino, D. *et al.* 2021. Million-year-old DNA sheds light on the genomic history of mammoths. *Nature*, **591**, 265–269. doi.org/10.1038/s41586-021-03224-9.
9. Palkopoulou, E., Lipson, M., Mallick, S. *et al.* 2018. A comprehensive genomic history of extinct and living elephants. *Proceedings of the National Academy of Sciences of the United States of America*, **111**, E2566–E2574. doi.org/10.1073/pnas.1720554115.
10. Mayr, E. 1942. *Systematics and the Origin of Species from the Viewpoint of a Zoologist*. New York: Columbia University Press.
11. Fontsere, C., de Manuel, M., Marques-Bonet, T. *et al.* 2019. Admixture in mammals and how to understand its functional significance: on the abundance of gene flow in mammalian species, its impact on the genome, and roads into a functional understanding. *Bioessays*, **41**, e1900123. doi.org/10.1002/bies.201900123.
12. Lister, A.M. 2013. The role of behaviour in adaptive morphological evolution of African proboscideans. *Nature*, **500**, 331–334. doi.org/10.1038/nature12275.
13. Saarinen, J. and Lister, A.M. 2023. Fluctuating climate and dietary innovation drove ratcheted evolution of proboscidean dental traits. *Nature Ecology and Evolution*, **7**, 1490–1502. doi.org/10.1038/s41559-023-02151-4.
14. Liem, K.F. 1980. Adaptive significance of intra- and interspecific differences in the feeding repertoires of cichlid fishes. *American Zoologist*, **20**, 295–314.
15. Baker, J., Meade, A., Pagel, M. *et al.* 2014. Adaptive evolution towards larger size in mammals. *Proceedings of the National Academy of Sciences of the United States of America*, **112**, 5093–5098. doi.org/10.1073/pnas.1419823112
16. Philipps, P.K. and Heath, J.E. 1992. Heat exchange by the pinna of the African elephant (*Loxodonta africana*). *Comparative Biochemistry and Physiology Part I: Physiology*, **101**, 693–699.
17. Myhrvold, C.L., Stone, H.A. and Bou-Zeid, E. 2012. What is the use of elephant hair? *PLoS One*, **7**(10), e47018. doi.org/10.1371/journal.pone.0047018.
18. Peto, R. 1977. Epidemiology multistage models, and short-term mutagenicity tests. In:, H.H. Hialt, J.D. Watson and J.A. Winsten (eds), *The Origin of Human Cancer*. New York, Cold Harbour Conferences on Cell Proliferation, 4. Cold Spring Harbour Laboratory, pp. 1403–1428.

19. Abegglen, C.M., Caulin, A.F., Chan, A. *et al.* 2015. Potential mechanisms for cancer resistance in elephants and comparative cellular response to DNA damage in humans. *JAMA*, **314**, 1850–1860. doi.org/10.1001/jama.2015.13134.
20. Sulak, M., Fong, L., Mika, K. *et al.* 2016. *TP53* copy number expansion is associated with the evolution of increased body size and an enhanced DNA damage response in elephants. *eLife*, **5**, e11994. doi.org/10.7554/eLife.11994.
21. Mai, P.L., Best, A.F., Peters, J.A. *et al.* 2016. Risks of first and subsequent cancers among *TP53* mutations carriers in the National Cancer Institute Li-Fraumeni syndrome cohort. *Cancer*, **122**, 3673–3681. doi.org/10.1002/cncr.30248.
22. Preston, A.J., Rogers, A., Sharp, M. *et al.* 2023. Elephant *TP53-RETROGENE 9* induces transcription-independent apoptosis at the mitochondria. *Cell Death Discovery*, **16**, 66. doi.org/10.1038/s41420-023-01348-7.
23. Padariya, M., Jooste, M-L., Hupp, T. *et al.* 2022. The elephant evolved p53 isoforms that escape MDM2-mediated repression and cancer. *Molecular Biology and Evolution*, **39**, msac149. doi.org/10.1093/molbev/msac149.
24. Vazquez, J.M., Sulak, M., Chigurupati, S. *et al.* 2018. A zombie *LIF* gene in elephants is upregulated by TP53 to induce apoptosis in response to DNA damage. *Cell Reports*, **24**, 1765–1776. doi.org/10.1016/j.celrep.2018.07.042.
25. Vallrath, F. 2023. Uncoupling elephant *TP53* and cancer. *Trends in Ecology and Evolution*, **38**, 705–707. doi:10.1016/j.tree.2023.05.011.
26. Sharma, V., Lehmann, T., Stuckas, H. *et al.* 2018. Loss of *RXFP2* and *INSL2* genes in Afrotheria show that testicular descent is the ancestral condition in placental mammals. *PLoS Biology*, **16**(6), e2005293. doi.org/10.1371/journal.pbio.2005293.
27. Rogers, R.L. and Slatkin, M. 2017. Excess of genomic defects in a woolly mammoth on Wrangel island. *PLoS Genetics*, **13**(3), e1006601. doi.org/10.1371/journal.pgen.1006601.
28. Razeto-Barry, P., Díaz, J. and Vásquez, R.A. 2012. The nearly neutral and selection theories of molecular evolution under Fisher geometrical framework: substitution rates, population size, and complexity. *Genetics*, **191**, 523–534. doi.org/10.1534/genetics.112.138628.
29. Díez-del-Molino, D., Dehasque, M., Chacón-Duque, J.C. *et al.* 2023. Genomics of adaptive evolution in the woolly mammoth. *Current Biology*, **33**, 1753–1764, doi.org/10.1016/j.cub.2023.03.084.
30. Shapiro, B. 2020. *How to Clone a Mammoth: the Science of De-extinction*. Princeton: Princeton University Press, p.132.

Chapter 8: The Sloth's Story

1. Asher, R.J., Bennett, N. and Lehmann, T. 2009. The new framework for understanding placental mammal evolution. *BioEssays*, **31**, 853–864. doi:10.1002/bies.200900053.
2. Gibb, G.C., Condamine, F.L., Kuch, M. *et al.* 2016. Shotgun mitogenomics provides a reference phylogenetic framework and timescale for living xenarthrans. *Molecular Biology and Evolution*, **33**, 621–642. doi.org/10.1093/molbev/msv250.
3. Simpson, G. 1931. Metacheiromys and the Edentata. *Bulletin of the American Museum of Natural History*, **59**, 295–381.
4. McDonald, H.G. 2003. Xenarthran skeletal anatomy: primitive or derived? *Senckenbergiana Biologica*, **83**, 5–17.
5. Emerling, C.A. and Springer, M.S. 2015. Genomic evidence for rod monochromacy in sloths and armadillos suggests early subterranean history for Xenarthra. *Proceedings of the Royal Society B*, **282**, 20142192. doi.org/10.1098/rspb.2014.2192.
6. For rod monochromacy in other animals, see:
    Fish – Douglas, R.J., Partridge, J.H. and Hope, A.C. 1995. Visual and lenticular pigments in the eyes of deep-sea fishes. *Journal of Comparative Physiology*, **A177**, 111–122. doi.org/10.1007/BF00243403.
    Cetaceans – Meredith, R.W., Gatesy, J., Emerling, C.A. *et al.* 2013. Rod monochromacy and the coevolution of cetacean retinal opsins. *PLoS Genetics*, **9**(4), e1003432. doi.org/10.1371/journal.pgen.1003432.

Mammals – Emerling, C.A. and Springer, M.S. 2014. Eyes underground: regression of visual protein networks in subterranean mammals. *Molecular Phylogenetics and Evolution*, **78**, 260–270. doi.org/10.1016/j.ympev.2014.05.016.
7. Hautier, L., Rodrigues, H.G., Billet, G. *et al.* 2016. The hidden teeth of sloths: evolutionary vestiges and the development of a simplified dentition. *Scientific Reports*, **6**, 27763. doi.org/10.1038/srep27763.
8. Emerling, C.A., Gibb, G.C., Tilak, M-K. *et al.* 2023. Genomic data suggest parallel dental vestigialization within the xenarthran radiation. *Peer Community Journal*, **3**, e75. doi.org/10.24072/pcjournal.303.
9. Delsuc, F., Vizcaíno, S.F., and Douzery, E.J.P. 2004. Influence of Tertiary paleoenvironmental changes on the diversification of South American mammals: a relaxed molecular clock study within xenarthrans. *BMC Evolutionary Biology*, **4**, 11. doi.org.10.1186/1471-2148-4-11.
10. Casali, D.M., Martius-Sautos, E., Santos, A.L.Q. *et al.* 2017. Morphology of the tongue of Vermilingua (Xenarthra: Pilosa) and evolutionary considerations. *Journal of Morphology*, **278**, 1380–1399. doi.org/10.1002/jmor.20718.
11. Toledo, N., Bargo, M.S., Vizcaíno, S.F. *et al.* 2017. Evolution of body size in anteaters and sloths (Xenarthra, Pilosa): phylogeny, metabolism, diet and substrate preferences. *Earth and Environmental Science Transactions of the Royal Society of Edinburgh*, **106**, 289–301. doi:10.1017/S1755691016000177.
12. Presslee, S., Slater, G.J., Pujos, F. *et al.* 2019. Palaeoproteomics resolves sloth relationships. *Nature Ecology & Evolution*, **3**, 1121–1130. doi.org/10.1038/s41559-019-0909-z.
13. Delsuc, F., Kuch, M., Gibb, G.C. *et al.* 2019. Ancient mitogenomes reveal the evolutionary history and biogeography of sloths. *Current Biology*, **29**, 2031–2042, doi.org/10.1016/j.cub.2019.05.043.
14. Gaudin, T.J. 2004. Phylogenetic relationships among sloths (Mammalia, Xenarthra, Tardigrada): the craniodental evidence. *Zoological Journal of the Linnean Society*, **140**, 255–305. doi.org/10.1111/j.1096-3642.2003.00100.x.
15. Hautier, L., Weisbecker, V., Sánchez-Villagra, M.R. *et al.* 2010. Skeletal development in sloths and the evolution of mammalian vertebral patterning. *Proceedings of the National Academy of Sciences of the United States of America*, **107**, 18903–18908. doi.org/10.1073/pnas.1010335107.
16. Böhmer, C., Arnson, E., Arnold, P. *et al.* 2018. Homeotic transformations reflect departure from the 'rule of seven' cervical vertebrae in sloths: inferences on the *Hox* code and morphological modularity of the mammalian neck. *BMC Evolutionary Biology*, **18**, 84. doi.org/10.1186/s12862-018-1202-5.
17. Galis, F. 1999. Why do almost all mammals have seven cervical vertebrae? Developmental constraints, *Hox* genes, and cancer. *Journal of Experimental Zoology*, **285**, 19–26.
18. Merten, L.J.F., Manafzadeh, A.R., Herbst, E.C. *et al.* 2023. The functional significance of aberrant cervical counts in sloths: insights from automated exhaustive analysis of cervical range of movement. *Proceedings of the Royal Society B*, **290**, 20231592. doi.org/10.1098/rspb.2023.1592.
19. Buffon, G.L.L, Compte de (1749). *Histoire naturelle, générale et particulière*. À Paris de l'imprimerie royal.
20. Pauli, J.N., Peery, M.Z., Fountain, E.D. *et al.* 2016. Arboreal folivores limit their energetic output, all the way to slothfulness. *The American Naturalist*, **188**, 196–204. doi.org/10.1086/687032.
21. Cliffe, R.N., Avey-Arroyo, J.A., Arroyo, F.R. *et al.* 2024. Mitigating the squash effect: sloths breathe easily upside down. *Biology Letters*, **10**, 20140172. doi.org/10.1098/rsbl.2014.0172.
22. Tejada, J.V., Flynn, J.T., MacPhee, R. *et al.* 2021. Isotope data from amino acids indicate Darwin's ground sloth was not a herbivore. *Scientific Reports*, **11**, 18944. doi.org/10.1038/s41598-021-97996-9.
23. Delsuc, F., Gibb, G.C., Kuch, M. *et al.* 2016. The phylogenetic affinities of extinct glyptodonts. *Current Biology*, **26**, PR155–PR156. doi.org/10.1016/j.cub.2016.01.039.
24. Chen, I.H., Kiang, J.H., Correa, V. *et al.* 2011. Armadillo armor: mechanical testing and micro-structural evaluation. *Journal of the Mechanical Behaviour of Biomedical Materials*, **4**, 713–722. doi.org/10.1016/j.jmbbm.2010.12.013.

25. Superina, M. and Loughry, W. 2011. Life on the half-shell: consequences of a carapace in the evolution of armadillos (Xenarthra: Cingulata). *Journal of Mammalian Evolution*, **19**, 217–224. doi.org/10.1007/s10914-011-9166-x.
26. Superina, M. 2011. Husbandry of a pink fairy armadillo (*Chlamyphorus truncatus*): case study of a cryptic and little known species in captivity. *Zoo Biology*, **30**, 225–231. doi.org/10.1002/zoo.20334.
27. Delsuc, F., Superina, M., Tilak, M-K. *et al.* 2012. Molecular phylogenetics unveils the ancient origins of the enigmatic fairy armadillos. *Molecular Phylogenetics and Evolution*, **62**, 673–680. doi.org/10.1016/j.ympev.2011.11.008.
28. Prodöhl, P.A., Loughry, W.J., McDonough, C.M. *et al.* 1996. Molecular documentation of polyembryony and the micro-spatial dispersion of clonal sibships in the nine-banded armadillo *Dasypus novemcinctus*. *Proceedings of the Royal Society B*, **263**, 1643–1649. doi:10.1098/rspb.1996.0240.
29. Ballouz, S., Kawaguchi, R.K., Pena, M.T. *et al.* 2023. The transcriptional legacy of developmental stochasticity. *Nature Communications*, **14**, 7226. doi.org/10.1038/s41467-023-43024-5.

Chapter 9: The Solenodon's Story

1. Murphy, W.J., Eizirik, E., O'Brien, S.J. *et al.* 2001. Resolution of the early placental mammal radiation using Bayesian phylogenetics. *Science*, **294**, 2348–2351. doi.org/10.1126/science.1067179.
2. Fostowicz-Frelik, L., Ge, D. and Ruf, I. 2021. Advances in the evolution of Euarchontoglires. *Frontiers in Genetics*, **12**. doi.org/10.3389/fgene.2021.773789.
3. Douady, C.J., Chatellier, P.I., Madsen, O. *et al.* 2002. Molecular phylogenetic evidence confirming the Eulipotyphla concept and in support of hedgehogs as the sister group to shrews. *Molecular Phylogenetics and Evolution*, **25**, 200–209. doi.org/10.1016/S1055-7903(02)00232-4.
4. Foley, N.M., Mason, V.C., Harris, A.J. *et al.* 2023. A genomic timescale for placental mammal evolution. *Science*, **380**, eabl8189. doi.org/10.1126/science.abl8189.
5. Sato, J.J., Ohdachi, S.D., Echenique-Diaz, L.M. *et al.* 2016. Molecular phylogenetic analysis of nuclear genes suggests a Cenozoic over-water dispersal origin for the Cuban solenodon. *Scientific Reports*, **6**, 31173; doi.org/10.1038/srep31173.
6. Springer, M.S., Murphy, W.J. and Roca, A.L. 2018. Appropriate fossil calibrations and tree constraints uphold the Mesozoic divergence of solenodons from other extant mammals. *Molecular Phylogenetics and Evolution*, **121**, 158–165. doi.org/10.1016/j.ympev.2018.01.007.
7. Monroe, J.G., Srikant, T., Carbonell-Bejerano, P. *et al.* 2022. Mutation bias reflects natural selection in *Arabidopsis thaliana*. *Nature*, **602**, 101–105. doi.org/10.1038/s41586-021-04269-6.
8. Gatesy, J. and Springer, M.S. 2017. Phylogenomic red flags: homology errors and zombie lineages in the evolutionary diversification of placental mammals. *Proceedings of the National Academy of Sciences of the United Sates of America*, **114**, 45, E9431–E9432. doi.org/10.1073/pnas.1715318114.
9. Sato, J.J., Bradford, T.M., Armstrong, K.N. *et al.* 2019. Post K–Pg diversification of the mammalian order Eulipotyphla as suggested by phylogenomic analyses of ultra-conserved elements. *Molecular Phylogenetics and Evolution*, **141**. doi.org/10.1016/j.ympev.2019.106605.
10. Turvey, S. 2009. *Witness to Extinction: How We Failed to Save the Yangtze River Dolphin*. Oxford: Oxford University Press.
11. Roca, A.L., Bar-Gal, G.K., Eizirik, E. *et al.* 2004. Mesozoic origin for West Indian insectivores. *Nature*, **429**, 649–651. doi.org/10.1038/nature02597.
12. Casewell, N.R., Petras, D., Card, D.C. *et al.* 2019. Solenodon genome reveals convergent evolution of venom in eulipotyphan mammals. *Proceedings of the National Academy of Sciences of the United States of America*, **116**, 25746–25755. doi.org/10.1073/pnas.1906117116.
13. Depalma, R.A., Oleinik, A.A., Gurche, L.P. *et al.* 2021. Seasonal calibration of the end-cretaceous Chicxulub impact event. *Scientific Reports*, **11**, 23704. doi.org/10.1038/s41598-021-03232-9.
14. Collins, G.S., Patel, N., Davison, T.M. *et al.* 2020. A steeply-inclined trajectory for the Chicxulub impact. *Nature Communications*, **11**. doi.org/10.1038/s41467-020-15269-x.

15. Sluijs, A., Bowen, G.J., Brinkhuis, H. *et al*. 2007. The Palaeocene–Eocene Thermal Maximum super greenhouse – biotic and geochemical models and mechanisms of global change. *Geological Society Special Publications*, **1**, 323–249.
16. Kosintsev, P., Mitchell, K.J., Devièse, T. *et al*. 2019. Evolution and extinction of the giant rhinoceros *Elasmotherium sibiricum* sheds light on late Quaternary megafaunal extinctions. *Nature Ecology & Evolution*, **3**, 31–38. doi.org/10.1038/s41559-018-0722-0.
17. Morelle, R. 2007. The cave of bones: a story of solenodon survival. www.bbc.co.uk/news (accessed July 2025).
18. Gradstein, F.M., Ogg, J.G., Schmitz, M. *et al*. (eds). 2012. *The Geologic Time Scale 2012*. Amsterdam: Elsevier.
19. Alvarez, L.W., Alvarez, W., Asaro, F. *et al*. 1980. Extraterrestrial cause for the Cretaceous–Tertiary extinction. *Science*, **208**, 1095–1108. doi:10.1126/science.208.4448.1095.
20. Goderis, S., Sato, H., Ferrière, L. *et al*. 2021. Globally distributed iridium layer preserved within Chicxulub impact crater. *Science Advances*, **7**, eabe3647. doi.org/10.1126/sciadv.abe3647.
21. Nesvorný, D., Bottke, W.F. and Marchi, S. 2021. Dark primitive asteroids account for a large share of K/Pg-scale impacts on the Earth. *Icarus*, **368**, 114621. doi.org/10.1016/j.icarus.2021.114621.
22. Nicholson, V., Bray, V.J., Gulick, S.P.S. *et al*. 2022. The Nadir Crater offshore West Africa: a candidate Cretaceous–Paleogene impact structure. *Science Advances*, **8**, eabn3096. doi.org/10.1126/sciadv.abn3096.
23. Morgan, J.V., Bralower, T.J., Brugger, J. *et al*. 2022. The Chicxulub impact and its environmental consequences. *Nature Reviews Earth and Environment*, **3**, 338–354. doi.org/10.1038/s43017-022-00283-y.
24. Schulte, P., Alegert, L., Arenillas, I. *et al*. 2010. The Chicxulub asteroid impact and mass extinctions at the Cretaceous–Paleogene boundary. *Science*, **327**, 1214–1218. doi:10.1026/science.1177265.
25. Range, M.M., Arbic, B.K., Johnson, B.C. *et al*. 2022. The Chicxulub impact produced a powerful global tsunami. *AGU Advances*, **3**, e2021AV000627. doi.org/10.1029/2021AV000627.
26. Richards, M.A., Alvarez, W., Self, S. *et al*. 2015. Triggering of the largest Deccan eruptions by the Chicxulub impact. *GSA Bulletin*, **127**, 1507–1520. doi.org/10.1130/B31167.1.
27. Junium, C.K., Zerkle, A.L., Witts, J.D. *et al*. 2022. Massive perturbations to atmospheric sulfur in the aftermath of the Chicxulub impact. *Proceedings of the National Academy of Sciences of the United States of America*, **119**, 14, e2119194119. doi.org/10.1073/pnas.2119194119.
28. Wagnall, P.B. 2001. Large igneous provinces and mass extinctions. *Earth-Science Reviews*, **53**, 1–33. doi.org/10.1016/S0012-8252(00)00037-4.
29. Robertson, D.S., Lewis, W.M., Sheehan, P.M. *et al*. 2013. K–Pg extinction: revaluation of the heat-fire hypothesis. *Journal of Geophysical Research: Biogeosciences*, **118**, 329–336. doi.org/10.1002/jgrg.20018.
30. Frank, T.D., Fielding, C.R., Winguth, A.M.E. *et al*. 2021. Pace, magnitude, and nature of terrestrial climate change though the end-Permian extinction in southeastern Gondwana. *Geology*, **49**, 1089–1095. doi.org/10.1130/G48795.1.
31. Alvarez, S.A., Gibbs, S.J., Bown, P.R. *et al*. 2019. Diversity decoupled from ecosystem function and resilience during mass extinction recovery. *Nature*, **574**, 242–245. doi.org/10.1038/s41586-1590-8.
32. Geiser, F. and Turbill, C. 2009. Hibernation and daily torpor minimize mammalian extinctions. *Naturwissenschaften*, **96**, 1235–1240. doi.org/10.1007/s00114-009-0583-0.
33. Lovegrove, B.G., Lobban, K.D. and Levesque, D.L. 2014. Mammal survival at the Cretaceous–Palaeogene boundary: metabolic homeostasis in prolonged tropical hibernation in tenrecs. *Proceedings of the Royal Society B*, **281**, 20141304. doi.org/10.1098/rspb.2014.1304.
34. Lee, A.H. and Wernig, S. 2008. Sexual maturity in growing dinosaurs does not fit reptilian growth models. *Proceedings of the National Academy of Sciences of the United States of America*, **105**, 582–587. doi.org/10.1073/pnas.0708903105.
35. Hughes, J.J., Berv, J.S., Chester, S.G.B. *et al*. 2021. Ecological selectivity and the evolution of mammalian substrate preferences across the K–Pg boundary. *Ecology and Evolution*, **21**, 14540–14554. doi.org/10.1002/ece3.8114.

36. García-Girón, J., Chiarenza, A.A., Alahuhta, J. et al. 2022. Shifts in food webs and niche stability shaped survivorship and extinction at the end-Cretaceous. *Science Advances*, **8**, eadd5040. doi.org/10.1126/sciadv.add5040.
37. Liow, L.H., Fortelius, M., Lintulaakso, K. et al. 2009. Lower extinction risk on sleep-or-hide mammals. *The American Naturalist*, **173**, 264–272. doi.org/10.1086/595756.
38. Springer, M.S., Foley, N.M., Brady, P.L, et al. 2019. Evolutionary models for the diversification of placental mammals across the KPg boundary. *Frontiers in Genetics*, **10**, 1241. doi.org/10.3389/fgene.2019.01241.
39. Cunningham, J.A., Liu, A.G., Bengtson, S. et al. 2016. The origin of animals: can molecular clocks and the fossil record be reconciled? *BioEssays*, **39**, 1. doi.org/10.1002/bies.201600120.
40. Halliday, T.J.D., dos Reis, M., Tamuri, A.U. et al. 2019. Rapid morphological evolution in placental mammals post-dates the origin of the crown group. *Proceedings of the Royal Society B*, **286**, 20182418. doi.org/10.1098/rspb.2018.2418.
41. Meredith, R.W., Janěcka, J.E., Gatesy, J. et al. 2011. Impacts of the Cretaceous Terrestrial Revolution and KPg extinctions on mammal diversification. *Science*, **334**, 521–524. doi.org/10.1126/science.1211028.
42. Wu, S., Rheindt, F.E., Zhang, J. et al. 2024. Genomes, fossils, and the concurrent rise of modern birds and flowering plants in the Late Cretaceous. *Proceeding of the National Academy of Sciences of the United States of America*, **121**, e2319696121. doi.org/10.1073/pnas.2319696121
43. Carlisle, E., Janis, C.M., Pisani, D. et al. 2023. A timescale for placental mammal diversification based on Bayesian modelling of the fossil record. *Current Biology*, **33**, 3073–3082. doi.org/10.1016/j.cub.2023.06.016.
44. Bininda-Emonds, O.R., Cardillo, M., Jones, K.E. et al. 2007. The delayed rise of present-day mammals. *Nature*, **446**, 507–512. doi.org/10.1038/nature05634.

Chapter 10: The Camel's Story

1. Rybczynski, N., Gosse, J.C., Harington, C.R. et al. 2013. Mid-Pliocene warm-period deposits in the High Arctic yield insight into camel evolution. *Nature Communications*, **4**, 1550. doi.org/10.1038/ncomms2516.
2. Grossnickle, D.M. and Polly, P.D. 2013. Mammal disparity decreases during the Cretaceous angiosperm radiation. *Proceedings of the Royal Society B*, **280**. doi.org/10.1098/rspb.2013.2110.
3. Doronina, L., Hughes, G.M., Moreno-Santillan, D. et al. 2022. Contradictory phylogenetic signals in the Laurasiatheria anomaly zone. *Genes*, **15**, 5. doi.org/10.3390/genes13050766.
4. Pascual-Rico, R., Morales-Reyes, Z., Aguilera-Alcalá, N. et al. 2021. Usually hated, sometimes loved: a review of wild ungulates contributions to people. *Science of the Total Environment*, **801**, 149652. doi.org/10.1016/j.scitotenv.2021.149652; Halliday, T.J.D., Upchurch, P. and Goswani, A. Resolving the relationship of Paleocene placental mammals. *Biological Reviews*, **92**, 521–550. doi.org/10.1111/brv.12242; Zurano, J.P., Magalhães, F.M., Agato, A.E. et al. 2019. Cetartiodactyla: updating a time-calibrated molecular phylogeny. *Molecular Phylogenetics and Evolution*, **133**, 256–262. doi.org/10.1016/j.ympev.2018.12.015; Steiner, C.C. and Ryder, O.A. 2011. Molecular phylogeny and evolution of the Perissodactyla. *Zoological Journal of the Linnean Society*, **163**, 1289–1363. doi.org/10.1111/j.1096-3642.2011.00752.x.
5. Price, S.A., Bininda-Emonds, O.R.P. and Gittleman, J.L. 2005. A complete phylogeny of the whales, dolphins and even-toed hoofed mammals (Cetartiodactyla). *Biological Reviews of the Cambridge Philosophical Society*, **80**, 445–473. doi.org/10.1017/s1464793105006743.
6. Cui, P., Ji, R., Ding, F. et al. 2007. A complete mitochondrial genome sequence of the wild two-humped camel (*Camelus bactrianus ferus*): an evolutionary history of Camelidae. *BMC Genomics*, **8**, 241. doi.org/10.1186/1471-2164-8-241.
7. Mitchell, W.T., Rybczynski, N., Schröder-Adams, C. et al. 2016. Stratigraphic and Palaeoenvironmental reconstruction of a Mid-Pliocene fossil site in the High Arctic (Ellesmere Island, Nunavut): evidence of an ancient peatland with beaver activity. *Arctic*, **69**, 121–223. doi.org/10.14430/arctic4567.
8. Wu, H., Guang, X., Al-Fageeh, M.B. et al. 2014. Camelid genomes reveal evolution and adaption to desert environments. *Nature Communications*, **5**, 5188. doi.org/10.1038/ncomms6188.

9. Mohandesan, E., Fitak, R.R., Corander, J. *et al.* 2017. Mitogenome sequencing in the genus *Camelus* reveals evidence for purifying selection and long-term divergence between wild and domestic Bactrian camels. *Science Reports*, **7**, 9970. doi.org/10.1038/s41598-017-08995-8.
10. Balmus, G., Trifonov, V.A., Biltueva, L.S. *et al.* 2007. Cross-species chromosome painting among camel, cattle, pig and human: further insights into the putative Cetartiodactyla ancestral karyotype. *Chromosome Research*, **15**, 498–515. doi.org/10.1007/s10577-007-1154-x.
11. Almathen, F., Charruau, P., Mohandesan, E. *et al.* 2016. Ancient and modern DNA reveal dynamics of domestication and cross-continental dispersal of the dromedary. *Proceedings of the National Academy of Sciences of the United States of America*, **113**, 6707–6712. doi.org/10.1073/pnas.1519508113.
12. Ali, A., Baby, B. and Vijayan, R. 2019. From desert to medicine: a review of camel genomics and therapeutic products. *Frontiers in Genetics*, **10**, 17. doi.org/10.3389/fgene.2019.00017.
13. The Bactrian Camels Genome Sequencing and Analysis Consortium. 2012. Genome sequences of wild and domesticated Bactrian camels. *Nature Communications*, **3**, 1202. doi.org/10.1038/ncomms2192.
14. Carrasco, T.S., Scherer, C.S. Ribeiro, A.M. *et al.* 2022. Paleodiet of Lamini camelids (Mammalia: Artiodactyla) from the Pleistocene of southern Brazil: insights from stable isotope analysis ($\delta^{13}C$, $\delta^{18}O$). *Paleobiology*, **48**, 513–526. doi.org/10.1017/pab.2022.10.
15. Storz, J.F. 2007. Haemoglobin and physiological adaptation to hypoxia in high-altitude mammals. *Journal of Mammalogy*, **88**, 24–31. doi.org/10.1644/06-MAMM-S-199R1.1.
16. Marcovitz, A., Turakhia, Y., Chen, H.I. *et al.* 2019. A functional enrichment test for molecular evolution finds a clear protein-coding signal in echolocating bats and whales. *Proceedings of the National Academy of Sciences of the United States of America*, **116**, 21094–21103. doi.org/10.1073/pnas.1818532116.
17. Díaz-Lameiro, A.M., Kennedy, J.G.L., Craig, S. *et al.* 2022. Ancient DNA confirms cross-breeding of domestic South American camelids in two pre-conquest archaeological sites. *Journal of Archaeological Science*, **141**, 105593. doi.org/10.1016/j.jas.2022.105593.

Chapter 11: The Whale's Story

1. Gatesy, J., Hayashi, C., Cronin, M.A. *et al.* 1996. Evidence from milk casein genes that cetaceans are close relatives of hippopotamid artiodactyls. *Molecular Biology and Evolution*, **13**, 954–963. doi:10.1093/oxfordjournals.molbev.a025663.
2. Gingerich, P.D., Haq, M.U., Zalmout, I.S. *et al.* 2001. Origin of whales from early artiodactyls: hands and feet of Eocene Protocetidae from Pakistan. *Science*, **293**, 2239–1142. doi.org/10.1126/science.1063902.
3. Thewissen, J.G.M., Cooper, L.N., Clementz, M.T. *et al.* 2007. Whales originated from aquatic artiodactyls in the Eocene epoch of India. *Nature*, **450**, 1190–1194. doi.org/10.1038/nature06343; Thewissen, H. 2014. *The Walking Whales: From Land to Water in Eight Million Years*. Oakland: University of California Press.
4. Gingerich, P.D. 2015. Evolution of whales from land to sea. In: K.P Dial, N. Shubin and E.L. Brainerd (eds). *Great Transformations in Vertebrate Evolution*. Chicago: University of Chicago Press, pp. 239–256.
5. For publications covering the early fossil history of cetaceans see: Prothero, D.R. 2007. *Evolution: What the Fossils Say and Why it Matters*. New York: Columbia University Press; Uhen, M.D. 2010. The origin(s) of whales. *Annual Review of Earth Planetary Sciences*, **38**, 189–219. doi.org/10.1146/annurev-earth-o40809-152453; Thewissen 2014 (see note 3); Pyenson, N.D. 2017. The ecological rise of whales chronicled by the fossil record. *Current Biology*, **27**, R558–R564. doi.org/10.1016.j.cub.2017.05.001.
6. Lambert, O., Bianucci, G., Salas-Gismondi, R. *et al.* 2019. An amphibious whale from the Middle Eocene of Peru reveals early South Pacific dispersal of quadrupedal cetaceans. *Current Biology*, **29**, 1352–1359.e3. doi.org/10.1016/j.cub.2019.02.050.
7. Dines, J.P., Otárola-Castillo, E., Ralph, P. *et al.* 2014. Sexual selection targets cetacean pelvic bones. *Evolution*, **68**, 3296–3306. doi.org/10.1111/evo.12516.
8. Churchill, M., Geisler, J.H., Beatty, B.L. *et al.* 2018. Evolution of cranial telescoping in echolocating whales (Cetacea: Odontoceti). *Evolution*, **72**, 1092–1108. doi.org/10.1111/evo.13480.

9. Coombs, E.J., Clavel, J., Park, T. 2020. Wonky whales: the evolution of cranial asymmetry in cetaceans. *BMC Biology*, **18**, 86. doi.org/10.1186/s12915-020-00805-4.
10. Peredo, C.M., Pyenson, N.D., Marshall, C.D. *et al.* 2018. Tooth loss precedes the origin of baleen in whales. *Current Biology*, **28**, 3992–4000. doi.org/10.1016/j.cub.2018.10.047.
11. Geisler, J.H., Boessenecker, R.W., Brown, M. *et al*. 2017. The origin of filter feeding in whales. *Current Biology*, **27**, 2036–2042. doi.org/10.1016/j.cub.2017.06.003.
12. Fitzgerald, E.M.G. 2010. The morphology and systematics of *Mammalodon colliveri* (Cetacea: Mysticeti), a toothed mysticete from the Oligocene of Australia. *Zoological Journal of the Linnean Society*, **158**, 367–476. doi.org/10.1111/j.1096-3642.2009.00572.x
13. Ekdale, E.G. and Deméré, T.A. 2022. Neurovascular evidence for a co-occurrence of teeth and baleen in an Oligocene mysticete and the transition to filter-feeding in the baleen whales. *Zoological Journal of the Linnean Society*, **194**, 395–415. doi.org/10.1093/zoolinnean/zlab017; Deméré, T.A., McGowen, M.R., Berta, A. *et al.* 2008. Morphological and molecular evidence for a stepwise evolutionary transition from teeth to baleen in mysticete whales. *Systematic Biology*, **57**, 15–37. doi.org/10.1080/10635150701884632.
14. Slater, G.J., Goldbogen, J.A. and Pyenson, N.D. 2007. Independent evolution of baleen whale gigantism linked to Plio-Pleistocene ocean dynamics. *Proceedings of the Royal Society B*, **284**, 20170546. doi.org/10.1098/rspb.2017.0546.
15. Hamner, W.M., Hamner, P.P., Strand, S.W. *et al.* 1983. Behavior of Antarctic krill, *Euphansia superba*: chemoreception, feeding, schooling, and molting. *Science*, **220**, 433–435. doi.org/10.1126/science.220.4595.433.
16. Nery, M.F., Arroyo, J.I. and Opazo, J.C. 2014. Increased rate of hair keratin gene loss in the cetacean lineage. *BMC Genomics*, **15**, 869. doi.org/10.1186/1471-2164-15-869.
17. Themudo, G.E., Alves, L.Q., Machado, A.M. *et al.* 2020. *Losing* genes: the evolutionary remodelling of Cetacea skin. *Frontiers in Marine Science*, **7**. doi.org/10.3389/fmars.2020.592375.
18. Springer, M.S., Guerrero-Juarez, G.F., Huelsmann, M. *et al.* 2021. Genomic and anatomical comparisons of skin support independent adaptation to life in water by cetaceans and hippos. *Current Biology*, **31**, 2124–2139. doi.org/10.1016/j.cub.2021.02.057.
19. Deméré *et al.* 2008 (see note 13); Meredith, R.W., Gatesy, J., Murphy, W.J. *et al.* 2009. Molecular decay of the tooth gene enamelin (*ENAM*) mirrors the loss of enamel in the fossil record of placental mammals. *PlosS Genetics*, **5**(9), 1–12. doi.org/10.1371/journal.pgen.1000634.
20. Meredith, R. W., Gatesy, J., Cheng, J. et al. (2011). Pseudogenization of the tooth gene enamelysin (MMP20) in the common ancestor of extant baleen whales. Proceedings of the Royal Society B, **278**, 993–1002. doi.org/10.1098/rspb.2010.1280.
21. Sharma, V., Hecker, N., Roscito, J.G. *et al.* 2018. A genomics approach reveals insights into the importance of gene losses for mammalian adaptation. *Nature Communications*, **9**, 1215. doi.org/10.1038/s41467-018-03667-1.
22. Huelsmann, M., Hecker, N., Springer, M.S. *et al.* (2019). Genes lost during the transition from land to water in cetaceans highlighting genomic changes associated with aquatic adaptations. *Science Advances*, **5**, eaaw6671. doi.org/10.1126/sciadv.aaw6671.
23. Lopes-Marques, M., Ruivo, R., Alves, L.Q. *et al.* 2019. The singularity of Cetacea behavior parallels the complete inactivation of melatonin gene modules. *Genes*, **10**, 121. doi.org/10.3390/genes10020121.
24. Olsen, M.V. 1999. When less is more: gene loss as an engine of evolutionary change. *American Journal of Human Genetics*, **64**, 18–23. doi:10.1086/302219.
25. Kishida, T., Thewissen, J.G.M., Hayakawa, T. *et al.* 2015. Aquatic adaptation and evolution of smell and taste in whales. *Zoological Letters*, **1**, 9. doi.org/10.1186/s40851-041-0002-z.
26. Ryan, C., Martins, M.C.I., Healy, K. *et al.* 2024. Morphology of nares associated with stereo-olfaction in baleen whales. *Biology Letters*, **20**, 20230479. doi.org/10.1098/rsbl.2023.0479.
27. Zhu, K., Zhou, X. and Xu, S. *et al.* 2014. The loss of taste genes in cetaceans. *BMC Evolutionary Biology*, **14**, 218. doi.org/10.1186/s12862-014-0218-8.
28. Huelsmann *et al.* 2019 (see note 22).
29. Rule, J.P., Duncan, R.J., Marx, F.G. *et al.* 2023. Giant baleen whales emerged from a cold southern cradle. *Proceedings of the Royal Society B*, **290**, 20232177. doi.org/10.1098/rspb.2023.2177.

30. Silva, F.A., Souza, E.M.S., Ramos, E. *et al.* 2023. The molecular evolution of genes previously associated with large sizes reveals possible pathways to cetacean gigantism. *Scientific Reports*, 13, 67. doi.org/10.1038/s41598-022-24529-3.
31. Tollis, M., Robbins, J., Webb, A.E. et al. 2019. Return to the sea, get huge, beat cancer: an analysis of cetacean genomes including an assembly for the Humpback whale (Megaptera novaeangliae). Molecular Biology and Evolution, 36, 1746–1763. doi.org/10.1093/molbev/msz099.

Chapter 12: The Buffalo's Story

1. Geist, V. 1999. *Deer of the World: Their Evolution, Behaviour and Ecology.* Shrewsbury: Swan Hill Press.
2. Yang, C., Xiang, C., Qi, W. *et al.* 2013. Phylogenetic analyses and improved resolution of the family Bovidae based on complete mitochondrial genomes. *Biochemical Systematics and Ecology*, 48, 136–143. doi.org/10.1016/j.bse.2012.12.005.
3. Calamari, Z.T. 2021. Total evidence phylogenetic analysis supports new morphological synapomorphies for Bovidae (Mammalia, Artiodactyla). *American Museum Novitates*, 3970, 1–38. doi.org/10.1206/3970.1.
4. Clauss, M., Frey, R., Kiefer, B. *et al.* 2003. The maximum attainable body size of herbivorous mammals: morphophysiological constraints on foregut, and adaptations of hindgut fermenters. *Oecologia*, 136, 14–27. doi.org/10.1007/s00442-003-1254-z.
5. Barts, N., Bhatt, R.H., Toner, C. *et al.* 2024. Functional convergence in gastric lysozymes of foregut-fermenting rodents, ruminants, and primates is not attributed to convergent molecular evolution. *Comparative Biochemistry and Physiology Part B: Biochemistry and Molecular Biology*, 271, 110949. doi.org/10.1016/j.cbpb.2024.110949.
6. Valerio, S.O., Hummel, J., Codron, D. *et al.* 2022. The ruminant sorting mechanism protects teeth from abrasives. *Proceedings of the National Academy of Sciences of the United States of America*, 119, e2212447119. doi.org/10.1073/pnas.2212447119.
7. Pečnerová, P., Lord, E., Garcia-Erill, G. *et al.* 2024. Population genomics of the muskox' resilience in the near absence of genetic variation. *Molecular Ecology*, 33, e17205. doi.org/10.1111/mec.17205.
8. University of Copenhagen. 2023. The genomic secrets to how the muskox mastered living on the edge. phys.org/news/2023-11-genomic-secrets-muskox-mastered-edge.html (accessed July 2025).
9. Hempel, E., Tyler Faith, J., Preick, M. *et al.* 2024. Colonial-driven extinction of the blue antelope despite genomic adaptation to low population size. *Current Biology*, 34, 2020–2029. doi.org/10.1016/j.cub.2024.03.051.
10. Signore, A.V. and Storz, J.F. 2020. Biochemical pedomorphosis and genetic assimilation in the hypoxia adaptation of Tibetan antelope. *Science Advances*, 6, eabb5447. doi.org/10.1126/sciadv.abb5447.
11. Abraham, J.O., Upham, N.S., Damian-Serrano, A. *et al.* 2022. Evolutionary causes and consequences of ungulate migration. *Nature Ecology and Evolution*, 6, 998–1006. doi.org/10.1038/s41559-022-01749-4.
12. Green, G. 2024. Migration of 6m antelope in South Sudan dwarfs previous records for world's biggest, aerial study reveals. *The Guardian*, 25 June 2024. www.theguardian.com/environment/article/2024/jun/25/south-sudan-antelope-migration-worlds-largest-aoe (accessed July 2025).
13. Dutta, P., Talenti, A., Young, R. *et al.* 2020. Whole genome analysis of water buffalo and global cattle breeds highlights convergent signatures of domestication. *Nature Communications*, 11, 4739. doi.org/10.1038/s41467-020-18550-1.
14. Chan, E.K.F., Nagaraj, S.H. and Reverter, A. 2020. The evolution of tropical adaptation: comparing taurine and zebu cattle. *Animal Genetics*, 41, 467–477. doi.org/10.1111/j.1365-2052.2010.02053.x.
15. Alberto, F.J., Boyer, F., Orozco-terWengel, P. *et al.* 2018. Convergent genomic signatures of domestication in sheep and goats. *Nature Communications*, 8, 813. doi.org/10.1038/s41467-018-03206-y.

16. Robinson, T.J., Cernohorska, H., Kubickova, S. *et al.* 2021. Chromosomal evolution in *Raphicerus* antelope suggests divergent X chromosomes may drive speciation through females, rather than males, contrary to Haldane's rule. *Science Reports*, **11**, 3152. doi.org/10.1038/s41598-021-82859-0.

Chapter 13: The Giraffe's Story

1. Lamarck, J-B. 1809. *Philosophie Zoologique*. Paris, Dentu.
2. Gould, S.J. 1991. *Bully for Brontosaurus: Reflections in Natural History*. London: Hutchinson Radius, p.166.
3. Simmonds, R.E. and Scheepers, L. 1996. Winning by a neck: sexual selection in the evolution of giraffes. *The American Naturalist*, **148**, 771–786.
4. Mitchell, G., Sittert, J.V. and Skinner, J.D. 2009. Sexual selection is not the origin of long necks in giraffes. *Journal of Zoology*, **278**, 281–286. doi.org/10.1111/j.1469-7998.2009.00573.x.
5. Wang, S-Q., Ye, J., Meng, J. *et al.* 2022. Sexual selection promotes giraffoid head–neck evolution and ecological adaptation. *Science*, **376**, eabl8316. doi.org/10.1126/science.abl8316.
6. Williams, E.M. 2016. Giraffe stature and neck elongation: vigilance as an evolutionary mechanism. *Biology*, **5**, 35. doi.org/10.3390/biology5030035.
7. Danowitz, M., Vasilyev, A., Kortlandt, V. *et al.* 2015. Fossil evidence and stages of elongation of the *Giraffa camelopardalis* neck. *Royal Society Open Science*, **2**, 150393. doi.org/10.1098/rsos.150393.
8. Danowitz, M., Domalski, R. and Solounias, N. 2015. The cervical anatomy of *Samotherium*, and intermediate-necked giraffe. *Royal Society Open Science*, **2**, 150521. doi.org/10.1098/rsos.150521.
9. Badlangana, N.L., Adams, J.W. and Manger, P.R. (2009). The giraffe (*Giraffa camelopardalis*) cervical vertebral column: a heuristic example in understanding evolutionary processes? *Zoological Journal of the Linnean Society*, **155**, 736–757. doi.org/10.1111/j.1096-3642.2008.00458.x
10. Gunji, M. and Endo, H. 2016. Functional cervicothoracic boundary modified by anatomical shifts in the neck of giraffes. *Royal Society Open Science*, **3**, 150604. doi.org/10.1098/rsos.150604.
11. Galis, F. 1999. Why do almost all mammals have seven cervical vertebrae? Developmental constraints, *Hox* genes, and cancer. *Journal of Experimental Zoology*, **285**, 19–26.
12. Nikura, A., Nabae, H., Endo, G. *et al.* 2022. Giraffe neck robot: first step toward a powerful and flexible robot prototyping based on giraffe anatomy. *IEEE Robotics and Automation Letters*, **7**, 3539–3546. doi.org/10.1109/LRA.2022.3146611.
13. Aalkjær, C. and Wang, T. 2021. The remarkable cardiovascular system of giraffes. *Annual Review of Physiology*, **83**, 1–15. doi.org/10.1146/annurev-physiol-031620-094629.
14. Binder, P-M. and Taylor, D.L. 2015. How giraffes drink. *The Physics Teacher*, **53**, 518–520. doi.org/10.1119/1.4935758.
15. Liu, C., Gao, J., Cui, X. *et al.* (2021). A towering genome: experimentally validated adaptations to high pressure and extreme stature in the giraffe. *Science Advances*, **7**, eabe9459. doi.org/10.1126/sciadv.abe9459.
16. Eldredge, N. and Gould, S.J. 1972. Punctuated equilibria: an alternative to phyletic gradualism. In: T.J.M. Schopf (ed.), *Models in Paleobiology*. San Francisco: Freeman Cooper, pp. 82–115.
17. Agaba, M., Ishengoma, E., Miller, B.C. *et al.* 2016. Giraffe genome sequence reveals clues to its unique morphology and physiology. *Nature Communications*, **7**, 11519. doi.org/10.1038/ncomms11519.
18. Augliere, B. 2016. Genome reveals why giraffes have long necks. *Nature*. doi.org/10.1038/nature.2016.19931.
19. Fennessy, J., Bidon, T., Reuss, F. *et al.* 2016. Multi-locus analyses reveal four giraffe species instead of one. *Current Biology*, **26**, P2543–P2549. doi.org/10.1016/j.cub.2016.07.036; Coimbra, R.T.F., Winter, S., Kumar, V. *et al.* 2021. Whole-genome analysis of giraffe supports four distinct species. *Current Biology*, **12**, 2929–2938. doi.org/10.1016/j.cub.2021.04.033.
20. Mayr, E. 1942. *Systematics and the Origin of Species from the Viewpoint of a Zoologist*. New York: Columbia University Press.

21. Wang, X., he, Z., Shi, S. *et al.* 2019. Genes and speciation: is it time to abandon the biological species concept? *National Science Reviews*, **7**, 1387–1397. doi.org/10.1093/nsr/nwz220.
22. Baker, R.J. and Bradley, R.D. 2009. Speciation in mammals and the genetic species concept. *Journal of Mammalogy*, **87**, 643–662. doi.org/10.1644/06-MAMM-F-038R2-1.
23. Bertola, L.D., Quinn, L., Haughøj, K. *et al.* 2024. Giraffe lineages are shaped by major ancient admixture events. *Current Biology*, **34**, 1576–1586. doi.org/10.1016/j.cub.2024.02.051.
24. Stanton, D.W.G., Hart, J., Galbusera, P. *et al.* 2014. Distinct and diverse: range-wide phylogeography reveals ancient lineages and high genetic variation in the endangered okapi (*Okapi johnstoni*). *PLoS One*, **9**(7), e101081. doi.org/10.1371/journal.pone.0101081.

Chapter 14: The Horse's Story

1. Rose, K.D. and Archibald, J.D. (eds). 2005. *The Rise of Placental Mammals: Origin and Relationships of the Major Clades*. Baltimore: Johns Hopkins University Press.
2. Rose, K.D., Holbrook, L.T., Rana, R.S. *et al.* 2024. Early Eocene fossils suggest that the mammalian order Perissodactyla originated in India. *Nature Communications*, **5**, 5570. doi.org/10.1038/ncomms6570.
3. Bai, B., Wang, Y-Q. and Meng, J. 2018. The divergence and dispersal of early perissodactyls as evidenced by early Eocene equids from Asia. *Communications Biology*, **1**, 115. doi.org/10.1038/s42003-018-0116-5.
4. Forrest, S. 2016. *The Age of the Horse: An Equine Journey through Human History*. New York: Atlantic Books.
5. Gould, S.J. 1991. *Bully for Brontosaurus: Reflections in Natural History*. London: Hutchinson Radius, pp. 168–181.
6. 'Iconography of an expectation' is the title of chapter 1 of Gould's seminal work *Wonderful Life* (1990), in which he debunks that great warhorse of tradition, the evolutionary ladder of horses.
7. Franzen, J.L. 2010. *The Rise of Horses: 55 million Years of Evolution*. Baltimore: Johns Hopkins University Press.
8. Janis, C.M. and Bernor, R.L. 2019. The evolution of equid monodactyly: a review including a new hypothesis. *Frontiers in Ecology and Evolution*, **7**. doi.org/10.3389/fevo.2019.00119.
9. Librado, S. and Orlando, L. 2021. Genomics and the evolutionary history of equids. *Annual Review of Animal Biosciences*, **9**, 81–101. doi.org/10.1146/annurev-animal-061220-023118.
10. Buck, C.E. and Bard, E. 2007. A calendar chronology for Pleistocene mammoth and horse extinction in North America based on Bayesian radiocarbon calibration. *Quaternary Science Reviews*, **26**, 2031–2035. doi.org/10.1016/quascirev.2007.06.013.
11. Taylor, W.T.T., Librado, P., Icu, M.H.T. *et al.* 2023. Early dispersal of domestic horses into the Great Plains and northern Rockies. *Science*, **379**, 1316–1323. doi.org/10.1126/science.adc9691.
12. Vershinina. A.O., Heintzman, P.D., Froese, D.G. *et al.* 2021. Ancient horse genomes reveal the timing and extent of dispersals across the Bering Land Bridge. *Molecular Ecology*, **30**, 6144–6161. doi.org/10.1111/mec.15977.
13. Gaunitz, C., Fages, A., Haughøj, K. *et al.* 2018. Ancient genomes revisit the ancestry of domestic and Przewalski's horses. *Science*, **360**, 111–114. doi.org/10.1126/science.aao3297.
14. Taylor, W.T.T. and Barrón-Ortiz, C.I. 2021. Rethinking the evidence for early horse domestication at Botai. *Scientific Reports*, **11**, 7440. doi.org/10.1038/s41598-021-86832-9.
15. Orlando, L. 2020. Ancient genomes reveal unexpected horse domestication and management dynamics. *BioEssays*, **42**, 1900164. doi.org/10.1002/bies.201900164.
16. Cardinali, I., Giontella, A., Tommasi, T. *et al.* 2022. Unlocking horse Y chromosome diversity. *Genes*, **13**, 2272. doi.org/10.3390/genes13122272.
17. Wang, Y., Hua, X., Shi, X. *et al.* 2022. Origin, evolution, and research development of donkeys. *Genes*, **13**, 1945. doi.org/10.3390/genes13111945.

18. Jónsson, H., Schbert, M., Seguin-Orlando. *et al.* 2014. Speciation with gene flow in equids despite extensive chromosomal plasticity. *Proceedings of the National Academy of Sciences of the United States of America*, **111**, 18655–18660. doi.org/10.1073/pnas.1412627111.
19. How, M.J. and Zanker, J.M. 2014. Motion camouflage induced by zebra stripes. *Zoology*, **117**, 163–170. doi.org/10.1016/j.zool.2013.10.004.
20. Caro, T. 2016. *Zebra Stripes*. Chicago: University of Chicago Press.
21. Turing, A. 1952. The chemical basis of morphogenesis. *Philosophical Transactions of the Royal Society B*, **237**, 37–72. doi.org/10.1098/rstb.1952.0012.
22. Meinhardt, H. 1982. *Models of Biological Pattern Formation*. London: Academic Press; Murray, J.D. 1990. *Mathematical Biology*, Berlin: Springer Verlag.
23. Alessio, B.M. and Gupta, A. 2023. Diffusiophoresis-enhanced Turing patterns. *Science Advances*, **9**, eadj2457. doi.org/10.1126/sciadv.adj2457.
24. Glover, J.D., Sudderick, Z.R., Shih, B.B-J. *et al.* 2023. The developmental basis of fingerprint pattern formation and variation. *Cell*, **186**, 940–956. doi.org/10.1016/j.cell.2023.01.015.
25. Wallace, A.R. 1889. *Darwinism: An Exposition of the Theory of Natural Selection and Some of its Applications*. London: Macmillan.
26. Eberle, J.J. 2005. A new 'tapir' from Ellesmere Island, Arctic Canada – implications for northern high latitude palaeobiogeography and tapir palaeobiology. *Palaeogeography, Palaeoclimatology, Palaeoecology*, **227**, 311–322. doi.org/10.1016/j.palaeo.2005.06.008.
27. de Thoisy, B., Gonçalves da Silva, A., Ruiz-García, M. *et al.* 2010. Population history, phylogeography, and conservation genetics of the last Neotropical mega-herbivore, the lowland tapir (*Tapirus terrestris*). *BMC Evolutionary Biology*, **10**, 278. doi.org/10.1186/1471-2148-10-278.
28. DeSantis, L.R.G. and MacFadden, B. 2007. Identifying forested environments in deep time using fossil tapirs: evidence from evolutionary morphology and stable isotopes. *Courier Forschungsinstit Senckenberg*, **258**, 147–157.
29. Cerdeño, E. 1998. Diversity and evolutionary trends of the family Rhinocerotidae (Perissodactyla). *Palaeogeography, Palaeoclimatology, Palaeoecology*, **141**, 13–34.
30. Bai, B., Meng, J., Zhang, C. *et al.* 2020. The origin of Rhinocerotoidea and phylogeny of Ceratomorpha (Mammalia, Perissodactyla). *Communications Biology*, **3**, 509. doi.org/10.1038/s42003-020-01205-8.
31. Liu, S., Westbury, M.V., Dussex, N. *et al.* 2021. Ancient and modern genomes unravel the evolutionary history of the rhinoceros family. *Cell*, **184**, 4874–4885. doi.org/10.1016/j.cell.2021.07.032.
32. Brownstein, C.D., MacGuigan, D.J., Kim, D. *et al.* 2024. The genomic signatures of evolutionary stasis. *Evolution*, **78**, 821–834. doi.org/10.1093/evolut/qpae028.

Chapter 15: The Bear's Story

1. Hassanin, A., Veran, G., Ropiquet, A. *et al.* 2021. Evolving history of Carnivora (Mammalia, Laurasiatheria) inferred from mitochondrial genomes. *PLoS One*, **16**(2), e0240770. doi.org/10.1371/journal.pone.0240770.
2. Krause, J., Unger, T., Noçon, A. *et al.* 2008. Mitochondrial genomes reveal an explosive radiation of extinct and extant bears near the Miocene-Pliocene boundary. *BMC Evolutionary Biology*, **8**, 220. doi.org/10.1186/1471-2148-8-220.
3. Jiangzuo, Q. and Spasser, N. 2022. A late Turolian giant panda from Bulgaria and the early evolution and dispersal of the panda lineage. *Journal of Vertebrate Paleontology*, **42**, e2054718. doi.org/10.1080/02724634.2021.2054718.
4. Wang, X., Su, D.F., Jablonski, N.G. *et al.* 2022. Earliest giant panda false thumb suggests conflicting demands for locomotion and feeding. *Scientific Reports*, **12**, 10538. doi.org/10.1038/s41598-022-13402-y.
5. Hu, Y., Wu, Q., Ma, S. *et al.* 2017. Comparative genomics reveals convergent evolution between bamboo-eating giant and red pandas. *Proceedings of the National Academy of Sciences of the United States of America*, **114**, 1081–1086. doi.org/10.1073/pnas.1613870114.
6. Wang, A., Zhan, M. and Pei, E. 2021. Succession of intestinal microbial structure of giant pandas (*Ailuropoda melanoleuca*) during different developmental stages and its correlation with cellulose activity. *Animals*, **11**, 2358. doi.org/10.3390/ani11082358.

7. Salis, A.T., Gower, G., Schubert, B.W. et al. 2021. Ancient genomes reveal hybridization between extinct short-faced bears and the extant spectacled bear (*Tremarctos ornatus*). *BioRxiv.* doi.org/10.1101/2021.02.05.429853.
8. Zou, T., Kuang, W., Yin, T. et al. 2022. Uncovering the enigmatic evolution of bears in greater depth: the hybrid origins of the Asiatic bear. *Proceedings of the National Academy of Sciences of the United States of America*, **119**, e2120307119. doi.org/10.1073/pnas.2120307119.
9. Fisher, R. A. 1930. *The Genetical Theory of Natural Selection.* Oxford: Clarendon Press, p. 130.
10. Mavárez, J. and Linares, M. 2008. Homoploid speciation in animals. *Molecular Ecology*, **17**, 4181–4185. doi.org/10.1111/j.1365-294x.2008.03898.
11. Homoploid hybrid speciation (HHS):
    Bats – Larsen, P.A., Marchán-Rivadeneira, M.R. and Baker, R.J. 2010. Natural hybridization generates mammalian lineage with species characteristics. *Proceedings of the National Academy of Sciences of the United States of America*, **107**, 11447–11452. doi.org/10.1073/pnas.1000133107; Andriollo, T., Ashrafi, S., Arlettaz, R. et al. 2019. Porous barriers? Assessment of gene flow within and among sympatric long-eared bat species. *Ecology and Evolution*, **8**, 12841–12854. doi.org/10.1002/ece3.4714.
    Fur seals – Lopes, F., Oliveira, L.R., Beux, Y. et al. 2023. Genomic evidence for homoploid hybrid speciation in a marine mammal apex predator. *Science Advances*, **9**, eadf6601. doi.org/10.1126/sciadv.adf6601.
12. Schumer, M., Cui, R., Rosenthal, G.G. et al. 2015. Reproductive isolation of hybrid populations driven by genetic incompatibilities. *PloS Genetics*, 11(3), e1005041. doi.org/10.1371/journal.pgen.1005041.
13. Barlow, A., Cahill, J.A., Hartmann, S. et al. 2018. Partial genomic survival of cave bears in living brown bears. *Nature Ecology and Evolution*, **10**, 1563–1570. doi.org/10.1038/s41559-018-0654-8.
14. Lindqvist, C., Schuster, S.C., Sun, Y. et al. 2010. Complete mitochondrial genome of a Pleistocene jawbone unveils the origin of polar bears. *Proceedings of the National Academy of Sciences of the United States of America*, **107**, 5053–5057. doi.org/10.1073/pnas.0914266107.
15. Castruita, J.A.S., Westbury, M.V. and Lorenzen, E.D. 2020. Analyses of key genes involved in Arctic adaptation in polar bears suggest selection on both standing variation and de novo mutations played an important role. *BMC Genomics*, **21**, 543. doi.org/10.1186/s12864-020-06940-0.
16. Sun, Y., Lorenzen, E.D. and Westbury, M.V. 2023. Late Pleistocene polar bear genomes reveal the timing of allele fixation in key genes associated with Arctic adaptation. *BioRxiv.* doi.org/10.1101/2023.11.30.569368.
17. Castruita *et al.* 2020 (see note 15).
18. Welch, A.J., Bedoya-Reina, O.C., Carretero-Paulet, L. et al. 2014. Polar bears genome-wide signatures of bioenergetic adaptation to life in the Arctic environment. *Genome Biology and Evolution*, **6**, 433–450. doi.org/10.1093/gbe/evu025.
19. Pongracz, J.D., Paetkau, D., Branigan, M. et al. 2017. Recent hybridization between a polar bear and grizzly bears in the Canadian Arctic. *Arctic*, **70**, 151–160. doi.org/10.14430/arctic4643.
20. Cahill, J.A., Stirling, I., Kistler, L. et al. 2015. Genomic evidence of geographically widespread effect of gene flow from polar bears into brown bears. *Molecular Ecology*, **24**, 1205–1217. doi.org/10.1111/mec.13038.
21. Wang, M-S., Murray, G.G.R., Mann, D. et al. 2022. A polar bear paleogenome reveals extensive ancient gene flow from polar bears into brown bears. *Nature Ecology and Evolution*, **6**, 936–944. doi.org/10.1038/s41559-022-01753-8.
22. Cahill, J.A., Heintzman, P.D., Harris, K. et al. 2018. Genomic evidence of widespread admixtures from polar bears into brown bears during the Last Ice Age. *Molecular Biology and Evolution*, **35**, 1120–1129. doi.org/10.1093/molbev/msy018.
23. Kumar, V., Lammers, F., Bidon, T. et al. 2017. The evolutionary history of bears is characterized by gene flow across species. *Scientific Reports*, **7**, 46487. doi.org/10.1038/srep46487.
24. Wu, C-I. 2001. The genic view of the process of speciation. *Journal of Evolutionary Biology*, **14**, 851–865. doi.org/10.1046/j.1420-9101.2001.00335.x.

Chapter 16: The Cat's Story

1. Johnson, W.E., Eizirik, E., Pecon-Slattery, J. *et al.* 2006. The Late Miocene radiation of modern Felidae: a genetic assessment. *Science*, **311**, 73–77. doi.org/10.1126/science.1122277.
2. Kitchener, A.C., Breitenmoser-Würsten, C., Eizirik, E. *et al.* 2017. A revised taxonomy of the Felidae. The final report of the Cat Classification Task Force of the IUCN/SSC Cat Specialist Group. *Cat News* Special Issue 11, 80 pp.
3. Li, X., Li, W., Wang, H. *et al.* 2005. Pseudogenization of a sweet-receptor gene accounts for cats indifference towards sugar. *PLoS Genetics*, **1**(1), e3. doi.org/10.1371/journal.pgen.0010003.
4. Jiang, P., Josue, J., Li, X. *et al.* 2012. Major taste loss in carnivorous mammals. *Proceedings of the National Academy of Sciences of the United States of America*, **109**, 4956–4961. doi.org/10.1073/pnas.1118360109.
5. Bredemeyer, K.R., Hillier, L., Harris, A.J. *et al.* 2023. Single-haplotype comparative genomics provides insights into lineage-specific structural variation during cat evolution. *Nature Genomics*, **55**, 1953–1963. doi.org/10.1038/s41588-023-01548-y.
6. Figueríó, H.V., Li, G., Trindade, F.J. *et al.* 2017. Genome-wide signature of complex introgression and adaptive evolution in the big cats. *Science Advances*, **3**, e1700299. doi.org/10.1126/sciadv.1700299.
7. Li, G., Davis, B.W., Eizirik, E. *et al.* 2016. Phylogenomic evidence for ancient hybridization in the genomes of living cats (Felidae). *Genome Research*, **26**, 1–11. doi.org/10.1101/gr.186668.114.
8. Dures, S.G., Carbone, C., Loveridge, A.J. *et al.* 2019. A century of decline: loss of genetic diversity in a southern African lion conservation stronghold. *Diversity and Distributions*, **25**, 870–879. doi.org/10.1111/ddi.12905; de Manuel. M., Barnett, R., Sandoval-Velasco, M. *et al.* 2019. The evolutionary history of extinct and living lions. *Proceedings of the National Academy of Sciences of the United States of America*, **117**, 10927–10934. doi.org/10.1073/pnas.1919423117.
9. Srigyan, M., Schubert, B.W., Bushell, M. *et al.* 2023. Mitogenomic analysis of a late Pleistocene jaguar from North America. *Journal of Heredity*, **115**, 424–431. doi.org/10.1093/jhered/esad082.
10. Hayward, M.W., Kamler, J.F., Montgomery, R.A. *et al.* 2016. Prey preferences of the jaguar *Panthera onca* reflect the post-Pleistocene demise of large prey. *Frontiers in Ecology and Evolution*, **3**, 148. doi.org/10.3389/fevo.2015.00148.
11. Figueríó *et al.* 2017 (see note 6).
12. Bredemeyer *et al.* 2023 (see note 5).
13. Armstrong, E.E., Khan, A., Taylor, R.W. *et al.* 2021. Recent evolutionary history of tigers highlights contrasting roles of genetic drift and selection. *Molecular Biology and Evolution*, **38**, 2366–2379. doi.org/10.1093/molbev/msab032.
14. Smith, H.F., Townsend, K.E.B., Adrian, B. *et al.* 2021. Functional adaptations in the forelimb of the snow leopard (*Panthera uncia*). *Integrative and Comparative Biology*, **61**, 1852–1866. doi.org/10.1093/icb/icab018.
15. Cho., Y.S., Hu, L., Hou, H. *et al.* 2013. The tiger and comparative analysis with lion and snow leopard genomes. *Nature Communications*, **4**, 2433. doi.org/10.1038/ncomms3433.
16. Wurster-Hill, D.H. and Centerwall, W.R. 1982. The interrelationships of chromosome banding patterns in canids, mustelids, hyena, and felids. *Cytogenetics and Cell Genetics*, **34**, 178–192. doi.org/10.1159/000131806.
17. Lescroart, J., Bouilla-Sánchez, A., Napolitana, C. *et al.* 2023. Extensive phylogenomic discordance and the complex evolutionary history of the neotropical cat genus *Leopardus*. *Molecular Biology and Evolution*, **40**, msad255. doi.org/10.1093/molbev/msad255.
18. O'Brien, S., Roelke, M.E., Marker, L. *et al.* 1985. Genetic basis for species vulnerability in the cheetah. *Science*, **227**, 1428–1434. doi.org/10.1126/science.2983425.
19. Dobrynin, P., Liu, S., Tamazian, G. *et al.* 2015. Genomic legacy of the African cheetah, *Acinonyx jubatus*. *Genome Biology*, **16**, 277. doi.org/10.1186/s13059-051-0837-4.

20. Driscoll, C.A., Menotti-Raymond, M., Roca, A.L. et al. 2017. The Near Eastern origin of cat domestication. *Science*, 317, 519–523. doi.org/10.1126/science.1139518.
21. Vigne, J-D., Guilaine, J., Debue, K. et al. 2004. Early taming of the cat in Cyprus. *Science*, 304, 259. doi.org/10.1126/science.1095335.
22. Montague, M.J., Li, G., Gandolfi, B. et al. 2014. Comparative analysis of the domestic cat genome reveals genetic signatures underlying feline biology and domestication. *Proceedings of the National Academy of Sciences of the United Sates of America*, 111, 17230–17235. doi.org/10.1073/pnas.1410083111.

Chapter 17: The Bat's Story

1. Simmons, N.B. and Cirranello, A.L. 2024. Bat species of the world: a taxonomic and geographical database. Version 1.8.1. batnames.org (accessed July 2025).
2. Rietbergen, T.B., van den Hoek Ostende, L.W., Aase, A. et al. 2023. The oldest known bat skeletons and their implications for Eocene chiropteran diversification. *PLoS One* 18(4), e0283505. doi.org/10.1371/journal.pone.0283505.
3. Hand, S.J., Maugoust, J., Beck, R.M.D. et al. 2023. A 50-million-year-old, three-dimensionally preserved bat skull supports an early origin for modern bat echolocation. *Current Biology*, 33, 4624–4640. doi.org/10.1016/j.cub.2023.09.043.
4. Colleary, C., Dolocan, A., Gardner, J. et al. 2015. Chemical, experimental, and morphological evidence for diagenetically altered melanin in exceptionally preserved fossils. *Proceedings of the National Academy of Sciences of the United States of America*, 112, 12592–12597. doi.org/10.1073/pnas.1509831112.
5. Lei, M. and Dong, D. 2016. Phylogenomic analyses of bat subordinal relationships based on transcriptome data. *Scientific Reports*, 6, 27726. doi.org/10.1038/srep27726.
6. Jones, M.F., Li, Q., Ni, X. et al. 2021. The earliest Asian bats (Mammalia: Chiroptera) address major gaps in bat evolution. *Biology Letters*, 6, 20210185. doi.org/10.1098/rsbl.2021.0185.
7. Weatherbee, S.D., Behringer, R.R., Rasweiller, J.J. et al. 2006. Interdigital webbing retention in bat wings illustrates genetic changes underlying amniote limb diversification. *Proceedings of the National Academy of Sciences of the United States of America*, 103, 15103–15107. doi.org/10.1073/pnas.0604934103.
8. Anthwal, N., Urban, D.J., Sadier, A. et al. 2023. Insights into the formation and diversification of a novel chiropteran wing membrane from embryonic development. *BMC Biology*, 21, 101. doi.org/10.1186/s12915-023-01598-y.
9. Cretekos, C.J., Wang, Y., Green, E.D. et al. 2008. Regulatory divergence modifies limb length between mammals. *Genetics and Development*, 22, 141–151. doi.org/10.1101/gad.1620408.
10. Anderson, S.C. and Ruxton, G.D. 2020. The evolution of flight in bats: a novel hypothesis. *Mammal Review*, 50, 426–439. doi.org/10.1111/mam.12211.
11. Zhang, G., Cowled, C., Shi, Z. et al. 2022. Comparative analysis of bats genomes provide insight into the evolution of flight and immunity. *Science*, 339, 456–460. doi.org/10.1126/science.1230835.
12. Liu, Z., Chen, P., Li, Y-Y. et al. Cochlea hair cells of echolocating bats are immune to intense noise. *Journal of Genetics and Genomics*, 48, 984–993. doi.org/10.1016/j.jgg.2021.06.007.
13. Lei and Dong 2016 (see note 5).
14. Liu, Z., Chen, P., Xu, D-M. et al. 2022. Molecular convergence and transgenic evidence suggesting single origin of laryngeal echolocation in bats. *iScience*, 25, 104114. doi.org/10.1016/j.isci.2022.104114.
15. Gracheva, E.O., Cordero-Morales, J.F., Gonzáles-Caracía, J.A. et al. 2011. Ganglion-specific splicing of TRPV1 underlies infrared sensation in vampire bats. *Nature*, 476, 88–91. doi.org/10.1038/nature10245.
16. Zepeda-Mendoza, M.L., Xiong, Z., Escalera-Zamudio, M. et al. 2018. Hologenomic adaptations underlying the evolution of sanguivory in the common vampire bat. *Nature Ecology and Evolution*, 4, 659–668. doi.org/10.1038/s41559-018-0476-8.

Chapter 18: The Rat's Story

1. Guiry, E., Kennedy, R., Orton, D. et al. 2024. The ratting of North America: a 350-year retrospective on *Rattus* species composition and competition. *Science Advances*, **10**, eadm6755. doi.org/10.1126/sciadv.adm6755.
2. Harpak, A., Garud, N., Rosenberg, N.A. et al. 2021. Genetic adaptation in New York City rats. *Genome Biology and Evolution*, **13**, evaa247. doi.org/10.1093/gbe/evaa247.
3. Krijger, I.M., Strating, M., von Gent-Pelzer, M. et al. 2023. Large-scale identification of rodenticide resistance in *Rattus norvegicus* and *Mus musculus* in the Netherlands based on *Vkorv1* codon 139 mutations. *Pest Management Science*, **79**, 989–995. doi.org/10.1002/ps.7261.
4. Murphy, W.J., Eizirik, E., O'Brien, J. et al. 2001. Resolution of the early placental mammal radiation using Bayesian phylogenetics. *Science*, **294**, 2348–2351. doi.org/10.1126/science.1067179.
5. Doronina, L., Reising, O., Clawson, H. et al. 2022. Euarchontoglires challenged by incomplete lineage sorting. *Genes*, **13**, 774. doi.org/10.3390/genes13050774.
6. Wang, X.Y., Liang, D., Jin, W. et al. 2020. Out of Tibet: genomic perspectives on the evolutionary history of extant pikas. *Molecular Biology and Evolution*, **37**, 1577–1592. doi.org/10.1093/molbev/msaa026.
7. Ferreira, M.S., Jones, M.R., Callahan, C.M. et al. 2021. The legacy of recurrent introgression during the radiation of hares. *Systematic Biology*, **70**, 593–607. doi.org/10.1093/sysbio/syaa088.
8. Irving-Pease, E.K., Frantz, L.A.F., Sykes, N. et al. 2018. Rabbits and the specious origins of domestication. *Trends in Ecology and Evolution*, **33**, 149–152. doi.org/10.1016/j.tree.2017.12.009.
9. Amori, G., Gippoliti, S. and Helgen, K.M. 2008. Diversity, distribution, and conservation of endemic island rodents. *Quaternary International*, **182**, 6–15. doi.org/10.1016/j.quaint.2007.05.014.
10. Terray, L., Denys, C., Goodman, S.M. et al. 2022. Skull morphological evolution in Malagasy endemic Neomyinae rodents. *PLoS One*, **17(2)**, e0263045. doi.org/10.1371/journal.pone.0263045.
11. Swanson, M.T., Oliveros, C.H. and Esselstyn, J.A. 2019. A phylogenomic rodent tree reveals the repeated evolution of masseter architectures. *Proceedings of the Royal Society B*, **286**, 20190672. doi.org/10.1098/rspb.2019.0672.
12. Esselstyn, J.A., Achmadi, A.S. and Rowe, K.C. 2012. Evolutionary novelty in a rat with no molars. *Biology Letters*, **8**, 990–993. doi.org/10.1098/rsbl.2012.0574.
13. Thybert, D., Roller, M., Navarro, F.C.P. et al. 2018. Repeat associated mechanisms of genome evolution and function revealed by the *Mus caroli* and *Mus pahari* genomes. *Genome Research*, **28**, 448–459. doi.org/10.1101/gr.234096.117.
14. Simkin, J., Aloysius, A., Adam, M. et al. 2024. Tissue-resident macrophages specifically express *Lactotransferrin* and *Vegfc* during ear pinna regeneration in spiny mice. *Developmental Cell*, **59**, 496–516. doi.org/10.1016/j.devcel.2023.12.017.
15. Maden, M., Polvadore, T., Polanco, A. et al. 2023. Osteoderms in a mammal the spiny mouse *Acomys* and the independent evolution of dermal armor. *iScience*, **26**, 106779. doi.org/10.1016/j.isci.2023.106779.
16. Kingdon, J., Agwanda, B., Kinnaird, M. et al. 2012. A poisonous surprise under the coat of the African crested rat. *Proceedings of the Royal Society B*, **279**, 675–680. doi.org/10.1098/rspb.2011.1169.
17. Tigano, A., Colella, J.P. and MacManes, M.D. 2020. Comparative and population genomics approaches reveal the basis of adaptation to deserts in a small rodent. *Molecular Ecology*, **29**, 1300–1314. doi.org/10.1111/mec.15401.
18. Smith, E.S.J., Park, T.J. and Lewin, G.R. 2020. Independent evolution of pain insensitivity in African mole rats: origin and mechanisms. *Journal of Comparative Physiology A*, **260**, 313–325. doi.org/10.1007/s00359-020-01414-w.
19. Rowe, A.H., Xiao, Y., Rowe, M.P. et al. 2013. Voltage-gated sodium channel in grasshopper mice defends against bark scorpion toxin. *Science*, **342**, 441–446. doi.org/10.1126/science.1236451.

20. Wang, Y., Qiao, Z., Mao, L. *et al.* 2022. Sympatric speciation of the spiny mouse from Evolution Canyon in Israel substantiated genomically and methylomically. *Proceedings of the National Academy of Sciences of the United States of America*, **119**, e2121822119. doi.org/10.1073/pnas.2121822119.
21. Terao, M., Ogawa, Y., Takada, S. *et al.* 2022. Turnover of mammal sex chromosomes in the *Sry*-deficient Amami spiny rat is due to the male-specific upregulation of *Sox9*. *Proceedings of the National Academy of Sciences of the United States of America*, **119**, e2211574119. doi.org/10.1073/pnas.2211574119.
22. Zalasiewicz, J. 2008. *The Earth After Us: What Legacy will Humans Leave in the Rocks?* Oxford: Oxford University Press.

Chapter 19: The Lemur's Story

1. Gimlette. J. 2021. *The Gardens of Mars: Madagascar, an Island Story*. London: Head of Zeus.
2. De Gregorio, C., Maiolini, M., Raimondi, T. *et al.* 2024. Isochrony as ancestral condition to call and song in a primate. *Annals of the New York Academy of Sciences*, **1537**, 41–50. doi.org/10.1111/nyas.15151.
3. Mason, V.C., Li, G., Minx, P. *et al.* 2016. Genomic analysis reveals hidden biodiversity within colugos, the sister group to primates. *Science Advances*, **2**, e1600633. doi.org/10.1126/sciadv.1600633.
4. Morse, P.E., Chester, S.G.B., Boyer, D.M. *et al.* 2019. New fossils, systematics, and biogeography of the oldest known crown primate *Teilhardina* from the earliest Eocene of Asia, Europe, and North America. *Journal of Human Evolution*, **128**, 103–131. doi.org/10.1016/j.jhevol.2018.08.005.
5. Roos, C., Schmitz, J. and Zischler, H. 2004. Primate jumping genes elucidate strepsirrhine phylogeny. *Proceedings of the National Academy of Sciences of the United States of America*, **101**, 10650–10654. doi.org/10.1073/pnas.0403852101.
6. Everson, K.M., Pozzi, L., Barrett, M.A. *et al.* 2025. Multiple bursts of speciation in Madagascar's endangered lemurs. *BioRxiv*. doi.org/10.1101/2023.04.26.537867.
7. Dewar, R.E. and Richard, A.F. 2007. Evolution in the hypervariable environment of Madagascar. *Proceedings of the National Academy of Sciences of the United States of America*, **104**, 13723–13727. doi.org/10.1073/pnas.0704346104.
8. Campbell, C.R., Tiley, G.P., Poelstra, J.W. *et al.* 2021. Pedigree-based and phylogenetic methods support surprising pattern of mutation rate and spectrum in the gray mouse lemur. *Heredity*, **127**, 233–244. doi.org/10.1038/s41437-021-00446-5.
9. Kollikowski, A., Zimmermann, E. and Radespiel, U. 2019. First experimental evidence for olfactory species discrimination in two nocturnal primate species (*Microcebus lehilahytsara* and *M. murinus*). *Scientific Reports*, **9**, 20386. doi.org/10.1038/s41598-019-56893-y.
10. Gunnell, G.F., Boyer, D.M., Friscia, A.R. *et al.* 2018. Fossil lemurs from Egypt and Kenya suggest an African origin for Madagascar's aye-aye. *Nature Communications*, **9**, 3193. doi.org/10.1038/s41467-018-05648-w.
11. Morris, P.J.R., Cobb, S.N.F., Cox, P.G. 2018. Convergent evolution in the Euarchontoglires. *Biological Letters*, **14**. 20180366. doi.org/10.1098/rsbl.2018.0366.
12. 12;   Ezran, C., Karanewsky, C.J., Pendleton, J.L. *et al.* 2017. The mouse lemur, a genetic model organism for primate biology, behavior, and health. *Genetics*, **206**, 651–664. doi.org/10.1534/genetics.116.199448.
13. Merritt, D.J., MacLean, E.L., Crawford, J.C. *et al.* 2011. Numerical rule-learning in ring-tailed lemurs (*Lemur catta*). *Frontiers in Psychology*, **2**, 23. doi.org/10.3389/fpsyg.2011.00023.
14. Itoigawa, A., Fierro, F., Chaney, M.E. *et al.* 2021. Lowered sensitivity of bitter taste receptors to β-glucosides in bamboo lemurs: an instance of parallel and adaptive functional decline in TAS2R16? *Proceedings of the Royal Society B*, **288**, 20210346. doi.org/10.1098/rspb.2021.0346.
15. Chaney, M.E., Tosi, A.J. and Bergey, C.M. 2024. Hyper-specialized primates possess a reduced suite of xenobiotic-metabolizing cytochrome P450 genes. *BioRxiv*. doi.org/10.1101/2023.12.06.570463.

Chapter 20: The Tarsier's Story

1. Hartig, G., Churakov, G., Warren, W.C. et al. 2013. Retrophylogenomics place tarsiers on the evolutionary branches of anthropoids. *Scientific Reports*, **3**, 1756. doi.org/10.1038/srep01756.
2. Bandari, T. 2016. Decoding of tarsier genome reveals ties to humans: findings refine understanding of primate evolution. WashU Medicine news release. medicine.wustl.edu/news/decoding-tarsier-genome-reveals-ties-humans (accessed July 2025).
3. Schmitz, J., Noll, A., Raabe, C.A. et al. 2016. Genome sequence of the basal haplorrhine primate *Tarsius syrichta* reveals unusual insertions. *Nature Communications*, **7**, 12997. doi.org/10.1038/ncomms12997.
4. Melin, A.D., Matsushita, Y., Moritz, G.L. et al. 2013. Inferred L/M cone opsin polymorphism of ancestral tarsiers sheds dim light on the origin of anthropoid primates. *Proceedings of the Royal Society B*, **280**, 1759. doi.org/10.1098/rspb.2013.0189.
5. Mihola O, Trachtulec Z, Vlcek C et al. 2009. A mouse speciation gene encodes a meiotic histone H3 methyltransferase. *Science*, **323**, 373–375. doi.org/10.1126/science.1163601.
6. Davies, B., Hatton, E., Altemose, N. et al. 2016. Re-engineering the zinc finger of PRDM9 reverses hybrid sterility in mice. *Nature*, **530**, 171–176. doi.org/10.1038/nature16931.
7. Davies, B., Hinch, A.G., Cebrian-Serrano, A. et al. 2021. Altering the binding properties of PRDM9 partially restores fertility across the species boundary. *Molecular Biology and Evolution*, **38**, 5555–5562. doi.org/10.1093/molbev/msab269.
8. AbuAlia, K.F.N., Damm, E., Ullrich, K.K. et al. 2024. Natural variation in the zinc-finger-encoding exon of *Prdm9* affects hybrid sterility phenotypes in mice. *Genetics*, **226**, iyae004. doi.org/10.1093/genetics/iyae004.
9. Heerschop, S., Zischer, H., Merker, S. et al. 2016. The pioneering role of *PRDM9* indel mutations in tarsier evolution. *Scientific Reports*, **6**, 34618. doi.org/10.1038/srep34618.

Chapter 21: The Howler Monkey's Story

1. Bond, M., Tejedor, M.F., Campbell, E.K. et al. 2015. Eocene primates of South America and the African origins of New World monkeys. *Nature*, **520**, 538–541. doi.org/10.1038/nature14120.
2. Singer, S.S., Schmitz, J., Schweigk, C. et al. 2002. Molecular cladistic markers in New World monkey phylogeny (Platyrrhini, Primates). *Molecular Phylogenetics and Evolution*, **26**, 490–501. doi.org/10.1016/S1055-7903(02)00312-3.
3. Silvestro, D., Tejedor, M., Serrano-Serrano, M.L. et al. 2019. Early arrival and climatically-linked geographic expansion of New World monkeys from tiny ancestors. *Systematic Biology*, **68**, 78–92. doi.org/10.1093/sysbio/syy046.
4. Wang, X., Lim, B.K., Ting, N. et al. 2018. Reconstructing the phylogeny of New World monkeys (Platyrrhini): evidence from multiple non-coding loci. *Current Zoology*, **65**, 579–588. doi.org/10.1093/cz/zoy072.
5. Silva, F.E., Luna, L.W., Batista, R. et al. 2023. The impact of Quaternary Amazonian river dynamics on patterns and process of diversification in uakari monkeys (genus *Cacajao*). *BioRxiv*. doi.org/10.1101/2023.06.23.546215.
6. Martins-Junior, A.M.G., Sampaio, I., Silva, A. et al. 2022. Out of the shadows: multilocus systematics and biogeography of night monkeys suggest a Central Amazonian origin and a very recent widespread southwestward expansion in South America. *Molecular Phylogenetics and Evolution*, **170**, 107426. doi.org/10.1016/j.ympev.2022.107426.
7. Lima, M.G.M., Buckner, J.C., Silva-Júnior, J.S. et al. 2017. Capuchin monkey biogeography: understanding *Sapajus* Pleistocene range expansion and the current sympatry between *Cebus* and *Sapajus*. *Journal of Biogeography*, **44**, 810–820. doi.org/10.1111/jbi.12945.
8. Orkin, J.D., Montague, M.J., Tejada-Martinez, D. et al. 2021. The genomics of ecological flexibility, large brains, and long lives in capuchin monkeys revealed by fecal FACS. *Proceedings of the National Academy of Sciences of the United States of America*, **118**, e2010632118. doi.org/10.1073/pnas.2010632118.
9. Dunn, J.C., Halenar, L.B., Davies, T.G. et al. 2015. Evolutionary trade-off between vocal tract and testes dimensions in howler monkeys. *Current Biology*, **25**, 2839–2844. doi.org/10.1016/j.cub.2015.09.029.

10. Baiz, M.D., Tucker, P.K. and Cortés-Ortiz, L. 2018. Multiple forms of selection shape reproductive isolation in a primate hybrid zone. *Molecular Ecology*, **28**, 1056–1069. doi.org/10.1111/mec.14966.
11. Veilleux, C.C. and Hessy, C.P. 2023. Visual system of the only nocturnal anthropoid. *Aotus*: the owl monkey. In: Fernandez-Duque, E. (ed.), *Owl Monkeys. Developments in Primatology: Progress and Prospects*. Cham: Springer.
12. Dyer, M.A., Martins, R., da Silva Filho, M. *et al.* 2009. Developmental sources of conservation and variation in the evolution of the primate eye. *Proceedings of the National Academy of Sciences of the United States of America*, **106**, 8963–8968. doi.org/10.1073/pnas.0901484106.

Chapter 22: The Gibbon's Story

1. Catalini, L. and Fedder, J. 2020. Characteristics of the endothelium in menstruating species: lessons learned from the animal kingdom. *Biology of Reproduction*, **102**, 1160–1169. doi.org/10.1093/biolre/ioaa029.
2. Rasmussen, D.T., Friscia, A.R., Gutierrez, M. *et al.* 2019. Primitive Old World monkey from the earliest Miocene of Kenya and the evolution of cercopithecoid bilophodonty. *Proceedings of the National Academy of Sciences of the United States of America*, **116**, 6051–6056. doi.org/10.1073/pnas.1815423116.
3. Stevens, N.J., Seiffert, E.R., O'Conner, P.M. *et al.* 2013. Palaeontological evidence for an Oligocene divergence between Old World monkeys and apes. *Nature*, **497**, 611–614. doi.org/10.1038/nature12161.
4. Xia, B., Zhang, W., Zhao, G. *et al.* 2024. On the genetic basis of tail-loss evolution in humans and apes. *Nature*, **626**, 1042–1048. doi.org/10.1038/s41586-024-07095-8.
5. Sorek, R. 2007. The birth of new exons: mechanisms and evolutionary consequences. *RNA*, **13**, 1603–1608. doi.org/10.1261/rna.682507.
6. Grabowski, M. and Jungers, W.L. 2017. Evidence of a chimpanzee-sized ancestor of humans but a gibbon-sized ancestor of apes. *Nature Communications*, **8**, 880. doi.org/10.1038/s41467-017-00997-4.
7. Carbone, L., Harris, R.A., Gnerre, S. *et al.* 2014. Gibbon genome and the fast karyotype evolution of small apes. *Nature*, **513**, 195–201. doi.org/10.1038/nature13679.
8. Carbone *et al.* 2014 (see note 7).
9. Mattle-Greminger, M.P., Bilgin Sonay, T.B., Nater, A. *et al.* 2018. Genomes reveal marked differences in the adaptive evolution between orangutan species. *Genome Biology*, **19**, 193. doi.org/10.1186/s13059-018-1562-6.
10. Amorim, C.E.G., Daub, J.T., Salzano, F.M. *et al.* 2015. Detection of convergent genome-wide signals of adaptation to tropical forests in humans. *PLoS One*, **10**(4), e0121557. doi.org/10.1371/journal.pone.0121557.

Chapter 23: The Gorilla's Story

1. Owen, R. 1865. *Memoir on the Gorilla (*Troglodytes Gorilla, *Savage)*. London: Taylor and Francis.
2. Desmond, A. and Moore, J. 1991. *Darwin*. London, Michael Joseph, p. 555.
3. Darwin, C. 1871. *The Descent of Man, and Selection in Relation to Sex*. London: John Murray, Vol. 1, p. 199.
4. Katoh, S., Beyene, Y., Haya, T. *et al.* 2016. New geological and palaeontological age constraint for the gorilla–human split. *Nature*, **530**, 7589. doi.org/10.1038/nature16510.
5. Morrison, R.E., Groenenberg, M., Breuer, T. *et al.* 2019. Hierarchical social modularity in gorillas. *Proceedings of the Royal Society B*, **286**, 20190681. doi.org/10.1098/rspb.2019.0681.
6. van der Valk, T., Jensen, A., Caillaud, D. *et al.* 2024. Comparative genomic analyses provide new insights into evolutionary history and conservation genomics of gorillas. *BMC Ecology and Evolution*, **24**, 14. doi.org/10.1186/s12862-023-02195-x.
7. Thalmann, O., Wegmann, D., Arandjelovic, M. *et al.* 2011. Historical sampling reveals dramatic demographic changes in western gorilla populations. *BMC Evolutionary Biology*, **11**, 85. doi.org/10.1186/1471-2148-11-85.

8. Scally, A., Dutheil, J.Y., Hillier, L.W. *et al.* 2012. Insights into hominid evolution from the gorilla genome sequence. *Nature*, **483**, 169–175. doi.org/10.1038/nature10842.
9. Skirble, R. 2012. Genome shows humans more gorilla-like than thought. *VOA News*. www.voanews.com/a/genome-shows-humans-more-gorilla-like-than-thought-143978626/180675.html (accessed July 2025).
10. Bowman, J.D., Silva, N., Schüftan, E. *et al.* 2024. Pervasive relaxed selection on spermatogenesis genes coincident with the evolution of polygyny in gorillas. *eLife*, **13**, RP94563. doi.org/10.7554/eLife.94563.1.
11. Pawar, H., Rymbekova, A., Cuadros-Espinoza, S. *et al.* 2023. Ghost admixtures in eastern gorillas. *Nature Ecology and Evolution*, **7**, 1503–1514. doi.org/10.1038/s41559-023-02145-2.
12. Hawks, J. 2023. Tracing the genetic histories of ghost apes. johnhawks.net/p/genetic-histories-of-ghost-apes (accessed July 2025).
13. Ottenburghs, J. 2020. Ghost introgression: spooky gene flow in the distant past. *BioEssays*, **42**, 2000012. doi.org/10.1002/bies.202000012.

Chapter 24: The Bonobo's Story

1. Herzfeld, C. 2007. L'invention du bonobo. *Bulletin d'Histoire et d'Épistémologie des Sciences de la Vie*, **14**, 139–162.
2. Bjork, A., Liu, W., Wertheim, J.O. *et al.* 2010. Evolutionary history of chimpanzees inferred from complete mitochondrial genes. *Molecular Biology and Evolution*, **28**, 615–623. doi.org/10.1093/molbev/msq227.
3. Pawar, H., Ostridge, H.J., Schmidt, J.M. *et al.* 2022. Genetic adaptations to SIV across chimpanzee populations. *PLoS Genetics*, **18**(8), e1008485. doi.org/10.1371/journal.pgen.1010337.
4. Gunasekaram, C., Battison, F., Sadekar, O. *et al.* 2024. Population connectivity explains the distribution and complexity of chimpanzee cumulative culture. *Science*, **386**, 920–925. doi.org/10.1126/science.adk3381.
5. Takemoto, H., Kawamoto, Y. and Furuichi, T. 2015. How did bonobos come to range south of the Congo River? Reconsiderations of the divergence of *Pan paniscus* from other *Pan* populations. *Evolutionary Anthropology*, **5**, 170–184. doi.org/10.1002/evan.21456.
6. de Waal, F.B.M. 2006. Bonobo sex and society. *Scientific American*, **282**, 82–88. doi.org/10.1038/scientificamerican0395-82.
7. Diogo, R., Molnar, J.L. and Wood, B. 2017. Bonobo anatomy reveals stasis and mosaicism in chimpanzee evolution, and supports bonobos as the most appropriate extant model for the common ancestor of chimpanzees and humans. *Scientific Reports*, **7**, 608. doi.org/10.1038/s41598-017-00548-3.
8. Moscovice, L.R., Surbeck, M., Fruth, B. *et al.* 2019. The cooperative sex: sexual interactions among female bonobos are linked to increases in oxytocin, proximity and coalitions. *Hormones and Behavior*, **116**, 104581. doi.org/10.1016/j.yhbeh.2019.104581.
9. Wilson, M.L., Boesch, C., Fruth, B. *et al.* 2014. Lethal aggression in *Pan* is better explained by adaptive strategies than human impacts. *Nature*, **513**, 414–417. doi.org/10.1038/nature13727.
10. Behncke, I. 2015. Play in the Peter Pan ape. *Current Biology*, **25**, R24–R27. doi.org/10.1016/j.cub.2014.11.020.
11. Hare, B., Wobber, V. and Wrangham, R. 2012. The self-domestication hypothesis: evolution of bonobo psychology is due to selection against aggression. *Animal Behaviour*, **83**, 573–585. doi.org/10.1016/j.anbehav.2011.12.007.
12. Rilling, J.K., Scholz, J., Preuss, T.M. *et al.* 2011. Differences between chimpanzees and bonobos in neural systems supporting social cognition. *Social Cognition and Affective Neuroscience*, **7**, 369–379. doi.org/10.1093/scan/nsr017.
13. Kovalaskas, S., Rilling, J.K. and Lindo, J. 2020. Comparative analyses of the Pan lineage reveal selection on gene pathways associated with diet and sociality in bonobos. *Genes, Brain and Behavior*, **20**, e12715. doi.org/10.1111/gbb.12715.

14. Brand, C.M., Kuang, S., Gilbert, E.N. *et al.* 2024. Sequence-based machine learning reveals 3D genome differences between bonobos and chimpanzees. *Genome Biology and Evolution*, 16, evae210. doi.org/10.1093/gbe/evae210.
15. Kuhlwilm, M., Han, S., Sousa, V.C. *et al.* 2019. Ancient admixture from an extinct ape lineage into bonobos. *Nature Ecology and Evolution*, 3, 957–965. doi.org/10.1038/s41559-019-0881-7.
16. Scally, A., Dutheil, J.Y., Hillier, L.W. *et al.* 2012. Insights into hominid evolution from the gorilla genome sequence. *Nature*, 483, 169–175. doi.org/10.1038/nature10842.

Chapter 25: The Human Story

1. Hu, W., Hao, Z., Du, P. *et al.* 2023. Genomic inference of a severe bottleneck during the Early to Middle Pleistocene transition. *Science*, 381, 979–984. doi.org/10.1126/science.abq7487.
2. Drummond-Clarke, R.C., Kivell, T.L., Sarringhaus, L. *et al.* 2022. Wild chimpanzee behavior suggests that a savanna-mosaic habitat did not support the emergence of hominin terrestrial bipedalism. *Science Advances*, 8, eadd9752. doi.org/10.1126/sciadv.add9752.
3. Young, M., Richard, D., Grabowski, M. *et al.* 2022. The development impacts of natural selection on human pelvic morphology. *Science Advances*, 8, eabq4884. doi.org/10.1126/sciadv.abq4884. Senevirathne, G., Fernandopulle, S.C., Richard, D. *et al.* 2025. The evolution of hominin bipedalism in two steps. *Nature*, 645, 952–963. doi.org/10.1038/s41586-025-09399-9.
4. Yegian, A.K., Heymsfield, S.B., Castilla, E.R. *et al.* 2024. Metabolic scaling, energy allocation tradeoffs, and the evolution of humans' unique metabolism. *Proceedings of the National Academy of Sciences of the United Sates of America*, 121, e2409674121. doi.org/10.1073/pnas.2409674121.
5. Zhang, Y., Wang, J., Yi, C. *et al.* 2025. An ancient regulatory variant of *ACSF3* influences the coevolution of increased human height and basal metabolic rate via metabolic homeostasis. *Cell Genomics*, 5, 100855. doi.org/10.1016/j.xgen.2025.100855.
6. Aldea, D., Atsuta, Y., Kokalari, B. *et al.* 2021. Repeated mutation of a developmental enhancer contributed to human thermoregulatory evolution. *Proceedings of the National Academy of Sciences of the United States of America*, 118, e2021722118. doi.org/10.1073/pnas.2021722118.
7. Kowalczyk, A., Chikina, M. and Clark, N. 2022. Complementary evolution of coding and noncoding sequence underlies mammalian hairlessness. *eLife*, 11, e76911. doi.org/10.7554/eLife.76911.
8. Püschel, T.A., Nicholson, S.L., Baker, J. *et al.* 2024. Hominin brain size increase has emerged from within-species encephalization. *Proceedings of the National Academy of Sciences of the United States of America*, 121, e2409542121. doi.org/10.1073/pnas.2409542121.
9. Vallender, E.J., Mekel-Bobrov, N. and Lahn, B.T. 2009. Genetic basis of human brain development. *Trends in Neuroscience*, 31, 627–644. doi.org/10.1016/j.tins.2008.08.010.
10. An, N.A., Zhang, J., Mo, F. *et al.* 2023. De novo genes with an lncRNA origin encode unique human brain development functionality. *Nature Ecology and Evolution*, 7, 264–278. doi.org/10.1038/s41559-022-01925-6.
11. Libé-Philippot, B., Iwata, R., Recupero, A.J. *et al.* 2024. Synaptic neoteny of human cortical neurons requires species-specific balancing of SRGAP2–SYNGAP1 cross-inhibition. *Neuron*, 112, P3602–P3617. doi.org/10.1016/j.neuron.2024.08.021.
12. Mitteroecker, P. and Fischer, B. 2024. Evolution of the human birth canal. *American Journal of Obstetrics and Gynecology*, 230, S841–S855. doi.org/10.1016/j.ajog.2022.09.010.
13. Lacruz, R.S., Bromage, T.G., O'Higgins, P. *et al.* 2015. Ontogeny of the maxilla in Neanderthals and their ancestors. *Nature Communications*, 6, 8996. doi.org/10.1038/ncomms9996.
14. Cann, R.L., Stoneking, M. and Wilson, A.C. 1987. Mitochondrial DNA and human evolution. *Nature*, 325, 31–36. doi.org/10.1038/325031a0.
15. Gusareva, E.S., Ghosh, A.G., Kharkov, V.N. et *al.* 2025. From North Asia to South America: Tracing the longest human migration through genomic sequencing. *Science*, 388, eadk5081. doi:10.1126/science.adk5081.

16. Green, R.E., Krause, J., Briggs, A.W. et al. 2010. A draft sequence of the Neandertal genome. *Science*, 328, 710–722. doi.org/10.1126/science1188021.
17. Vernot, B. and Akey, J.M. 2014. Resurrecting surviving Neandertal lineages from modern human genomes. *Science*, 343, 1017–1021. doi.org/10.1126/science.1245938.
18. Krause, J., Fu, Q., Good, J.M. et al. 2010. The complete mitochondrial DNA genome of an unknown hominin from southern Siberia. *Nature*, 464, 894–897. doi:10.1038/nature08976.
19. Tsutaya, T., Sawafuji, R., Taurozzi, A.J. et al. 2025. A male Denisovan mandible from Pleistocene Taiwan. *Science*, 388, 176–180. doi.org/10.1126/science.ads3888.
20. Zeberg, H., Jakobsson, M. and Pääbo, S. 2024. The genetic changes that shaped Neandertals, Denisovans, and modern humans. *Cell*, 187, 1047–1058. doi.org/10.1016/j.cell.2023.12.029.
21. Iasi, L.N.M., Chintalapati, M., Skov, L. et al. 2024. Neanderthal ancestry through time: insights from genomes of ancient and present-day humans. *Science*, 386, eadq3010. doi.org/10.1126/science.adq3010.
22. Racimo, F., Gokhman, D., Fumagalli, M. et al. 2017. Archaic adaptive introgression in *TBX15/WARS2*. *Molecular Biology and Evolution*, 34, 509–524. doi.org/10.1093/molbev/msw283.
23. Bolognini, D., Halgren, A., Long, R.N. et al. 2024. Recurrent evolution and selection shape structural diversity at the amylase locus. *Nature*, 634, 617–625. doi.org/10.1038/s41586-024-07911-1.
24. Ilardo, M.A., Moltke, I., Korneliussen, T.S. et al. 2018. Physiological and genetic adaptation to diving in Sea Nomads. *Cell*, 173, 569–580. doi.org/10.1016/j.cell.2018.03.054.
25. Aguilar-Gómez, D., Bejder, J., Graae, J. et al. 2025. Genetic and training adaptations in the Haenyeo divers of Jeju, Korea. *Cell Reports*, 44, 115577. doi.org/10.1016/j.celrep.2025.115577.
26. Zeberg, H., Jakobsson, M. and Pääbo, S. 2024. The genetic changes that shaped Neandertals, Denisovans, and modern humans. *Cell*, 187, 1047–1058. doi.org/10.1016/j.cell.2023.12.029.

Epilogue: The Descent of Mammals

1. Turvey, S.T. and Crees, J.J. 2019. Extinction in the Anthropocene. *Current Biology*, 29, R982–R986. doi.org/10.1016/j.cub.2019.07.040.
2. Andermann, T., Faurby, S., Turvey, S.T. et al. 2020. The past and future human impact on mammalian diversity. *Science Advances*, 6, eabb2313. doi.org/10.1126/sciadv.abb2313.
3. Cooke, R.S.C., Eigenbrod, F. and Bates, A.E. 2019. Projected losses of global mammal and bird ecological strategies. *Nature Communications*, 10, 2279. doi.org/10.1038/s41467-019-10284-z.
4. Farnsworth, A., Lo, Y.T.E., Valdes, P.J. et al. 2023. Climate extremes likely to drive land mammal extinction during the next supercontinent assembly. *Nature Geoscience*, 16, 901–908. doi.org/10.1038/s41561-023-01259-3.
5. Kaku, M. 2018. *The Future of Humanity*. London: Allen Lane.

Bibliography

Ashby, J. 2022. *Platypus: The Extraordinary Story of Australian Mammals.* London: William Collins.
Ball, P. 2024. *How Life Works: A User's Guide to the New Biology.* London: Picador.
Benton, M.J. 2023. *Extinctions: How Life Survives, Adapts and Evolves.* London: Thames & Hudson.
Buffon, G.L.L., compte de. 1749. *Histoire naturelle, générale et particulière.* À Paris de l'imprimerie royale.
Brusatte, S. 2022. *The Rise and Reign of the Mammals.* New York: Mariner Books.
Carlquist, S. 1965. *Island Life: A Natural History of the Islands of the World.* New York: Natural History Press.
Caro, T. 2016. *Zebra Stripes.* Chicago: University of Chicago Press.
Castello, T.R. 2016. *Bovids of the World: Antelopes, Gazelles, Cattle, Goats, Sheep, and Relatives.* Princeton and Oxford: Princeton University Press.
Cech, T.R. 2024. *The Catalyst: RNA and the Quest to Unlock Life's Deepest Secrets.* London: W.W. Norton & Company.
Conway Morris, S. 1988. *The Crucible of Life: The Burgess Shale and the Rise of Animals.* Oxford: Oxford University Press.
Conway Morris, S. 2003. *Life's Solutions: Inevitable Humans in a Lonely Universe.* Cambridge: Cambridge University Press.
Conway Morris, S. 2014. *The Runes of Evolution: How the Universe Became Self-Aware.* West Conshohocken: Templeton Press.
Coyne, J.A. 2009. *Why Evolution is True.* Oxford: Oxford University Press.
Crockford, S.J. 2023. *Polar Bear Evolution: A Model for How New Species Arise.* Ottawa: Library and Archives Canada.
Darwin, C. 1871. *The Descent of Man, and Selection in Relation to Sex.* London: John Murray.
Darwin, C. 1872. *The Origin of Species*, 6th edition. London: John Murray.
Dawkins, R. 2004. *The Ancestor's Tale: A Pilgrimage to the Dawn of Life.* London: Weidenfeld & Nicolson.
Dawkins, R. 2024. *The Genetic Book of the Dead: A Darwinian Reverie.* London: Head of Zeus.
De Waal, F. 2013. *The Bonobo and the Atheist: In Search of Humanism among Primates.* New York: W.W. Norton & Company.
De Waal, F and Lanting, F. 1997. *Bonobo: The Forgotten Ape.* University of California Press.
Desmond, A. and Moore, J. 1991. *Darwin.* London, Michael Joseph.
Dial, K.P., Shubin, N. and Brainerd, E.L. (eds). 2015. *Great Transformations in Vertebrate Evolution.* Chicago: University of Chicago Press.
Doudna, J. and Sternberg, S.H. 2018. *A Crack in Creation: Gene Editing and the Unthinkable Power to Control Evolution.* London: HarperCollins.
Drew, I. 2017. *Mammal: The Story of What Makes Us Mammals.* London: Bloomsbury.
Everett, D. 2017. *How Language Began: The Story of Humanity's Greatest Invention.* London: Profile Books.
Fenton, M.B. and Simmons, N.B. 2015. *Bats: A World of Science and Mystery.* Chicago: University of Chicago Press.
Fernandez-Dugue, E. (ed.) 2023. *Owl Monkeys: Biology, Adaptive Radiation, and Behavioral Ecology of the only Nocturnal Primate in the Americas.* (Developments in Primatology: Progress and Prospects). Cham: Springer.

Fisher, R.A. 1930. *The Genetical Theory of Natural Selection*. Oxford: Clarendon Press.
Flannery, T.F. 1995. *Mammals of New Guinea*. Chatswood, N.S.W: Reed; Australian Museum.
Forrest, S. 2016. *The Age of the Horse: An Equine Journey through Human History*. New York: Atlantic Books.
Franzen, J.L. 2010. *The Rise of Horses: 55 million Years of Evolution*. Baltimore: Johns Hopkins University Press.
Gimlette, J. 2021. *The Gardens of Mars: Madagascar, an Island Story*. London: Head of Zeus.
Goin, F.L., Woodburne, M.O. and Zimicz, A.N. 2016. *A Brief History of South American Metatherians. Evolutionary Contexts and Intercontinental Dispersals*. Dordrecht: Springer.
Goodman, S.M (ed.) 2022. *The New Natural History of Madagascar*. Princeton: Princeton University Press.
Gould, S.J. 1990. *Wonderful Life: The Burgess Shale and the Nature of History*. London: Hutchinson.
Gould, S.J. 1991. *Bully for Brontosaurus: Reflections in Natural History*. London: Hutchinson.
Gunnell, G.F. and Simmons, N.B. (eds) 2012. *Evolutionary History of Bats: Fossils, Molecules and Morphology*. Cambridge: Cambridge University Press.
Hallett, M. and Harris, J.M. 2020. *On the Prowl: In Search of Big Cats Origins*. New York: Columbia University Press.
Halliday, T. 2022. *Otherlands. A World in the Making*. London: Allen Lane.
Hartston, W. 2018. *Sloths: A Celebration of the World's Most Maligned Mammal*. New York: Atlantic Books.
Higham, T. 2021. *The World Before Us: How Science is revealing a New Story of our Human Origins*. New York: Viking.
Holmes, B. and Linnard, G. (eds) 2023. *Thylacine: The History, Ecology and Loss of the Tasmanian Tiger*. Clayton, Vic: CSIRO Publishing.
Huxley, J. 1942. *Evolution: The Modern Synthesis*. London: George Allen & Unwin.
Kaku, M. 2018. *The Future of Humanity*. London: Allen Lane
Kemp, T.S. 2004. *The Origin and Evolution of Mammals*. Oxford: Oxford University Press.
Kemp, T.S. 2017. *Mammals: A Very Short Introduction*. Oxford: Oxford University Press.
Kielan-Jaworowska, Z. 2013. *In Pursuit of Early Mammals*. Bloomington: Indiana University Press.
Kielan-Jaworowska, Z., Cifelli, R.L. and Luo, Z-X. 2004. *Mammals from the Age of Dinosaurs: Origin, Evolution, and Structure*. New York: Columbia University Press.
Kingdon, J. 1997. *The Kingdon Field Guide to African Mammals*. London: Academic Press.
Krajewski, C., Westermann, M. and Woolley, P.A. 2024. *The Evolution of Dasyurid Marsupials*. Clayton, Vic: CSIRO Publishing.
Kulijian, C. 2016. *Darwin's Hunch: Science Race and the Search for Human Origins*. Johannesburg: Jacana Media.
Lamarck, J-B. 1809. *Philosophie Zoologique*. Paris: Dentu.
Lent, P.C. 2000. *Muskoxen and their Hunters: A History*. Norman: University of Oklahoma Press.
Losos, J.B. 2017. *Improbable Destinies: How Predictable is Evolution?* London: Allen Lane.
Losos, J.B. 2023. *The Age of Cats: From the Savannah to your Sofa*. London: William Collins.
Mayr, E. 1942. *Systematics and the Origin of Species from the Perspective of a Zoologist*. New York: Columbia University Press.
Meinhardt, H. 1982. *Models of Biological Pattern Formation*. London: Academic Press.
Mithen, S. 2024. *The Language Puzzle: How We Talked Our Way Out of the Stone Age*. London: Profile Books.
Mivart, St G.J. 1871. *On the Genesis of Species*, 2nd edition. London and New York: Macmillan.
Mukherjee, S. 2022. *The Song of the Cell: An Exploration of Medicine and the New Human*. London: Bodley Head.
Murray, J.D. 1990. *Mathematical Biology*. Berlin: Springer-Verlag.
Ohno, S. 1970. *Evolution by Gene Duplication*. Berlin: Springer-Verlag.
Owen, D. 2004. *Tasmanian Tiger: The Tragic Tale of How the World Lost its Most Mysterious Predator*. Baltimore: Johns Hopkins University Press.
Owen, R. 1865. *Memoir on the Gorilla (*Troglodytes Gorilla, *Savage)*. London: Taylor and Francis.
Owen, D. and Pemberton, D. 2005. *Tasmanian Devil: A Unique and Threatened Animal*. Crows Nest: Allen & Unwin.

Pääbo, S. 2014. *Neanderthal Man: In Search of Lost Genomes*. New York: Basic Books.
Panciroli, E. 2021. *Beasts Before Us: The Untold Story of Mammal Origins and Evolution*. London: Bloomsbury.
Prothero, D.R. and Schoch, R.M. 2003. *Horns, Tusks, and Flippers: The Evolution of Hoofed Mammals*. Baltimore: Johns Hopkins University Press.
Prothero, D.R. 2007. *Evolution: What Fossils Say and Why it Matters*. New York: Columbia University Press.
Prothero, D.R. 2017. *Princeton Field Guide to Prehistoric Mammals*. Princeton: Princeton University Press.
Quammen, D. 1996. *The Song of the Dodo: Island Biogeography in an Age of Extinctions*. London: Hutchinson.
Quammen, D. 2018. *The Tangled Tree of Life: A Radical New History of Life*. London: William Collins.
Reich, D. 2018. *Who We Are and How We Got Here*. Oxford: Oxford University Press.
Reilly, J. 2018. *The Ascent of Birds: How Modern Science is Revealing their Story*. Exeter: Pelagic Publishing.
Rich, T.H. and Vickers-Rich, P. 2020. *Dinosaurs of Darkness*, 2nd edition. Bloomington: Indiana University Press.
Richard, A. 2022. *The Sloth Lemur's Song: Madagascar from the Deep Past to the Uncertain Present*. London: William Collins.
Rose, K.D. and Archibald, J.D. (eds.). 2005. *The Rise of Placental Mammals: Origin and Relationships of the Major Clades*. Baltimore: Johns Hopkins University Press.
Rose, K.D. 2006. *The Beginning of the Age of Mammals*. Baltimore: Johns Hopkins University Press.
Rosenberger, A.L. 2024. *Primates: An Introduction*. Oxford: Routledge.
Shapiro, B. 2015. *How to Clone a Mammoth: The Science of De-extinction*. Princeton: Princeton University Press.
Simpson, G.G. 1945. *The Principles of Classification and a Classification of Mammals*. New York: Bulletin of the American Museum of Natural History.
Simpson, G.G. 1996. *The Dechronization of Sam Magruder*. New York: St. Martin's Press.
Swartz, S.L. 2014. *Lagoon Time: A Guide to Gray Whales and the Natural History of San Ignacio Lagoon*. The Ocean Foundation.
Thewissen, H. 2014. *The Walking Whales: From Land to Water in Eight Million Years*. Oakland: University of California Press.
Turvey, S. 2009. *Witness to Extinction: How we failed to save the Yangtze River Dolphin*. Oxford: Oxford University Press.
Tyndale-Biscoe, H. 2005. *Life of Marsupials*. Collingwood: CSIRO Publishing.
Voss, R.S. and Jansa, S.A. 2021. *Opossums: An Adaptive Radiation of New World Marsupials*. Baltimore: Johns Hopkins University Press.
Waal, F.B.M. and Lanting, F. 1997. *Bonobo: The Forgotten Ape*. Berkeley: University of California Press.
Wagner, A. 2014. *Arrival of the Fittest: Solving Evolution's Greatest Puzzle*. London: Oneworld.
Wallace, A.R. 1853. *A Narrative of Travels on the Amazon and Rio Negro*. London: Reeve & Co.
Wallace, A.R. 1880. *Island Life: or, the Phenomena and Causes of Insular Faunas and Floras, including a Revision and Attempted Solution of the Problem of Geological Climates*. London: Macmillan.
Wallace, A.R. 1889. *Darwinism: An Exposition of the Theory of Natural Selection and Some of its Applications*. London: Macmillan.
Wallace, D.R. 2004. *Beasts of Eden: Walking Whales, Dawn Horses, and Other Enigmas of Mammal Evolution*. Berkeley: University of California Press.
Wray, B. 2017. *Rise of the Necrofauna: The Science, Ethics, and Risks of De-extinction*. Vancouver: Greystone Books.
Zalasiewicz, J. 2008. *The Earth after Us: What Legacy will Humans Leave in the Rocks?* Oxford, Oxford University Press.

Index

References to figures appear in *italic* type; those in **bold** type refer to tables.

aardvark, 39, 85, 87, 88, *88*, 89–91, **90**, Plate 9
aardvark cucumber (*Cucumis humifructus*), 91
abdominal sweat production, 21
Aboriginal Australians, 305
Acokanthera schimperi ('poison arrow tree'), 232
acquired characteristics, 172
adaptive convergent evolution, 77
adaptive introgressions, 304–6
adaptive radiations, 53, 89
    bamboo lemurs, 249
    early primates, 242
    lemurs, 244, 246
    mammals, 139
    sloths, *124*
    tarsiers, 257
admixtures (gene flows), 64–5, 108–9, 200, 204
Africa
    equatorial region, *285*
    gorilla–human split, 277–8
    Homo species, 302–3
    isolated fauna, 85, 87
    surface ocean currents, 95
    trans-Atlantic sweepstake dispersals, 228, *260*
African black rhinoceros, 189
African buffalo, Plate 19
African cheetah, 208
African elephant, 108, 111, 113
African forest elephant, 107, 109
African manatee, 102–3
African mouse-deer (water chevrotain), 151
African savanna elephant, 105, 107, 109, Plate 12
African water buffalo, 167
African white rhinoceros, 165, 189
African wild ass, 185–6
Afro-Arabia primate evolution, 243
Afroinsectiphilia, 88, *88*, 93

AfroSINEs, 88
afrotheria, 85–9, *86*, *88*, 97, 105, 114
'Age of Mammals', 143
'Age of Reptiles' (Middle Mesozoic), 31
Alberto, Florian, 170
Alcheringa (journal), 37
Alesso, Benjamin, 187
Alexander Humboldt National Park, Cuba, 129
Ali, Jason, 95
alleles, 56, 115, 126, 197–8, 208, 232, 245, 254, 266
allopatric speciation, 234–6, 245, 246, 257, 287
allopolyploidy, 196
allyl isothiocyanate (AICL), 234
alpaca, *145*, 149
alternative splicing, 221, 271
altricial births
    human, 301–2
    marsupials, 44, 45, 71–2
    monotremes, 19, 25
Alu elements, 258, 271–2
Alu-induced exons, 272
'Alvarez hypothesis', 135
Alvarez, Luis, 134, 135
Alvarez, Walter, 134, 135
Amami spiny rat, 237, Plate 28
Amazonia, 49, 102, 117, 259–63
Amazonian manatee, 103
Amazonians, 303–4
American bear, *196*
American bison, 161, 169
American cheetah, 205, 207–8
American lion, 204, 205, 209
amino acids, 21, 27, 104
amylase, 292, 307
ancestral apes, 287
ancestral bonobos, 287, 290, 294
ancestral chimpanzees, 290
ancestral tarsiers, 254–5, *255*
ancient sloths, 121
Andean white-eared opossum, 49

Andeans, 303–4
Andermann, Tobias, 311
Anderson, Sophia, 217
angiosperms, 24, 38, 242
Anguilla, 94
animal calls/songs, 239–40
animal husbandry, 70
Anolis lizards, 244
ant and termite diets (myrmecophagy), 8, 39, 91
Antarctica, 9, 48, *48*, 50–4, 69–70, 143, 228
anteater, 17, 77, 86, 89, **90**, 119, 120–1
antelope, *144*, 167
Anthropocene Defaunation, 310
Anthwal, Neal, 25
anticoagulants, 221
Antilopinae, 160, *161*
Aotidae, 260
    see also owl monkey (douroucoulis)
apes, 269–74
    brachiation locomotion, 273
    divergence, 270
    and human last common ancestor, 272–3
    see also great apes; New World monkeys; Old World monkeys; primates
apocrine glands, 20–1
apoptosis, 112, 114, 216
aquatic mammals, 101, 152
aquatic skin traits, 157
Arabia, 85, 303
Arabian camel (dromedary), 142, *145*, 146–7, Plate 15
arboreal marsupials, 67
arboreal theory, 241
archaeocetes, 152
    see also whales
archaic DNA insertions, 283
Archer, Michael, 58
Arfak mountain range, New Guinea, 31
Aristotle, 2–3
Arizona bark scorpion, 234
armadillo, 39, 97, 117, 119, 121, 124–7, 299
Artiocetus, 151
Artiodactyla (even-toed ungulates), *129*, 143–4, *144*, 150–2, 160, 178
Asher, Robert, 122
Asian elephant, 107, 116
Asian horse, 184, 185
Asian leopard cat, 209
Asian water buffalo, 167–8
Asiatic bear, *196*
Asiatic black bear, 194–7
Asiatic mouflon, 170
asses, 183, 185–6

asteroid, 134–9, 101, 132
astragalus, 151
Atelidae, 260
Atlantogenata, 85, *86*
aurochs, 167, 168
Ausktribosphenos nyktos, 36–7
Australia
    colonisation, 16
    monito del monte-like founder population, 54
    monotremes, 9, 30
    National Threatened Species Day, 79
    second wave of extinctions, 311
    South American marsupials, 50–1
    therian fossils, 37
    wealth of habitats, 71
Australian marsupials, *48*, 54–74
    bandicoot, 54, *55*, 59, 68-9, 71, 78
    carnivores, 59
    de-extinction, 75–81
    honey possum, 54, *55*, 66–7
    kangaroo, 45, *55*, 63–5
    koala, *55*, 56, 61–3, Plate 6
    mice (genus *Antechinus*), 59–60
    moles, 57–9
    as monophyletic, 53
    phylogeny, 54–5, *55*
    potoroo, *55*, 65–6
    speciation, 71–2
    thylacines, 75-81
    see also marsupials
aye-aye, **90**, 246–7

Bab al-Mandab strait, Arabia, 303
babakoto (indri), 239–40, 241, 247
back-cross breeding, 149, 195–6, 197, 199
Bactrian camel, *145*, 146–7, 148
Badlands, Nebraska, 145
baiji (Yangtze River dolphin), 310
Baiz, Marcella, 264, 265
Baja California, 1
Bajau people (Sea Nomads), 307–8
baleen whale (Mysticeti), 119, 153–6, 157, 158–9
bamboo, 51, 192–4, 249
bamboo lemur, 249–50
banded anteater (numbat), *55*, 59, 77, **90**
bandicoot, 54, *55*, 59, 68–9, 71, 78
Bantu agriculturalists, 280
Barlow, Axel, 197
basal eutherian, 44
basal mammaliaforms, 11
bats, 67, 85, *129*, 196, 211–22
    'big' bats (Megachiroptera), 218
    classifications, 218–19

detect thermal radiation, 221
diverse group, 212–13
echolocation, 213–14, 218–20
evolution, 214–15
fossils, 213–14
interdigital apoptosis, 216
interdigital webbing, 216–17
noise-cancelling proteins, 218
sanguivory, 220–2
'small' bats (Microchiroptera), 218
tidal breathing patterns, 217
TRPV1 nerve channels, 221
bats' wings, 211–12, Plate 24
evolution, 215–18
and human upper limb, 212, *212*
patagium, 212
plagiopatagium, *212*, 216, Plate 24
bay cat, 202, 207
BCL-2 proteins, 73
beaked whales, 153
bears, 128, *129*, 133, 146, 191–201, 301
adaptations, 191–2
gene flow, 200–1
speciation, 195
see also brown bear; polar bear; spectacled bear
Beer, Marc, 78
Behncke, Isabel, 290, 291
'the bends' (decompression sickness), 157–8
Benin tree hyrax, 99
Bennett's tree-kangaroo, 68
benthic suction feeders, 155
Bering Land Bridge (Beringia), 107, 146, 161, 169, 184, 205, 207, 208, 209, 227
Bering Strait, 303
Bernor, Raynor, 183
bettongs, 55, 65
Bezoar ibex, 170
Bi, Shundong, 44
'big' bats (Megachiroptera), 218
'big' cats (Pantherinae), 202, 203–7
bilby, 54, *55*
bilophodonty, 269
Binder, Phillipe, 176
'biological species concept', 178–9
bipedalism, 271, 272, 296–8
birds, 3, 11, 18, 23, 28, 43, 49, 60, 62, 70, 73, 96, 139, 141, 196, 214, 217, 256, 291
bison, 160, *161*, 169
bitter taste receptors, 26, 62, 77, 232, 249, 282
bitter tastes, 62, 282
black-and-white ruffed lemur, 246

black howler monkey, 264–5
black rat, 133, 223–4, 224
black-striped capuchin, 262
Blainville, Henri de, 17–18
'blood sweat', 157
blossom bat (*Syconycteris* sp.), 5
blue antelope, 165
blue whale, 2, 111, 154
blue wildebeest, 162, 166–7
body vibrissae (tactile hairs), 99, 104, 106
Bolivia, 46–7
Bond, Mariano, 258
bone morphogenetic proteins (BMPs), 187, 216
bonobo, 284–5, *285*, 287–95, Plate 34
ancestral river crossing, 287–8
bipedal, 288
genomic analyses, 292–3
infanticide, 289–90
introgressed DNA, 293–4
neoteny, 290–1
peaceful and egalitarian, 288
playfulness, 290–1
'pygmy chimpanzee', 284
recreational sex, 288–9
'self-domestication', 291–3
Boreoeutheria, 85, *86*, 128, 225, *226*
Borhyaena, 47
Bornean orangutan, 274–5
Bornean tarsiers, 255
Bougainville, 305
Bovidae, 160–2, *161*
bovids
diversification, 161–3
extreme adaptations, 164–6
foregut fermentation, 162–3
migration, 166
ungulates/ruminants, 160
Bovinae, 160, *161*
bowhead whale, 159
brachiation locomotion, 273
Bradypus (three-toed sloth), 121–2, 123
Bramble Cay mosaic-tailed rat, 310
Brasilodon, 10
Brawand, David, 21–2
breastfeeding, 4
see also lactation
bronze quoll, 60
brown bear, 194, *196*, 197–200, Plate 22
Brown, Emily, 214
brown fat, 306
brown lemur, 245
brown long-eared bat, 218
brown rat, 223–4, Plate 25
Brown-throated sloth, Plate 13

'Bruno' (juvenile polar bear skull), 200
Brusatte, Steve, 6
Buckley, Mike, 142
buffalo, 93, 98, 160, *161*, 167–8
Buffon, Georges, 122–3
bushbabies (galagos), 241
Bwindi Impenetrable Forest, Uganda, 278, 280

caballines (true horses), 183–5
Cactus mice, 232–3
Caesar, Julius, 168
Cahill, James, 200
Caldwell, William Hay, 18
Callitrichidae, 260
camelid, 144–6, *145*, 147
camels, 144–8, *144*
Cameroon highlands, West Africa, 278, 280
cancer, 61, 112, 113, 122, 159
canids (dogs, wolves and foxes), 76
Caniformia, 192
Cantalapiedra, Juan, 107
Cape grysbok, 170
capsaicin, 221, 233–4
capuchin monkey, 258, *260*, 261, 262–3
caracals, 202, 207
carbon dioxide, 53, 137, 162, 181, 233, 312
cardiotonic glycosides, 232
Caribbean, 46, 69, 94, 101, 130, 132, 224, 229, 311
Carlquist, Sherwin, 81
carnassial teeth, 202
Carnivora, 192
carnivores, 39, 46, 47, 49, 59, 93, 111, 130, 143, 192, 202, 203, 208
Caro, Tim, 186
Carrier's constraint, 39
Carstens, Bryan, 5, 227
Carter, John, 22
Casali, Daniel, 120–1
casein proteins, 22
Casewell, Nicholas, 131, 132
Catarrhini, 259
Cathariostachys madagascariensis (giant/woody bamboo), 249
cats (family Felidae), *129*, 202–10, 255, 283, 291
  'big' cats (Pantherinae), 202, 203–7
  domestication, 209–10
  DXZ4 gene, 203
  genetic diversity, 190, 203
  obligate carnivores, 203
  'small' cats (Felinae), 202, 207–9
  sweet tastes, 203
cattle, 86, 123, *144*, 147, 148, 152, 160, *161*
cave bear, 197
cave lion, 204
Cavener, Douglas, 178
caviomorph rodents, 228–9
Cebidae, 260
Cenozoic, 72, 87, 98, 228
Central American Seaway, 102
central chimpanzee, 285, *285*
Cephalostachyum perrieri (bamboo), 249
Ceratomorpha, 188
cerebral cortex, 300, 301
cervical vertebrae, 57, 122, 126, 173–5, 251
Cetacea, 143, *144*, 152
cetaceans, 104, 118, 150–3, 158, 159
Cetartiodactyla, 143, *144*, 152
Chaco Amerindians, 303–4
Chacoan pygmy opossum, 49
Chamois, *161*
character displacement, 263
cheetah, 205, 207–9
chemical diffusion gradients, 187
chickens, 21
Chicxulub impact, Yucatán Peninsula, 118, 132–7
Chile, 50, 51, 53, 207
chimpanzee, *241*, 284–95
  aggressive primates, 289–90
  differences with bonobos, 290–3
  distribution, 285–6, *285*
  divergence from bonobo, 287–8
  DNA, 293
  human-impact hypothesis, 289
  tool use, 286
China, 11, 43, 102, 147, 193, 223–4, 311
Chiroptera, 211, 218–19
chiru (Tibetan antelope), 165–6
chitinase proteins, 39
chorio-allantoic placentas, 59
Chorora Formation, Ethiopia, 277–8
Chororapithecus abyssinicus, 277–8
Christidis, Les, 37–8
Christmas Island pipistrelle, 310
chromosomes
  and gene flow, 64–5
  LCA, 10–11
  territories, 293
chuffling ('prusten'), 204
Church, George, 115
cichlids, 110, 244
Cingulata
  see armadillo
Cladosictis, 47
Clark, Nathan, 299
climate change, 200, 209, 246, 308, 310–11

climate-facilitated admixtures, 200
cloaca, 16, 17, 19, 96–7
cloning, 79, 81
clouded leopards, 204
cobras, 243
coccyx, 271
cochlea, 23, 40, 208, 219
cold tolerance, 306
Colobine monkeys, 162, 163
Colossal Biosciences, 79–80, 115, 116
colour-blindness, 267
colour-sensitive retinal pigments
    see opsins
colour vision, 265–8
colugos, 67, 128, 215, 225, *226*, 240
Columbian mammoth, 107, 108
Columbus, Christopher, 101
columella (stapes), 23–4, *24*
Commentarii de Bello Gallico (Julius
    Caesar), 168
common hippopotamus, 95, Plate 16
common spiny mouse, 235–6, 269
common tenrec, 96, 138
comparative genomics, 9, 41, 76, 96, 176–7
complex numerical abilities, 248–9
condylarths, 143
cone photoreceptors, 254–5, 262, 265–8
coney (European rabbit), 227
coney (hyrax), 87–8, *88*, 98–101
congenital ichthyosis, 104
Congo River, 180, 279, 284, 285, *285*,
    287–8, 290, 294
connectome, 44
convergent evolution, 71–3
    'ant-eating' mammals, 8, 89
    extreme, 76
    fingerprints, 62
    ILS genes, 57
    metatherians, 46–7
    monotremes, 26
    otter shrews, 93
    phenotypic traits, 87
    sifakas, 247
    sloths 122
    Steller's sea cows, 104
    T1R2, 203
    thylacine and canids, 76–7, Plate 7
Conway Morris, Simon, 72–3
Coolidge, Harold, 284
Coombs, Ellen, 153–4
cooperative breeding, 233
Cope's rule, 156
copy number variations, 237
cottontail rabbit, 227
cougar, 207–8

courtship songs, 100
Cox, Philip, 247
crabeater seal, 155
cranial telescoping, 153
crested rat, 232
Cretaceous Angiosperm Revolution, 141
Cretaceous–Palaeogene mass extinction
    see K–Pg extinction events
Cretaceous period, *xv*, 24, 30, 37–9
Cretaceous Terrestrial Revolution, 141
Crick, F.H.C., 7
CRISPR gene-editing technology, 80
Cross River gorillas, 278, 280
Crowley, Brooke, 95
crown carnivorans, 192
crown marsupials, 46
crown primate, 242
crown whales, 153
CSIRO, Australia, 23
Cuban solenodon, **90**, 129, 131
cube-square law, 110–11
Cucumis humifructus (aardvark
    cucumber), 91
cuscus, 35–6, 69–71
cyanides, 249–50
cyanogenic glycosides, 250
Cyclops Mountains, 31
Cyprus, 210
cytochrome P450, 148, 250
cytoskeletal genes, 232

dactylopatagium, *212*, 215–16
Dahomey Gap, 285
Damaraland mole-rat, 233
Darwin, Charles, 4
    African origins for humanity, 302
    The Descent of Man, 277
    development of giraffes' neck, 172, 174
    evolution of flight, 215
    evolutionary innovations, 21
    human origins and gorillas, 276–7
    hybridisation, 196
    mammary glands, 19–20
    The Origin of Species, 18, 20, 178,
        196, 276
    sympatric speciation, 234–5
    'Vertebrata' (*Encyclopaedia
        Britannica*), 277
'Darwinian demon', 264
Darwinian gradualism, 174
Darwin's finches, 244
Dasyuridae (thylacine family), 80
    see also thylacine (Tasmanian tiger)
Dasyuromorphia, 54, 59
    see also quolls (tiger cats)

Dasyurus spartacus (bronze quoll), 60–1
Davies, Ben, 256
de-extinction, 79–81, 115
De Gregorio, Chiara, 239–40
de Waal, Frans, 288
Deakin University, Australia, 22
Deccan Traps, India, 134
deciduous teeth (milk teeth), 10, 119
decompression sickness ('the bends'), 157–8
deer, 86, 123, 142, *144*, 160
defensin genes, 27–8
dehydration, 232
Delsuc, Frédéric, 120
Demartsev, Vlad, 100
dementia, 248
Democratic Republic of the Congo (DRC), 278, 280, 286
Den Boer, Wendy, 64
Denisova Cave, Altai Mountains, Siberia, *304*, 305
Denisovans, 303, *304*, 305–6
'dental filtration hypotheses', 155
dentary, 4, 24, *24*, 30, 36
  see also teeth
Dermoptera, 240
The Descent of Man (Darwin), 277
Deschner, Tobias, 289
desert adaptations, 232–4
devil facial tumour disease (DFTD), 61
dichromacy (two-colour vision), 39, 254, 255
Didelphimorphia (opossums), 43, 48–50, *48*
diffusiophoresis, 187
dik-diks, 161, *161*
dim light levels (mesopic), 255
dimethyl sulphide (DMS), 158
dingo, 76, Plate 7
dinosaurs, 30, 89, 111, 130, 132, 137–40, 214, 242, 291
Diogo, Rui, 288
diphyodonty, 10
Diprotodontia, 54, 66
Discokeryx xiezhi, 173–4
diurnal mammals, 260, 262
  cone and ganglion cells, 267
  haplorrhines, 122
  patterned coats, 49
  polychromatic trichromacy, 254, 255
  three-toed sloths, 122
  xenarthrans, 118
diving adaptations, 308
DNA, 6–9
  non-coding, 76, 87, 113, 203, 217, 220, 271, 300–1

DNA-based phylogeny (Springer), 89
DNA polymerase, 7
dogs, 128, *129*, 131, 143, 192, 241, 265, 291, 301
Dollo's law of irreversibility, 65
dolphins, *144*, 150, 153, 203, 219
domestic cat, 203, 209–10
donkey, 182, 185–6
dormice, 43
douroucoulis (owl monkey), 258, *260*, 262, 267–8
Driscoll, Carlos, 209
dromedary (Arabian camel), 142, *145*, 146–7, Plate 15
Dromiciops, 52, 53
  see also monito del monte
Drummond-Clarke, Rhianna, 297
Duchêne, David, 54–5
duck-billed platypus
  see platypus
dugong, 88, *88*, 98, 101, 102, 103, 104
duiker, 160, *161*, 162
Dunn, Jacob, 263
dunnart, 54, *55*, 77, 78, 80–1, Plate 8
Dyer, Michael, 267

early mammals, 36, 39, 85, 89, 138, 150
early monotremes, 11, 30, 31, 138
early primates, 242–3
early therians, *38*, 39, 40, 43
ears
  ectotympanic bones, 151
  embryonic development, 25
  gorillas and humans, 281
  middle ear ossicles, 4, 11, *18*, 23–5, *24*, 39–40, 93
  newborn monotremes, 25
  pinnae (ear flaps), 25, 111
The Earth After Us (Zalasiewicz), 237–8
Earth BioGenome Project, 6
earthquakes, 93, 136–7
eastern chimpanzee, 285–6, 279, *285*
eastern gorilla, 276, 278–80, 282
eastern lowland gorilla (Grauer's gorilla), 278, 280
eastern quoll, 61
Eberle, Jaelyn, 187–8
eccrine sweat glands, 299
echidna, 17, 18–19, 25–7, 30–2, 231, Plate 3
echidna antimicrobial protein (EchAMP), 23
echolocation, 153–4, 158, 212–20
ectotympanic bones, 151
EDAR protein, 187

EDGE scores, 90, **90**, 131
egg-laying
    see oviparity
Egyptian fruit bat, 220
Einstein, Albert, 210
Eldredge, N., 177
electroreception, 28–9
elephant, 85–8, *88*, 105–16
    constraints on size, 110–12
    dental evolution, 109–10
    genomes, 108
    heat dissipation, 111
    Hyracoidae and Sirenia, 98–104
    intelligence and memory, 99
    isolation and hybridisation, 109
    pinnae, 111
    Proboscidea, 98, 105–6
    TP53 retrogenes, 113–14
    trunks (proboscises), 104, 106
    tusks, 106–7
    undescended testicles, 114
    vibrissae, 99, 106
elephant shrew (jumping shrew, sengis), 85, 87, 88, *88*, 89, 91–2, 114, 269
Elephantidae, 105, 107
Ellesmere Island, Nunavut, 142–3, 145–6, *145*, 188
Emerling, Christopher, 118, 119–20
enamelysin (*MMP20*), 157
Encyclopaedia Britannica, 277
endolymph fluids, 12
endothermy, xiii, 12, 111
Eocene epoch, 228, 242, 257
Eocene–Oligocene cooling, 257
Eohippus (*Hyracotherium*), 182–3
Eomaia scansoria, 44
Eotragus, xiv, 160
equatorial Africa, 276, *285*, 296
Equidae (horses), 181–5
equids, 181, 182, 184, 186
equines, 183–4
Equus, 182, 183
Eritherium, 106
Esselstyn, Jake, 230
Euarchonta, 225, *226*
Euarchontoglires, *86*, 128, 138–9, 225, *226*
Eucalyptus, 62
eukaryote genomes, 87–8
Eulipotyphla, 129–30
eulipotyphlans, 131–2
Eurasian proboscideans, 107
European rabbit (coney), 227
European shrews, 91
eusocial mammals, 233
Eutheria (true beast), *18*, 42, *86*

eutherians, *38*, 40–7, 59, 69, 72, 81
even-toed ungulates (Artiodactyla), *129*, 142–4, *144*, 150–2, 160, 178
evolution
    auditory genes and language, 281
    comparative genomics, 9
    filling niches, 73
    of flight, 215, 217
    mammalian jaw joint and middle ear, 23–5, *24*
    mammalian placenta and middle ear, *18*
    predictability versus contingency, 72–4
    spontaneous mutations, 108
Evolution Canyon, Mount Carmel, Israel, 235–6
evolutionary arms race, 234, 249, 273
evolutionary distinctiveness (ED), 89–91, **90**, 131, Plate 9
evolutionary pleiotropy, 177
evolutionary trade-off hypotheses, 249–50
exaptation, 20, 21, 24, 68
exons, 220–1, 272
Explosive model, 140–1, *140*, 242
extinctions
    Anthropocene Defaunation, 310
    Australia, 75
    Chicxulub impact, 30, 136, 137
    climate change, 310–11
    'Great Dying', 137
    human-induced, 310
    human population, 311
    Quaternary Extinction event, 117, 146, 184
    second wave (2100), 311
    solenodons' resilience, 139
    South American marsupials, 49
eyes
    golden moles, 92
    retinal cone receptors, 118
    retinal rod-to-cone ratios, 267
    sizes, 254
    see also vision

fat-tailed dunnart, 77, 78, 80, Plate 8
feathertail glider (pygmy gliding possum), 67
Feigin, Thomas, 76, 77
Felidae, 190, 202–10
    big cats, 203–7
    domestication, 209–10
    genomes, 203
    small cats, 202, 207–9
Feliformia, 192
Felinae, 202
Felis silvestris catus (domestic cat), 209–10

Felis silvestris labia (wildcat), 202, 209
Felsōtárkány, Hungary, 211
Feng, Shaohong, 56
feral cats, 210, 238
fertile hybrids, 199, 294
fibroblast growth factor 8 (Fgf8), 216
Filipino Negritos, 305
filter-feeding mysticetes, 153–6
fingerprints, 62, 187–8
fish, 20, 29, 38, 49, 89, 93, 104, 110, 118, 132, 151, 154, 196
Fisher, Diane, 60
Fisher, Ronald, 196
fishing cat, 209
Flannery, Tim, 29–31, 37, Plate 4
flat-headed cat, 209
flight, 67, 68, 211–17
floating islands, 228
flotation bladders, 20
flying foxes, 217–18
Flynn, John, 37
Foja mountains, New Guinea, 5, 31
folivorous foraging, 268
foramen magnum, 297
foregut fermenters, 144, 162–3, 182
forest ox (kouprey), 310
forest refugia, 139, 180, 280, 294
fossil-lagerstätte, 213
four-horned antelope, 160
fovea centralis, 254, 267–8
Frankham, Greta, 65–6
free radicals, 217
fruit bats, 218–20
Fu, Qiaomei, 305

galagos (bushbabies), 241
Galápagos Islands, 228, 236, 244
Galileo Galilei, 110
Gatesy, John, 150
Gaudin, Timothy, 121–2
Gause's law, 263
gazelles, 160, *161*
Geisler, Jonathan, 155
gene duplication, 9, 27–8, 28, 77, 113, 268
gene expression, 88, 10, 41, 76, 126, 271, 293
gene flows (admixtures), 64–5, 108–9, 200, 204
gene loss, 10, 157, 250
genes
    ACSF3, 298
    AMPD3, 157
    APOB, 198
    BGTG1, 225
    BNC2, 306

CACNA1C, 224
casein, 22
CHAT, 224
CHST11, 225
COL1A1, 273
DXZ4, 203
DYNCZ2H1, 193–4
EGLN1, 207
EPAS1, 207, 306
ESPR1, 206
EVPL, 281
FGF, 253
FGF8, 216
FGF12, 224
FGFRL1, 177
GHSR, 159
HCRT, 178
hearing and speech, 281
homeobox, 178
homeotic genes (*Hox* genes), 21, 122
Hr, 157
IGFBP7, 159
IL1A, 253
ILS1, 218
KU80, 217
L-gulonolactone oxidase (*GULO*), 252
L/M genes, 254, 255, 266, 268
L/M opsin, 255, 267
LIF6, 113–14
M opsin, 255
MMP20 (enamelysin), 157
MTOR, 263
PAPSS2, 57
PARP1, 263
PCNT, 193–4
PDE10A, 308
PEG10, 43
PER1, 178
pleiotropic functions, 76–7
PRDM9, 256–7
Prestin, 219
RAD50, 217
REP15, 222
SOX9, 236, 237
sperm production, 281
SRY gene, 237
SSTR4, 206
T1R2, 203
tail development, 271
TAS2R, 62
TAS2R14, 282
TBX5, 273
TBX15, 306
TBXT, 271
TP53, 112–14

TPR2, 261
VIR, 62
vitellogenin (*VTG*), 21–2
Vkorc1, 225
WARS2, 306
wavelength of light, 254
WFIKKN1, 56
Wnt5a, 67–8, 216
ZNF804B, 293
  see also 'jumping genes' (retroposons); pseudogenes; transposons
genetic adaptations, 220–2, 225, 240, 286, 308
genetic convergence, 194
genetic divergence, 65–6
genetic diversity, 56, 78
genetic shuffling (recombination), 200
genetic time machine' (Migliano), 286
genetic variation, 61, 115, 165, 198, 203, 208, 280, 302
genetics and genetic code, 6
'genomic fossils', 158, 293, 294
genomic imprinting, 9, 40–1
genomic plasticity, 272–4
genomic reconstruction, 79–81
geological ages, *xv*
Gheerbrant, Emmanuel, 89
'ghost' ape species, 293–4
ghost introgressions, 283
giant anteater, 17, 120–1
giant armadillo, 124
giant camel, 145–6
giant ground sloth (*Megatherium*), 121
giant panda, 192–4
giant parrot (*Heracles*), 73
gibbons, 100, 123, 201, *241*, 243, 272–4
gigantism, 154, 159
Gilbert's potoroo, 65
Gillis, Jesse, 126
Gimlette, John, 239
Gingerich, Philip, 151
Gir forest sanctuary, Gujarat, 205
Giraffa camelopardalis, 178
giraffe, 85, 93, 100, 123, *144*, 172–80, Plate 20
  cardiovascular challenges, 175–6
  comparative genomics, 176–8
  Darwinian gradualism, 172–4
  drinking, 176
  endangered, 179
  eye development and vision, 177
  FGFRL1 gene, 177
  gene flow, 179–80
  horizon-scanning, 174, 177
  interbreeding, 180
  neck elongation theories, 172–6
  olfaction, 177
  sexual selection, 173–4
  sleep patterns, 177–8
  speciation, 178–80
  thoracic vertebrae, 174–5
giraffe-inspired robotic arm, 175
Giraffidae, 178
gliding species, 67–8, 215, 240
Glires (lagomorphs and rodents), 225, *226*
global iridium layer, 134
global warming
  see Middle Miocene Climatic Optimum (MMCO); Palaeocene–Eocene Thermal Maximum (PETM)
globally endangered species
  see EDGE scores
goats, 160, *161*, 162, 170
Goderis, Steven, 135
golden bamboo lemur, 249
golden-headed lion tamarin, Plate 31
golden moles, 58, 87, 88, *88*, 92–3
Gombe National Park, Tanzania, 289
gonadotropins, 292
Gondwana, 9, 30, 37, *38*, 43, 50–1, 85
Goodall, Jane, 289
Goodfellow's tree-kangaroo, 68
gorilla, *241*, 276–83
  Darwin's theories, 276–8
  digestive anatomy and physiology, 278–9
  diverging lineages, 285
  ear morphology, 281
  evolution and distribution, 279–81
  genome sequence, 280–1
  'ghost' lineages, 282
  hierarchical social modularity (HSM), 279
  knuckle-walking, 278
  mating strategies, 281
  nests, 278
  silverbacks, 278, 279, 281
  vocalisations, 279
Goswami, Anjali, 42, 71
Gould, Stephen Jay, 72–4, 86
  evolution of horses, 182–3
  evolution of humans, 296
  evolution of long necks, 172
  'exaptation', 20
  'Life's little joke', 182
  'punctuated equilibrium, 177
  'tape of life' metaphor, 72
  Wonderful Life, 72
Grabowski, Mark, 273
gracile capuchin (untufted, *Cebus*), 262–3

Grant's golden mole, 92
Grauer's gorilla (eastern lowland gorilla), 278, 280
gray mouse opossum, 49
Great American Biotic Interchange, 50, 117, 145, 189, 229
great apes, 263, 272, 274, 278, 284, 294–5
'Great Dying', 137
  see also extinctions
Great Nile Migration, 167
Greater Antilles, 46, 229, 259–60
greater bamboo lemur, 249
greater fairy armadillo (pichiciego), 125
greater hedgehog tenrec, 96
green iguana, 94
Green River Formation, Wyoming, 213
Greenland land bridge, 181
Greenlandic Inuit, 306
Grévy's zebra, 186
grey bamboo lemur, 249
grey ox (kouprey), 310
grey whale, 1–2, 154, 157, Plate 17
grolars (hybrid brown and polar bear), 199
grooming, 243
'grooming' claws, 251
grysbok, *161*, 170
guanaco, 145, *145*, 148–9
Gubbio, Italy, 134
Guernsey, Michael, 45
guinea pigs, 53, 252
Gujarat lions, 205
GULO pseudogene, 252
gulp-feeders, 156
Gupta, Ankur, 187
gustatory sensory system, 158

Haast's eagle, 73
haemoglobin, 104, 116, 149, 157, 164, 166, 307
Haenyeo divers, 308
Hainan gibbon, 310
Haiti, 131
Haldane's rule, 171
hand digits, 212, *212*
hand-wing, 215
hanging and swinging behaviours, 273
haplorrhine hypotheses, 252, 253
haplorrhines, 241, 253–4
'hard contingency' evolution, 73
Hare, Richard, 291
hares, 225–7, *226*
hartebeest, *161*
Hawkesbury, New South Wales, 15
Hawks, John, 282
Haywood, Matt, 205–6

hearing, 24–5, 281
  see also ears
hedgehogs, 73, 87, 96, 129–30, *129*, 131, 231
Heerschop, Sacha, 256–7
Helgen, Kristofer, 30–1
Helicobacter pylori infections, 306
Heller, Rasmus, 179–80
Henneberg, Maciej, 62
Heptodon, 188
Heracles (giant parrot), 73
herbivores, 39
  Eurasian proboscideans, 107
  foregut fermentation, 162–3
  generating too much heat, 111
  Miocene epoch, 161, 166
  New Zealand, 73
  Palaeocene epoch, 143
  plant toxins, 249–50
  prey for large cats, 205
Heritage, Stephen, 102
heterochromatin, 170
heterochrony, 267–8
heterogametic sex, 255–6
heterothermy, 123
heterozygotes, 307
hibernation, 138
hierarchical social modularity (HSM), 279
higher primates, 119, 243, 252, 291
highveld mole-rat, 234
Hildebrand, Alan, 135
hindgut fermenters, 144, 163, 182
hippocampus minor, 100, 276
hippopotamus, 3, 95, 144, *144*, 150–2, 157
Hispaniola, 130, 133, 259
Hispaniolan hutia, 133
Hispaniolan solenodon, 129, 133, Plate 14
Histoire Naturelle (Buffon), 122
hoatzin, 163
Holt, Carl, 22
Home, Everard, 16, 28
homeobox genes, 178
homeotic genes (*Hox* genes), 21, 122
Hominidae (great apes and humans), 272
hominids (great apes), 272, 288
hominins, 279, 293, 296–7, 301
  see also bonobo (*Pan paniscus*); chimpanzee; human (*Homo sapiens*)
Hominoidea (primates superfamily), 272
hominoids (apes), 272
Homo erectus, 303
Homo habilis, 303
homoploid hybrid speciation (HHS), 196–7
honey possum, 54, *55*, 66–7

hoofed foot (monodactyly), 183
hopping, 59, 63–5, 91, 248
'horizon-scanning', 174
horses (Equidae), 181–5
Houston Chronicle, 135
How, Martin, 186
howler monkey, 258, *260*, 261, 263–5, 268
Huber, Matthew, 95
human (*Homo sapiens*), 3, *241*, 294–309
   adaptive introgressions, 304–6
   altricial neonates, 301–2
   Alu elements, 271–2
   auditory genes and language, 281
   bipedalism, 271, 298
   brain, 300, 301
   cerebral cortex, 300
   Darwin's views of origins, 277
   Denisovans and Neanderthal DNA, 294, 304–6
   diets and digestion, 224–5, 307
   eccrine glands and sweating, 299
   emergence, 296–302
   environmental selection, 306–8
   facial plan, 302
   fetal tails, 270–1
   and gorillas, 277, 281
   hairlessness, 299
   hand & arm and bat wing, 212, *212*
   hippocampus minor, 276
   last common ancestor (LCA), 272–3
   metabolic rates, 298–9
   migrations, 303–4, *304*
   numerical abilities, 248–9
   Out of Africa model, 296, 302–4
   prognathism, 302
   rate of evolution of genes, 281
   similarities with bonobos and chimpanzees, 288
   social systems, 279
   survival advantage, 308–9
   tool use and cultural development, 286
human immunodeficiency virus (HIV), 286
Hunter, John, 15
Hurricane Marilyn, 94
Huxley, Thomas, 276
hybrid sterility, 255–6
hybridisation
   Asiatic black bear, 195, *196*
   back-crossing, 149, 195–6, 197, 199
   and gene flow (admixture), 108
   introgressive hybridisation, 197–201
   mantled and black howler monkeys, 264–5

reinforcement, 265
reticulated giraffe, 179–80
speciation, 196–7, 245
see also introgressions
Hylobatidae (gibbons), 272
hyper-specialisations, 249–50
hypertrophied malleus, 93
hypoxia, 149, 169, 207, 306
hypsodonty, 109, 110, 163
Hyrachyus, 189
Hyracotherium (*Eohippus*), 182–3
hyrax (Hyracoidae), 85, 87–8, *88*, 98–101

ibexes, *161*
impact craters, 135
impala, *161*
incomplete lineage sorting (ILS), 54–7, 141, 143, 260
incus, 24, *24*
India, 37
Indian Ocean, 95
Indian rhinoceros, 189
indicine cattle, 168–9
Indohyus, 151
indri (*babakoto*), 239–40, *241*, 247
infanticide, 289
infrared radiation, 220
Insectivora, 87
insectivores, 87, 124, 127, 129, 130, 132, 143
insects, 24, 38–9, 43, 50, 60, 78, 91, 96, 97, 120, 141, 196
insertion–deletion mutation, 256–7
insulin, 159
inter-species gene transfer, 10
interbreeding, 108, 180, 184, 198–9, 236, 256, 265, 280, 296, 303, 304, 305
interdigital apoptosis, 216
interdigital webbing, 216–17
interferon responses, 221
introgression-detection technologies, 283
introgressions, 195, 201, 245, 282–3, 293–4
   see also hybridisation
introgressive hybridisation, 197–201
introns, 220, 271
Inuit people, 306
iodine, 103
iridium, 133–4, 135
Irish brown bear, 200
Isaac, Nick, 90
ischial spines, 297
island hopping, 95
Island Life (Wallace), 94

Isthmus of Panama, 50, 72, 117, 120, 145, 229
  see also Great American Biotic Interchange; North America; South America
ivory poaching, 107

jackrabbit, 227
jaguar, 203–4, 205–6, Plate 23
jaguarundi, 207–8
Janis, Christine, 183
Javan rhinoceros, 189
jaws, 23–5, *24*
  masseter muscles, 230
  prognathism, 302
Jehol Biota formation, China, 43
Jeju Island, Korea, 308
'jumping gene' (LAVA element), 272, 273, Plate 32
'jumping genes' (retroposons), 52, 87–8, 243, 253, 271
  see also genes; transposons
jumping shrew
  see elephant shrew (jumping shrew, sengis)
Jungers, William, 273
'junk' DNA (non-coding DNA), 76, 300–1
  see also DNA
Juramaia sinensis, 44–5

K–Pg boundary, *xv*, 118, 133–7, 140
K–Pg extinction events, 30, 48, 85, 92, 132–3, 140
Kaku, Michio, 312
kallikreins, 132
Kamilah (female lowland gorilla), 280
kangaroo, 45, *55*, 63–5
Kealy, Shimona, 70
keratin, hair, 157
Kingdon, Jonathan, 91
kiwi, 73
knuckle walking, 278, 281, 298
koala, 8, *55*, 56, 61–3, Plate 6
kob, *161*, 162, 166
kouprey (grey/forest ox), 310
Kovalaskas, Sarah, 292
Krause, Johannes, 305
kudus, 160, *161*
Kuhlwilm, Martin, 282, 293–4
Kumar, Viktar, 200
Kuroiwa, Asato, 237

L-gulonolactone oxidase (*GULO*) gene, 252
La Meseta Formation, Seymour Island, 51

laboratory research animals, 248
lactase enzyme, 307
lactase persistence, 307
lactation, 10, 19, 20, 22–3, 31, 45
  see also breastfeeding
lagomorphs, 128, 225–7
Laguna San Ignacio, Baja California, 1–2
Lamarck, Jean-Baptiste, 18, 172
Lamini tribe, 145, 148
Lamm, Ben, 115
Lane, David, 112
Lankester, Edwin Ray, 277
Laotian rock rat, 227
Lar gibbon, Plate 32
large-brained fetuses, 301
large-headed capuchin, 263
large mammals, 93, 110–14, 123, 310
laryngeal echolocation, 219
last common ancestors (LCA), xiii, xiv, 10–12, 272–3
Late Quaternary ice age, 133
Laurasia, 36, *38*, 43, 46, 85, 127, 128, 242
Laurasiatheria, *86*, 128, *129*, 143, *226*
LAVA element ('jumping gene'), 272, 273, Plate 32
Leadbeater's possum, **90**
leaves, 99, 123, 144, 172, 183, 188, 247
Leidy, Joseph, 145
lemuriform lineages, 243
lemurs, 53, 94, 239–50, *241*
  adaptions, 246–50
  dementia, 248
  introgression, 245
  Madagascar, 239, 244, 245
  reproductive cycles, 245–6
  speciation, 244
  unpredictable weather, 245–6
leopard, 155, 202, 203–4, 206
leopard cat, 209
leopard seal, 155
lesser hedgehog tenrec, 96
leukaemia inhibitory factor 6 (*LIF6*), 113–14
Liaoning Province, China, 43, 44
Liem, Karel, 110
Liem's paradox, 109–10
'Life's little joke' (Gould), 182
limb development, 193–4, 298
Linnaeus, Carolus (Carl von Linné), 3–4, 6, 31, 92, 178, 209
lion, 203–4, 204–6
Lisbon University, 12
Lister, Adrian, 109–10
little brown bat, 5

little red kaluta, 60
live births (viviparity), 2–3, 4, 17, 40–3
llama, 145, *145*, 149
Long Fuse model, *140*, 141
long non-coding RNA (lncRNA), 300
long-nosed bandicoot, 59
long-nosed potoroo, 65, 66
long-to-middle wavelength genes (L/M genes), 254, 255, 266, 268
Lorenzen, Eline, 198
loris, 241, *241*, 243, 252, 253
lorisiforms, 243
loss of gene function, 156–8
Lovegrove, Barry, 138
low-density lipoprotein (LDL), 198
low-fitness hybrids, 265
lower jaws, 23–4, 153, 247, 302
lower primates (strepsirrhines), 241, 243, 245
lowland streaked tenrec, 97, Plate 10
Lumholtz's tree-kangaroo, 68
lunar philic, 255
lungs, 20, 39, 101, 102, 124, 169, 207
Luo, Zhe-Xi, 44
lynx, 202, 207
lysozyme, 163

M opsin genes, 255
macroevolution, 150, 183
macrophages, 216, 231
Macropodidae, 63–5
Macroscelidea, 91
Madagascar, 9, 37, 93–6, 138
　allopatric speciation, 245
　biological diversity, 93
　climate, 245–6
　habitats, 245, 248
　isolated refugia, 245
　rainforests, 247
　rodents, 228, 229
　tenrecs, 53, 88, *88*, 93–7
　see also lemurs
mainland clouded leopard, 204
Malagasy hippopotamus, 94–5
Malagasy rodent, 229
malaria, 266, 307
Malayan tapir, 188
malleus, 23–4, *24*, 93
mammal radiation, 140–1, *141*
Mammalia, 2–6, 85–6, *86*, 240
Mammalia (Aristotle), 2
Mammalia (Linnaeus), 3–4
mammaliamorphs (pre-mammals), 6, 11, 12, 20, 21, 23–5, 39, 40
mammalian XY hybrids, 256

mammals
　diversity, 4–6, 139, 140
　endothermic, 12, 111, 244
　extinctions, 61, 75, 101, 139, 257, 294, 310, 311
　and flight, 211–17
　jaws, 23–4, *24*
　middle ear, 23, *24*
　viviparity, 4, 17, 18, 40–3
mammary glands, 3, 19–22, 99
mammoth, 107, 108, 114–16, 133
manatees (sea cows), 88, *88*, 98, 101–4, 299
maned rat, **90**
mantled howler monkey, 264–5, Plate 31
manual dexterity, 241
marmoset, 258, *260*, 261
Marquès-Bonet, Tomàs, 293
marsupial cat, 60–1
marsupial mice (genus *Antechinus*), 59–60
marsupial moles (Notoryctemorphia), 54, 55, *55*, 57–8, 92
Marsupialiformes, 46
marsupials, 35–53, *86*
　altricial births, 44, 45, 71–2
　brain anatomy, 44
　brood pouch, 44, 45, 72
　constraint hypotheses, 71–2
　defining features, 45
　developmental strategy, 42
　dispersal journeys, 9, 50–1
　divergence, 43–6
　diversity, 71
　evolution, 42, 56–7
　neonates, 25, 42
　New Guinea, 35, 68–70
　phylogeny and biogeography, *48*
　reproduction, 41–2
　reproductive tracts, 44
　solitary behaviour, 47
　South America, 46–50
　stem cell biobanks, 81
　see also Australian marsupials
Masai giraffe, 178–80
mass extinctions
　see extinctions
masseter muscles, 230
Masters, Judith, 95
mastication, 24
Mayr, Ernst, 108, 179, 235
Mediterranean Sea, 193
Megachiroptera ('big' bats), 218
Melanesians, 70, 306
melatonin, 158
Melin, Amanda, 255
Memoir on the Gorilla (Owen), 276

Mendelian genetics, 40–1
Mendoza, Zepeda, 221
mermaids, 101
mesopic light levels, 255
Mesozoic epoch, *xv*, 31, 134, 139
messenger RNA (mRNA), 113, 220, 272, 300
metabolic rates, 4, 95, 103, 121–3, 226, 232, 296, 298–9
Metatheria, *18*, 42
metatherian cerebral connections ('connectome'), 44
metatherian–eutherian dichotomy, 44
metatherian milk, 45
metatherians, 42–8, *38*, 208
Microbiotheria, 48, *48*, 50, 52
Microchiroptera ('small' bats), 218
Microraptor, 43
Microtragulus, 47
Mid-Cretaceous period, 38
middle ear ossicles, 4, 11, *18*, 23–5, *24*, 39–40, 93
Middle Mesozoic ('Age of Reptiles'), 31
Middle Miocene, 49, 53, 125, 161, 226, 260, 269, 294
Middle Miocene Climatic Optimum (MMCO), xiv, 49, 53, 259–60
Migliano, Andrea, 286, 287
migrations, 68, 146, 166–7, 184, 207, 209, 224, 286, 296, 302–4
milk, 21–3, 150, 307
milk teeth (deciduous teeth), 10, 119
Minckley's cichlid, 110
miniature primates, 248
Miocene epoch, *xv*, 49, 243, 269–70
Miocene–Pliocene transition, 274, 297
Mirceta, Scott, 100
Mitchell, Graham, 173
Mitteroecker, Philipp, 25
Mivart, St George Jackson, 19–20, 277
molar teeth, 4, 230, 269
Mole National Park, Ghana, 105
molecular approaches, 6–10
molecular clocks, 9, 120, 130–1, 141, 209, 229
molecular convergence, 67–8, 131–2, 149, 219
'molecular fossils', 39, 52, 88
moles, 92, 128, 129, 130, 131
monito del monte, xiv, 48, *48*, 50, 51–3, 54, 56–7, Plate 5
monodactyl horses, 183
monophyletic groups, 53, 218, 226, 228, 229, 243, 258
Monotremata, 17

monotreme lactation protein (MLP), 23
monotremes, 9, 17–32, *18*, *86*
altricial births, 19, 25
edentulous, 26
evolutionary origins, 18–19
lactation, 22–3
mammary glands, 19
mass extinction, 30
middle-ear bones, 25
oesophagus, 26
oviparous, 17–19
as primitive, 32
sex chromosomes, 28
South America, 51
taste receptors, 26
VTG genes, 21
see also echidna; platypus
Montana, 45
Morales, Ariadna, 5
Morganucodon, 11, Plate 1
morganucodons, 11–12, 25, 36
Morocco, 89
morphogens, 187, 216
mountain gorilla, 115, 278, 280, Plate 33
mountain pygmy possum, 90, **90**
mountain zebra, 186
mouse-deer, *144*, 151
mouse lemur, 245, 246, 248
Mozambican Civil War, 106–7
Müllerian mimicry, 243
multituberculates, 42, 71
Muroidea, 228
Murphy, William, 203
music, 100, 239
muskox, 160, *161*, 164–5
muskoxen, 164–5, Plate 18
mustangs (wild horses), 184
myoglobin, 100
myrmecophagy (ant and termite diets), 8, 39, 91
mysticetes
see baleen whale (Mysticeti)

Nadir Crater, West Africa, 136
naked mole-rat, 233–4, 299, Plate 27
Nancy Ma's night monkey, Plate 31
Narins, Peter, 92–3
A Narrative of Travels on the Amazon and Rio Negro (Wallace), 261
Natal droptail ant, 234
Naturalist's Miscellany (Shaw), 17
Nature, 302
Neanderthal, 303–6, *304*, 307, 308
'near' mammals, 10–11

'necks for sex' hypotheses, 173
Necrolestes, 47
Neofelis, 204
Neogene period, *xv*, 259
neoteny (paedomorphism), 290–1, 301
Neotropical rodents, 229
Nery, Mariana, 159
Nesvorný, D., 135
Nevo, Eviatar, 235–6
New Guinea, 29–31, 60, 68–70, 305
New Ireland, 70
New South Wales, 15, 58, 90
New World monkeys, *241*, 243, 254, 258, 260, 261–2, 266, **266**, Plate 31
New World primates (Platyrrhini), 9, 53, 258–61, 267
New York City rats, 224
New Zealand, 70, 73, 228, 304, *304*
Newman, Janet, 23
Niger River, 285
Nigeria-Cameroon chimpanzee, 285, *285*
night monkey
  see owl monkey (douroucoulis)
night vision, 243, 248, 253–4, 267
nilgai, 160
Nilsson, Maria, 52
nine-banded armadillo, 124, 126–7
nitric oxide, 199
noise-cancelling proteins, 218
non-echolocating Old World fruit bats, 218
non-mammalian mammaliaforms, 11
non-prehensile tails, 270
non-volatile odorants, 62
North America
  bison, 169
  black rat, 224
  camelids, 145–6, *145*
  cave lion, 204
  Chicxulub impact, 136
  equines, 183–4
  Great American Biotic Interchange, 50, 145
  horse, 184
  human migrations, 303
  Marsupialiformes, 46
  Pleistocene–Holocene extinction, 164
  proboscideans, 107
  see also Isthmus of Panama
northern brown bandicoots, 59
northern common cuscus, 70
northern giraffe (Rothschild's giraffe), 178, Plate 20
northern naked-tailed armadillo, 124
northern tamandua, 120

Nothofagus (southern beech), 51, 53
Notoryctemorphia (marsupial moles), 54, 55, *55*, 57–8, 92
numbat (banded anteater), *55*, 59, 77, **90**
nyala, 160, *161*

obligate carnivores, 202, 203
O'Brien, Stephen, 202
ocean currents, 12, 69, 94, 95, 102, 154, 156
Oceania, 304
Oceanic dispersals, 259
ocelots, 202, 207, 208
Ocepeia, 89
odd-toed ungulates (perissodactyls), *129*, 143, 144, 181–3, 189, 190
Odontoceti (toothed whales), 153–4, 158, 219
Oftedal, Olav, 20, 21
Ohno, Susumu, 27
okapi, 143, *144*, 173, 174, 176, 178, 180
Old World monkeys, *241*, 243, 259, *260*, 269–70
olfaction, 158, 177, 240, 243
Oligocene, *xv*, 48–50, 102, 106, 120, 125, 145, 155, 160, 192, 243, 259
Oligocene–Miocene boundary, 270
Olson, Maynard, 158
On the Genesis of Species (Mivart), 19–20
onager, 185
'one gene, one protein' hypotheses, 220
OneZoom website, 5
opossums (Didelphimorphia), 43, 48–50, *48*
opsins, 39, 118, 254, 255, 265, *266*, 267
orangutans, *241*, 270, 274–5, 296
The Origin of Species (Darwin), 20, 178, 196, 276
oryx, *161*
osteoclasts, 302
osteoderms (bony plates), 231–2
Ottenburghs, Jente, 283
otter shrews, 88, 93, 96
otters, 47, 72, 203
ouabain, 232
Oulad Abdoun Basin, Morocco, 89
'Out of Africa' model, 296, 302–4
overheating, 111
oviparity
  mammary gland development, 21–2
  monotremes, 17–19
  and viviparity, 40–1
ovum meroblastic, 18
Owen, Richard, 18, 19, 276
owl monkey (douroucoulis), 258, *260*, 262, 267–8

oxygen deprivation, 233
oxytocin, 289, 292

p53 tumour suppressor protein, 112, 113
P450 enzyme, 250
Pääbo, Svante, 304–5, 308, Plate 35
paedomorphism (neoteny), 290–1, 301
Paenungulata, 88, *88*, 98, 99, 100–1, 104, 105, 106
Palaeocene–Eocene Thermal Maximum (PETM), xiv, 181, 213, 242
Paleocene–Eocene boundary, 242
Palkopoulou, Eleftheria, 108–9
Pallas's cat, 209
pampas cat, 207
Pan genus
    see chimpanzee
Pan paniscus ('little Pan')
    see bonobo
Panamanian white-faced capuchin, 263
Pancho's monito del monte, 53
panda, 192–4
Pangaea, xiii, 10, 11, 12, 43, 85
Pangaea Ultima, 312
pangolins, 4, 9, 26, 86, 89, 119, 121, *129*, 231
Panthera, xiv, 203–7
Pantherinae, 202–4, 207
Paracamelus, 145, 146
Paranaense Sea, 259
paraphyly, 226
parasagittal stance, 39
Pask, Andrew, 76, 77, 79–80, 81, 115
patagium, 67–8, 212, *212*, 215, 216–17, 240
Patagonia, 37, 149, 205, 207, 259, 260
Patagonians, 303–4
Paucituberculata (shrew opossum), 48, *48*, 50
Pauli, Jonathan, 123
Pebas mega-wetland, 102
peccary, *144*
Pečnerová, Patrícia, 164, 165
pelvis, 122, 152, 274, 297–8, 301
pen-tailed treeshrew, **90**
Peramelemorphia, 54, 59
    see also bandicoot
'percussive foraging', 246–7
Peredo, Carlos, 155
perioral vibrissae, 104
perissodactyls (odd-toed ungulates), 86, *129*, 143, 144, 149, 181, 182, 189, 190
Perupithecus, 258
Peto, Richard, 112
Peto's paradox, 112, 159

Pettigrew, Jack, 29
Pezosiren, 101
phenotypic traits, 10, 28, 76, 86–7, 111, 219
Philippine colugo, 240
Philippine tarsier, 251–2, 255, Plate 30
Philosophie Zoologique (Lamarck), 172
photoreceptive genes
    see opsins
photosynthesis, xiii, 30, 136–7
phylogenetic species concept, 178, 179
phylogenies, 5, 8, 130, 214
    Afrotheria, *88*
    Australian marsupials, 54–5, *55*
    Bovidae, *161*
    Cetartiodactyla, *144*
    Euarchontoglires, *226*
    Laurasiatheria, *129*
    Mammalia, *86*
    marsupials, *48*
    molecular, 9, 87, 214
    primates, *241*
pichiciego (greater fairy armadillo), 125
pig-footed bandicoot, 59
pigs, 86, 99, 128, 142–4, *144*, 159, 238
pikas, 225, 226, *226*
Pilosa, 4
pink fairy armadillo, 124
pinnae (ear flaps), 25, 111
    see also ears
Pitheciidae, 260
pizzlies (hybrid brown and polar bear), 199
placental mammals, *18*, 213, 236, 265
    Afrotheria, 88–9
    marsupial species, 71
    models for diversification, 140–1, *140*
    size and heterogeneity, 143
    skeleton at birth, 25
    viviparity, 40–3
placental moles, 58
plagiopatagium, *212*, 216, Plate 24
plains zebra, 186, Plate 21
plant speciation, 196
Plasmodium falciparum, 307
platypus, 15–32, **90**, Plate 2
    casein genes, 22
    electroreception, 28–9
    evolutionary origins, 18–19
    oviparity, 17–19, 22
    puggles suckling, 19
    reproductive organs and ureters, 16
    teeth, 26
    venom, 26–7
Platyrrhini–Catarrhini split, 259
Platyrrhini (New World primates), 9, 53, 258–61, 267

pleiotropy, 177
Pleistocene–Holocene extinction, 164
Pliocene, *xv*, 47, 50, 63, 117, 143, 154, 164, 167, 227, 279, 282, 297
Pliocene–Pleistocene climatic oscillations, 170
plunger pumps, 176
Poebrotherium, 145
'poison arrow tree' (*Acokanthera schimperi*), 232
polar bear, 191–2, *196*, 197–200, Plate 22
polychromatic trichromacy, 254
polymorphic trichromacy, 253–5
polypeptides, 103–4
porcupines, 229, 231
porpoises, 143, *144*, 150, 153, 299, 310
positive selection, 177, 240, 253, 273, 275, 282, 308
possums, 54, *55*, 66, 67, 148
post-facial vibrissae, 104
post-zygotic reproductive isolation, 256
potoroo, 54, *55*, 65–6
Potter, Sally, 64
pottos, 241, *241*
pre-mammals (mammaliamorphs), 6, 11, 12, 20, 21, 23, 39, 40
precursor trunk, 106
predator-pain problem, 234
prehensile tails, 49, 51, 69, 261, 270
Prestin, 219
primates, 128, 225, *226*, 230, 241–8, *241*
  Afro-Arabia, 243
  colour vision, 265–8
  DNA introgressions, 282–3
  fossils, 242
  genomes, 203
  grooming and food acquisition, 243
  size of ancient primates, 273
  tails, *241*, 270–2
  see also New World primates (Platyrrhini)
primitive eutherian, 44
The Principles of Classification and a Classification of Mammals (Simpson), 86
Proboscidea, 98, 105
Proboscidean hypsodonty, 110
proboscideans, xiv, 99, 100, 106, 113–14
proboscis, 105–6, 188
prognathism, 302
  see also jaws
promiscuity, 60
'prosimian monophyly hypotheses', 252
protein-coding DNA, 76, 148, 271, 301

protein evolution, 73–4
protein fingerprinting, 142
proteome (protein repertoire), 43
Prothero, Donald, 6
proto-tarsiers, 257
Prototheria
  see monotremes
Protylopus, 144–5
'prusten' (chuffling), 204
Prx1 protein, 216–17
Przewalski, Nikolai, 185
Przewalski's horse, 185
pseudogenes, 9–10, 22, 39, 156–8
  catarrhine *TPR2*, 261
  as 'genomic fossils', 158
  GULO, 252
  insectivorous ancestry, 39
  LIF6, 113
  regressive evolution, 119
  testicondy, 114
  umami taste receptor *TAS1R1*, 194
  xenarthrans, 118–20
  see also genes
Pucadelphys, 46–7
puma, 202, 207–8
'punctuated equilibrium' (Eldredge and Gould), 177
'pygmy chimpanzee'
  see bonobo
pygmy cuscus, 69
pygmy gliding possum (feathertail glider), 67
pygmy possum, *55*, 90, **90**
Pyrenean ibex, 79

Qiu, Qiang, 176–8
qiviut, 164
Quaternary Extinction event, 117, 146, 184
'quolls (tiger cats), xiv, *55*, 59, 60–1, 68–9, 78, 81

rabbits, 146, 225–7, *226*, 291
radial sesamoid, 193
rafts of vegetation, 95–6, 228, 259
random mutations, 92, 108, 198, 283, 293, 300
Ranomafana, Madagascar, 249
Raphicerus, 170–1
rapid evolution
  archaic DNA insertions, 283
  genetic variations, 61, 115, 177
  neoteny, 291
  polar bears, 198
rapid speciation
  see adaptive radiations

rat kangaroo, 55
Raterman, Denise, 69
rats, 133, 223–5, 229, 237
reactive oxygen species (ROS), 217
recombination (genetic shuffling), 131, 200, 224
recreational sex, 288–9
red blood cells, 148, 149, 166
red panda, 192, 193, 194
reedbucks, *161*
regenerating damaged tissue, 231
regressive evolution, 10, 118–19
regulatory elements, 77, 163, 299, 300
reinforcement, 245, 265
relict species, 53
reproductive isolation, 179, 186, 195–6, 201, 235–6, 255, 264–5, 282
reptiles, 11, 18–19, 22
    jaws, 23–4, *24*
    middle-ear bone (columella/stapes), 23
    teeth, 10
reticulated giraffe, 178, 179–80
retinal cells, 267–8
retinal cone receptors, 118
retinal progenitor cells (RPCs), 267
retinal rod-to-cone ratios, 267
retrogenes, 9, 112–14, 115
retroposons ('jumping genes'), 52, 87–8, 243, 253, 271
retroviral genes, 9, 42–3
rhinoceroses, 143, 163, 181, 189–90, 299
rice rat, 228
Rich, Thomas, 36–7
ring-tailed lemur, 248, Plate 29
ringtail possum, 55
riverine barrier hypotheses, 261
Riversleigh World Heritage fossil site, Queensland, 58
RNA (ribonucleic acid), 7–8, 113, 253, 271, 272
Robertson, Douglas, 137
robust capuchins (*Sapajus*), 262–3
rock badger (hyrax), 87–8, *88*, 98–101
rock hyrax, 99, Plate 11
rock-wallabies, 55, 64–5
rod monochromacy, 118
Rodentia, 225–6, 227, 229
rodenticides, 225
rodents, 128, *226*, 227–32
    dental arrangement, 229–30
    diversity, 227
    evolution, 228, 230
    locomotion, 224
    masseter muscle, 230
    opportunistic omnivores, 230

post-human world, 237–8
radiations, 228
Rodhocetus, 151
Rogers, Rebekah, 115
Rothschild's giraffe (northern giraffe), 178, Plate 20
Rowe, Timothy, 11
Rukwa Rift Basin, Tanzania, 270
ruminants, 162–3, 182
running and breathing, 39
rusty-spotted cat, 209
Ruxton, Graeme, 217
Rybczynski, Natalia, 142–3, 145
Ryukyu islands, Japan, 237

Saarinen, J., 109
sable antelope, *161*
'sabre-toothed tiger' (*Smilodon*), 47, 209
Sagan, Carl, 276, 310
Sahul Shelf, 69
Saint-Hilaire, Etienne Geoffroy, 17, 18
sakis, 258, *260*, 261
Samotherium, 174
Sanaga River, equatorial Africa, 285–6, *285*
sanguivores, 220–2
saola, 167
Sato, J.J., 130
Scally, Alywyn, 281, 294–5
Scheepers, Lue, 173
Schiebinger, Londa, 4
Schmitz, Jürgen, 251, 252–3
Schodde, Richard, 38
Schouteden, Henri, 284
Schwartz, Ernst, 284
scrotum, 44, 96, 99, 114
sea cows (manatees), 85, 88, *88*, 98, 101–4
sea lions, 70, 203
Sea Nomads (Bajau people), 307–8
seagrass, 102, 103
seals, 36, 70, 128, *129*, 155, 192
sebaceous gland secretions, 157
Seba's short-tailed bat (eutherian), 67
second wave of extinctions, 311
segmental duplications, 203
Seiffert, Erik, 102
semelparity (suicidal reproduction), 60
sengis
    see elephant shrew (jumping shrew, sengis)
'serial founder effect', 302
sesamoid, 193, 215
sex determination, 230, 236–8
sexual dimorphism, 47
sexual reproduction, 59–60, 256
sexual selection, 60, 173, 174, 197

Seymour Island, Antarctica, 51, 137
Shapiro, Beth, 80–1, 184, 200
Sharpe's grysbok, 170
Shaw, George, 16–17
sheep, 75, *161*, 162, 170
Shi, Peng, 218, 219
short-beaked echidna, 31, Plate 3
short-faced bear (Tremarctinae), 194–5
Short Fuse model, *140*, 141, 242
short-lived placentas, 44, 45
shrew opossum (Paucituberculata), 36, 48, *48*, 50
shrew-rat, 230
shrew tenrec, 96
shrews, 5, 26, 129–31, *129*, 131–2
Siberian tiger, 206
sickle cell disease, 266, 307
sifakas, 246, 247–8, 255
silky anteater, **90**, 120
silverback gorillas, 278, 279, 281
simian immunodeficiency virus (SIV), 286
Simmonds, Robert, 173
Simpson, George Gaylord, 36, 86–7, 93–5, 117, 244
Sinai Peninsula, 303
SINEs (short interspersed nuclear elements), 87–8
Sinodelphys szalayi, xiii, 44
Sir David's long-beaked echidna, 31
Sirenia (sirenians, sea cows), 98, 99, 101–4
sirenians (dugongs and manatees), 88, 97–105, 214
Sixth Mass Extinction, 310
skin pigmentation, 186–7, 306
Slater, Graham, 156
Slatkin, Montgomery, 115
sloth, 86, 117–24, 162, 174, Plate 13
    apparent inactivity, 122
    cervical vertebrae, 122
    dental regression, 119
    diet of leaves, 123–4
    hanging upside down, 124
    heterothermic, 123
    metabolic rate, 123–4
    use of energy, 123
slow lorises, 243
'small' bats (Microchiroptera), 218
'small' cats (Felinae), 202, 207–9
Smilodon ('sabre-toothed tiger'), 47
snow leopard, 165, 203–4, 206–7
solenodon, 128–33, *129*
    anthropogenic extinction pressures, 139
    'Goldilocks' species, 133
    Haiti, 131

K–Pg extinction event, 132–3, 139
'living fossils', 131
Long Fuse Model, 141
venom, 26, 131–2
'zombie lineage', 130
Solomon Islands, 5, 29, 70, 228
Solounias, Nikos, 174
somatic cell nuclear transfer, 79
songbirds, 38, 152, 240
Sorek, Rotem, 272
sound generation echolocation, 220
South Africa lion, 205
South America, *48*
    camelids, 147, 148–9
    cats, 207–8
    caviomorph rodents, 53, 228–9
    Gondwanan dispersal, 50–1
    humans, 303
    manatees, 102
    marsupials, 45, 46–50, 51
    metatherians, 48
    monotremes, 51
    natural experiment, 117
    reunited with North America, 50
    rodents, 228
    see also Isthmus of Panama
southern beech (*Nothofagus*), 51, 53
southern giraffe, 178
southern grasshopper mouse, 234
southern greater glider, 67
southern opossum, 50
southern tamandua, 120
Sox9 molecules, 216, 236, 237
Sparassodonta, 47
speciation gene (*PRDM9*), 256–7
spectacled bear, 194–5
spectacled cobra, 243
speech, xiv, 281
sperm, 28, 41, 60, 96, 114, 197, 205, 208, 237, 256, 263, 281
sperm whales, 153, 154, 157
spider monkey, 258, *260*, 261, 270
spiny mouse, 231, 235–6, Plate 26
spleens, 308
spliceosome (protein-and-RNA complex), 113
spontaneous mutations, 108, 114, 235, 307
'spooky gene flow' (Ottenburghs), 283
sportive lemur, 245, 247
spotted-tailed quoll, 78
springbok, *161*
Springer, Mark, 87–9, 92, 118, 130, 157
squirrel monkey, 258, *260*
stapes (columella), 23–4, *24*
steenbok, 170

Steller's sea cow, 102–3, 104
stem bats, 213
stem therians, xiii, 36–7, *38*
stenonines, 183–5
stereoscopic vision, 241
Stevens, Nancy, 270
stone tools, 286
straight-tusked elephant, 107, 108, 111, 112
Strathcona Fjord, Ellesmere Island, 142
strepsirrhines (lower primates), 241, 243, 245, 252
stridulation, 97
Suárez, Rodrigo, 44
subterranean adaptations, 232–4
sugar glider, *55*, 67
suicidal reproduction (semelparity), 59–60
Sulawesi bear cuscus, 69
Sulawesi tarsiers, 255
Sumatran orangutan, 274, 275
Sumatran rhinoceros, 189
Sumatran tiger, 206
sun bear, 194, 195, *196*
Sunda clouded leopard, 204
Sunda colugo, 240
sweat glands, 20–1, 26, 62, 157, 299
sweepstake dispersals, 69, 93–5, 228, 244, 259
SWS1 opsin protein, 254, 265, **266**
sympatric speciation, 234–6, 236, 263
syndactyly (fusion of fingers), 216
Synergistaceae, 62
'Syrian camel', 146
Systema Naturae (Linnaeus), 3

T-helper lymphocytes, 286
tactile hairs (body vibrissae), 99, 104, 106
tail-assisted hindlimb suspension, 261
tailless primates, 270
tailless tenrec, 138
tails and tail loss, 261, 270–2
Talahpithecus, 258
talpid mole, 58
tamandua, 120
tamarin, 258, *260*, 261, Plate 31
Tanzania, 166, 270, 286, 289
Tapanuli orangutan, 274
'tape of life' metaphor (Gould), 72
tapetum lucidum, 243, 253–4, 262
tapir (Tapiridae), 143, 181, 188–9
tarsier, *241*, 251–7, Plate 30
  dichromatic, 255
  evolution, 252
  gene polymorphism, 254–5
  GULO pseudogene, 252

haplorrhines, 241, 253
insertion–deletion mutation, 256–7
L/M polymorphic genes, 255
nocturnal lifestyle, 254–5
PRDM9, 256
trichromatic vision, 254–5
ultrasonic calls, 251
'tarsier-first hypothesis', 252
TAS1R1 umami taste receptor, 194
Tasmanian devil, *55*, 56–7, 59, 61, 78
Tasmanian pademelon, 78
Tasmanian Passage, 51
Tasmanian tiger
  see thylacine (Tasmanian tiger)
taste receptor genes, 62, 77, 158, 194, 282
taurine cattle, 168–9
Taylor, Dale, 176
teeth
  bilophodonty, 269
  carnassial molars, 202
  deciduous/milk teeth, 10, 119
  dentary, 4, 24, *24*, 30, 36
  diphyodonty, 10
  elephants, 109–10
  enamel, 109, 119, 155, 157, 229
  filter-feeding whales, 155
  grind and chew, 36
  hypsodonty, 109, 110, 163
  molar teeth, 4, 230, 269
  monotremes, 26
  non-functioning genes, 157
  regressive evolution, 118–19
  reptiles, 10
  tribosphenic, 36, 37, 43, 58, 117, 119, 127
  Xenarthra, 119
  zalambdodonty, 58–9
Teilhardina, 242
Teinolophos trusleri, 30
tenrec, 53, 87, 88, *88*, 93–7, 138
termite-fishing chimpanzees, 286
testes, 96, 99, 114, 236, 263, 264
testicondy, 114
Tethys Ocean, 38, *38*, 89, 151, 152
tetrachromatic system, 265
therians, 36–43, *38*
  Cretaceous, 39
  ear ossicles, 39–40
  Gondwana, 43
  Laurasia, 43
  Mid-Cretaceous period, 38
  placenta, 21, 42
  stem, xiii, 66–8
  tribosphenic molars, 36

venom, 39
X and Y chromosomes, 236
Thewissen, Hans, 151
thoracic vertebrae, 174–5
three-toed sloth (*Bradypus*), 121–2, 123
Thybert, David, 230
thylacine (Tasmanian tiger), 55, 59, 75–81, Plate 7
   brain development, 77
   convergence with canids, 76
   de-extinction, 80–1
   extinction, 75–6
thylacine–wolf accelerated regions (TWARs), 76–7
Thylacosmilus, 47
thyroid hormones, 103
tiang, 167
Tibetan antelope (chiru), 165–6
Tibetan kiang, 185
Tibetans, 306
tiger, xiv, 203–4, 206, 256
tiger cat (quoll), xiv, 55, 59, 60–1, 69, 78
'Tiger' pet cat, 210
tiger quoll, 61
TIGRR Lab (Thylacine Integrated Genetic Restoration Research), 79–80
timelines, xiii–xiv
tissue patterning, 216
titi monkey, xiv, 258, *260*, 261
Tobin, George, 16–17
'toilet' claws, 251
Tokyo Institute of Technology, 175
Tollis, Marc, 159
tongue-clicking, 220
tongues, 89, 120–1, 154, 202
tools, 262, 263, 286–7, 297
tooth comb, 243
toothed whales (Odontoceti), 153–4, 157, 158, 219
Topi, *161*, 167
Torres Strait, 68
Touan short-tailed opossum, 49
toxic food, 158, 249–50
TP53 retrogenes, 112–14
trans-Antarctic dispersal, 50–1
trans-Atlantic sweepstake dispersals, *260*
transient marine corridors, 95
transitional fossils, 89, 214
transoceanic sweepstake dispersals, 53, 93–4
transposons, 221, 230, 253, 272
   see also genes; 'jumping genes' (retroposons)
Treacher Collins syndrome, 25
tree-kangaroo, xiv, 55, 63, 65, 68

treeshrew, 43, 128, 225, *226*, 240
Tremarctinae (short-faced bears), 194–5
Triassic period, 10, 12, 21, 85
tribosphenic molars, 36, 37
trichromatic vision, 254–5, 268
tridactyl horses, 183
Triossauria, 4
tropical animals, 242
trotting, 183
TRPV1 nerve channels, 221
true lemur, 245, 247
trunks, 99, 103, 105–6
Tubulidentata, 89–90
tufted capuchin (*Sapajus*), 262–3
tumour suppressors, 112, 113, 159
Turing, Alan, 186–7
Turing model, 187–8, 216
Turvey, Samuel, 131, 132–3
tusks, 99, 105, 106–7, 160
two-colour vision (dichromacy), 39, 254, 255
two-toed sloths (*Choloepus*), 121–2
Tyrannosaurus rex, 138

uakari monkey, 258, *260*, 261
Ubangi River, equatorial Africa, 285, *285*
ulnar sesamoid, 215
ultrasonic calls, 251, Plate 30
umami taste receptor *TAS1R1*, 158, 194
unfilled ecological niches, 245
ungulates, 86, 95, *129*, 147, 151–2, 160, 166–7
United States, 184
University of Austin in Texas, 11
untufted capuchin (gracile, *Cebus*), 262–3
urban rats, 224–5
Ursinae, 194–5

Valerio, S.O., 163
Vallrath, Fritz, 114
vampire bat, 220–2
van Dijk, M.A., 87
Vangunu giant rat, 5
vaquita porpoise, 310
vascular endothelial growth factor c (VEGFC), 231
vasopressin pathways, 292
venom, 16, 26–7, 39, 131–2, 234, 243
venomous primates, 243
Vermilingua
   see anteater
'vertebrata' (*Encyclopaedia Britannica*), 277
vibrissae, 99, 104, 106
vicariant speciation
   see allopatric speciation

Vickers-Rich, Patricia, 36–7
vicuña, *145*, 148–9
viral RNA, 42
Virginia opossum, 50
Virunga gorilla population, 278, 280, Plate 33
vision
    colour-blind, 118, 267
    dichromacy, 39, 254, 255
    night vision, 243, 248, 253–4, 267
    nocturnal animals, 243
    rod monochromacy, 118
    trichromatic vision, 254–5, 268
    xenarthrans, 118
    see also eyes
visual predation theory, 241
vitamin C, 252
vitellogenin (*VTG*) genes, 21–2
viviparity (live births), 2–3, 17, 40–3
Vogelkop peninsula, New Guinea, 31, 69
volatile fatty acids (VFAs), 162
volcanic activity, 136–7
vomeronasal organ, 243, 261
Vrba, Elisabeth, 20

Wagner, Andreas, 27
Waigeo cuscus, 35, 69
Waigeo, New Guinea, 35
Wallace, Alfred Russel, 4, 94, 186, 188, 261
warm-blooded animals, 4, 10, 12, 21, 111, 114, 220
Warren, Wesley, 210
wasabi, 234
water chevrotain (African mouse-deer), 151
water opossum, 72, 76
waterbuck, *161*
Watson, J.D., 7
Watson, Lyall, 105
Weisrock, David, 244, 245
West Africa, 102, 136, 161, 280
West Indian manatee, 103
western chimpanzee, 285, *285*
western gorilla, 276, 278–9, 280, 282
western lowland gorilla, 278, 280
western quoll, 61
whales, *129*, *144*, 150–9
    aquatic skin traits, 157
    evolution, 152–6, 159
    gigantism, 159
    gustatory sensory system, 158
    and hippos, 150
    sleeping patterns, 158
    taste receptor genes, 158
whiskers, 11, 104, 106, 202
whispering bats, 218
white-eared kob, 162, 166–7
white-fronted capuchin, 263
white-tailed deer, 151–2
Whittington, Camilla, 27
wild Bactrian camel, 146–7
wild cattle, 167
wild horses, 184–5
wild yaks, 169
wildcat, 202, 209
wildebeest, 98, 152, *161*, 162, 166–7
Williams, Edgar, 174
Wilson's bird-of-paradise, 35
Wnt protein, 187, 216
wolves, and thylacine, 76–7
wombat, 54, *55*, 62, 63
Wonderful Life (Gould), 72
wood-boring beetle larvae, 246–7
woodrat, 163
woolly mammoth, 107, 115–16
woolly monkey, xiv, 258, *260*, 261
Wrangel Island, Arctic, 115, 164
Wyoming, 45

X chromosomes, 28, 107, 126, 170–1, 199, 203, 236–7, 254, 266–8, 282
X-linked L/M opsin alleles, 265–6, **266**
Xenarthra, *86*, 117, 119
xenarthran, 117–20
Xia, Bo, 271
Xie, Victoria, 73–4

Y chromosomes, 236–7, 255
Yangochiroptera, 214, 219
Yangtze River, 168, 311
Yangtze River dolphin (baiji), 131, 310
Yegian, Andrew, 298, 299
Yinpterochiroptera, 214, 219
Yucatán Peninsula impact, 118, 132–7

zalambdodonty, 58–9
Zalasiewicz, Jan, 237–8
Zambezi river, 244
Zanker, Johannes, 186
zebra, 93, 183, 185–7, Plate 21
zebu cattle, 168–9
Zischler, Hans, 258
Zoonomia Consortium, 6
zooplankton, 158
Zou, Tiantian, 195

Plate 1. The 205-million-year-old *Morganucodon* – An insectivorous, nocturnal, shrew-like early mammal with venomous spurs on its hind feet like the modern platypus. (FunkMonk-Michael B.H.)

Plate 2. Platypus – This egg-laying mammal, endemic to eastern Australia and Tasmania, offers unique insights into the reptile–mammal transition. (Pepper Bush Adventures)

Plate 3. Short-beaked echidna – Originally evolved in New Guinea and later dispersed to Australia when sea levels fell. (Gunjan Pandey, CC BY-SA 4.0)

Plate 4. Tim Flannery, dubbed the 'Indiana Jones of mammalogy', with a fossil of a platypus-like monotreme from the Early Cretaceous (*Steropodon galmani*). Flannery proposes that stem therians evolved in Gondwana and echidnas in New Guinea. (Matt McCurry)

Plate 5. Monito del monte – This species evolved 59.7 million years ago in South American Gondwana, where it has remained. (José Luis Bartheld, CC BY-SA 4.0)

Plate 6. Koala – the marsupial equivalent of the placental sloth. Koalas have evolved a diverse set of proteins that help them detect and select the least toxic *Eucalyptus* leaves. (Diliff, CC BY-SA 3.0)

Plate 7. A comparison of the thylacine (A) and the dingo (B), highlighting their morphological similarities due to convergent evolution. (A: Tasmanian Museum and Art Gallery; B: Charles J. Sharp, CC BY-SA 4.0)

Plate 8. Fat-tailed dunnart – The Tasmanian tiger's closest living relative and a key species in the thylacine de-extinction project. (David Paul, Copyright Museums Victoria, CC BY-NC)

Plate 9. Aardvark – The oldest of the afrotherian orders, with the highest evolutionary distinctiveness (ED) score among mammals, having been genetically isolated for over 70 million years. (Theo Kruse, Burgers' Zoo, CC BY-SA 4.0)

Plate 10. Lowland streaked tenrec – A Madagascan endemic species, one of around 30 tenrec species whose last common ancestor floated across the Mozambique Channel from Africa approximately 50 million years ago. (Frank Vassen, CC BY 2.0)

Plate 11. Rock hyrax – A male hyrax singing; individuals with a more uniform song rhythm achieve higher reproductive success. (Amiyaal Ilany)

Plate 12. African savanna elephants at a waterhole in Mole, Ghana. These largest terrestrial animals, belonging to the superorder Afrotheria, paradoxically experience very low cancer rates despite their size. (John Reilly)

Plate 13. Brown-throated sloth – One of four three-toed sloth species that evolved their arboreal lifestyle independently of the two-toed sloths. (Daniella Maraschiello, CC BY-SA 4.0)

Plate 14. Hispaniolan solenodon – A venomous, burrowing, nocturnal, insectivorous mammal that evolved alongside the dinosaurs. (Joe Nunez-Mino, The Last Survivors Project)

Plate 15. Arabian camel (Dromedary) – Many of the camel's distinctive features (large eyes, humps, and wide, flat feet) originally evolved in the harsh Arctic environment. (Shreyas Shyam)

Plate 16. Common hippopotamus – The closest living relatives of whales, sharing a common ancestor that lived approximately 55 million years ago. (Pixabay, Herbert Bieser)

Plate 17. Grey whale calves – Whale gigantism evolved rapidly during the Pleistocene, approximately 4.5 million years ago. (John Reilly)

Plate 18. Muskoxen – One of the few species of Pleistocene megafauna to survive the Pleistocene–Holocene extinction event. (Hannes Grobe, CC BY-SA 2.5)

Plate 19. African buffalo – Diverged from the Asian water buffalo around 8 million years ago. (John Reilly)

Plate 20. Northern (Rothschild's) giraffe is native to Uganda and South Sudan. Despite having only seven cervical vertebrae like most mammals, the giraffe's extreme elongation of the neck has required numerous evolutionary adaptations. (John Reilly)

Plate 21. Plains zebra evolved approximately 1.3 million years ago in Africa. Zebra stripes are best explained by the 'Turing effect' and likely evolved to thwart the attacks of blood-sucking flies. (John Reilly)

Plate 22. The polar bear (A) evolved from the brown bear (B) approximately 400,000 years ago. (A: Alan Wilson, CC BY-SA 3.0; B: Ben Bluhm)

Plate 23. Jaguar – its powerful jaws evolved to penetrate the skulls of capybaras and the thick skin of caimans. (John Reilly)

Plate 24. A bat's wing – A highly modified forelimb featuring the plagiopatagium, a wing membrane extending from the last digit to the hindlimb. (Salix, CC BY-SA 3.0)

Plate 25. Brown rat – Genetic studies indicate that the brown and black rats emerged in East Asia around 2 million years ago. (Zeynel Cebeci, CC BY-SA 4.0)

Plate 26. Spiny mouse – A genus that offers insights into the mechanisms of speciation and tissue repair. (Marcel Burkhard, CC BY-SA-2.0)

Plate 27. Naked mole-rat – An underground-dwelling species that wins the prize for extreme adaptations. (Roman Klementschitz, CC BY-SA-3.0)

Plate 28. Amani spiny rat – One of only two mammal species that lack the male Y chromosome. (Asato Kuroiwa)

Plate 29. Ring-tailed lemur – One of approximately 110 species of lemur. They all evolved from a single ancestral species that reached Madagascar around 63 million years ago by rafting on a mat of vegetation. (John Reilly)

Plate 30. Western tarsier – Emits ultrasonic calls to help locate prey, like other tarsiers. (Pete Morris)

Plate 31. New World Monkeys – Includes the mantled howler monkey (A), Nancy Ma's night monkey (B), and golden-headed lion tamarin (C). All 170+ species of New World monkeys evolved from a single founder species that rafted from Africa to South America, approximately 40 million years ago. (A: Ariel Rodriguez-Vargas, CC-BY-4.0; B: Whaldener Endo, CC CCO-1.0; C: Pixabay, Garethdoc)

Plate 32. Lar gibbon – The gibbon lineage diverged from apes around 17 million years ago. A unique 'jumping gene' (LAVA) is believed to have driven their rapid speciation. (Paignton Zoo, Wild Planet Trust)

Plate 33. Mountain gorilla – Its ancestors inhabited tropical Africa during the Pliocene, around 4 million years ago. (John Reilly)

Plate 34. Bonobo – Diverged from chimpanzees approximately 2 million years ago, a vicariant event brought about by the mighty Congo River. (Licensed by Natataek, CC BY-SA 3.0)

Plate 35. Nobel Laureate Svante Pääbo, of the Max Planck Institute for Evolutionary Anthropology in Leipzig, Germany, holding a reconstructed Neanderthal skull. A pioneer in ancient DNA sequencing, Pääbo led the Neanderthal Genome Project, and, in 2010, identified the Denisovans, a novel archaic hominin. (Svante Pääbo)